VIRUSES

VIRUSES

. . . .

AGENTS *of* EVOLUTIONARY INVENTION

Michael G. Cordingley

Harvard University Press

Cambridge, Massachusetts · London, England

2017

Printed in the United States of America

First printing

Library of Congress Cataloging-in-Publication Data
Names: Cordingley, Michael G., 1958-
Title: Viruses : agents of evolutionary invention / Michael G. Cordingley.
Description: Cambridge, Massachusetts : Harvard University Press, 2017. |
Includes bibliographical references and index.
Identifiers: LCCN 2016056261 | ISBN 9780674972087 (hardcover)
Subjects: LCSH: Viruses. | Evolution (Biology)
Classification: LCC QR370 .C67 2017 | DDC 579.2--dc23
LC record available at https://lccn.loc.gov/2016056261

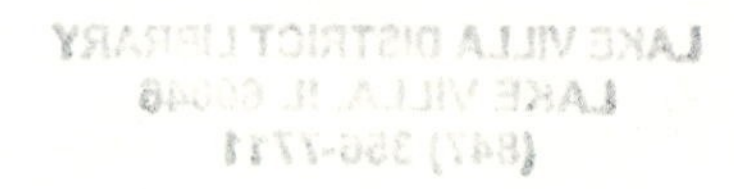

CONTENTS

VIRUSES

INTRODUCTION

Following the discovery of viruses in the last decade of the nineteenth century, science and society focused on the association of viruses with diseases. This remains the case, as reflected in the modern press and lay literature—we are most preoccupied with viruses as pathogens. This perspective is entirely appropriate because the viral diseases we encounter in our daily lives and as news features are manifestations of the viral world that impact our personal experience. Invisible to us, however, is the influence of viruses in every ecosystem on the planet. Viruses are obligate parasites of living cells; every living organism in the biosphere suffers viral parasitism, and its genome bears the indelible imprints of virus infection. In this publication, you will be introduced to viruses as agents of evolutionary invention. They are remarkable engines of genetic variation that powers their own adaptive evolution and catalyzes evolutionary change in their hosts.

It is a complex topic. Separating the evolution of viruses from the evolution of their hosts is impossible: every host is locked in evolutionary conflict with its viruses. This conflict will be evident as we follow the science that has contributed to our understanding of viruses and their place in our world, our diseases, and even our own evolution. Our journey of discovery will, by necessity, take a winding road through varied territory. At each destination on our route we will focus on a particular aspect of viral evolutionary invention. It is, of course, necessary to begin with the discovery of viruses and the nature of their physical composition. We will learn that today viruses can be conceptually understood as infectious egotistical genetic information that replicates and evolves in living systems. Their success is recorded in the novel genetic information they have created—the viral metagenome—the sheer volume and diversity of which overshadows all other genetic information in the biosphere.

In individual chapters we will explore often underappreciated aspects of virus biology, always with a focus on the "whys" and the underlying evolutionary principles. Early chapters will examine the viruses of microbes their role in shaping global ecosystems and in fueling microbial evolution and diversity. We will then look at how viruses potentiate many of our most feared bacterial diseases and influence bacterial pathogenesis and antibiotic drug resistance. Turning to viruses that infect higher organisms, we will discover how the viruses of today evolved to be successful and reveal the very distinct mechanisms that fueled their evolution. We ignore viral diseases at our peril, so substantial attention is dedicated to the pathobiology and evolutionary pressures at work in epidemic and zoonotic virus infections, as well as the evolution of pandemic viruses.

The human relationship with viruses is central to the closing chapters. Our unique cultured species can wield knowledge and ingenuity against viral diseases. In this regard, humans are a distinctively privileged species. Such is our ascendancy that viruses themselves are now becoming part of the medical armamentarium, used to prevent viral disease and treat a variety of illnesses. Nevertheless, new viruses are emerging and continue to challenge us. The evolutionary invention of viruses in our changing global reality commands our diligence and respect to prevent them from becoming an untenable health burden to future society.

It is an ambitious project for which I have relied on the writings of many inspirational experts and the publication of research results in a multitude of scientific and medical disciplines. By necessity, I will climb upon the shoulders of these giants to borrow their hard-won insights and develop a compelling narrative. I will state explicitly that viruses are just matter with informational content and do not meet criteria to be considered living organisms. That these entities can trigger and regulate such complex cascades of molecular events is wont to be compared to human qualities, but strictly speaking viruses have no motives, needs, or strategies. This imbues them with a life that they do not have. Nevertheless, I will occasionally make excursions into this dangerous anthropomorphic territory. I do so consciously believing it useful representational language to effectively communicate complex and nuanced concepts and context while maintaining the momentum of the narrative form. Finally, the dialogue I undertake is not intended to be a comprehensive text on viral evolution and should not be construed as such. There are many excellent source textbooks that already serve as gold standards in this regard and I recommend them as reference materials (Knipe and Howley 2013; Flint et al. 2015). I have chosen just a few destinations to explore intensively and to best illustrate viruses as agents of evolutionary invention. The important concepts illustrated in this work can be easily extended to understand other viruses and they will place the reader in a position to approach further reading on these fascinating topics with an appetite and a prepared mind. At the end of the journey the reader will have a new appreciation of viruses as life's greatest assets and most feared predators.

· 1 ·

OBLIGATE PARASITES *of* CELLS

THE STORY OF VIRUSES begins in 1879 at the Agricultural Experiment Station in Wageningen in the Netherlands. In the mid-part of the nineteenth century, a disease ravaged the tobacco crop. It was so severe in some regions that it "caused the cultivation of tobacco to be given up entirely" (Zaitlin 1998). Adolf Mayer, christened the disease tobacco mosaic disease, as it manifest in darkened patches on the leaves of the tobacco plants. Mayer was looking for the cause of the disease when he observed that the juice extracted by grinding up the leaves of a diseased plant could pass on the disease to a healthy plant. He rightly concluded that a transmissible infectious agent was responsible for the disease of the tobacco crop. However, his experimental results did not suggest to him that the agent was anything other than a microbe.

In 1892 in St. Petersburg, Russia, Professor Dimitri Ivanowski demonstrated that the same transmissible agent could pass through a porcelain filter. The filter, invented by Louis Pasteur and Charles Chamberland, was designed to have a pore size that retained bacteria, since it permitted only particles smaller than 0.5–1.0 microns in diameter to pass through. Ivanowski's results ruled out bacteria but he concluded that the disease-causing agent was most likely a bacterial by-product or a toxin. A few years elapsed before Martinus Willem Beijerinck, a Dutch scientist, refined the concept of the infectious principle. It was certainly not a bacterium. It could not be coaxed to grow in the laboratory in a nutritional medium that typically supported bacterial growth. He proposed that the infectious agent required close association with the metabolism of living plant cells for propagation. The infectious principle evidently depended on them for growth. He described the clear infectious filtrate as *contagium vivum fluidum*—a "contagious living fluid" (Bos 1999).

At the turn of the century, scientists had no tool other than the Chamberland filter to describe the physical nature of viruses. They were infectious entities small enough to pass through its pores, defined only by their diminutive size. It would take another forty years before tobacco mosaic virus particles themselves would be isolated and described as an "enzyme-like protein," and later characterized as a nucleoprotein, a particle containing both protein and nucleic acids.

Some twenty years after these first observations of a virus infecting a plant, another tandem effort of scientific discovery revealed viruses that infect prokaryotic cells. The English doctor Frederick Twort was studying the bacterium *Staphylococcus* because it was a frequent contaminant of cowpox lesions that he collected for use in the preparation of smallpox vaccine. While examining the bacterium in culture, he observed clear patches on the surface of the small bacterial colonies growing on his culture plates. He interpreted them, quite correctly, to be the result of the destruction of bacterial cells and hence a disease of the microorganism. He found that the "disease" could be passed from one colony to another and that the agent passed through a filter, just as Beijerinck had observed for the infectious agent of tobacco mosaic disease. Although Twort believed the disease-causing principle which destroyed bacterial cells was probably an enzyme or toxin, the key properties of a virus were met (Twort 1915).

Perhaps Twort did not recognize the real significance of his observations, but Félix d'Hérelle, a Québecois scientist working at the Pasteur Institute in Paris, soon did. He advanced the discovery of bacteria infecting viruses one step further. He observed a filterable "antagonistic microbe" that killed *Shigella dysenteriae*, rendering the bacterial cultures clear. D'Hérelle wrote, "The disappearance of the dysentery bacilli is coincident with the appearance of an invisible microbe . . . [it] is an obligate bacteriophage" (D'Hérelle 1917). This was the first use of the term "bacteriophage," which means "bacteria eating." He had discovered what we now know to be the group of viruses that make up the vast majority of the virosphere. They are parasites of prokaryotes, the organisms that comprise the ancient bacterial and archaeal domains of life.

Although Ivanowski, Beijerinck, and Twort grappled with the nature of the infectious agent—a bacterium, a toxin, or an enzyme—today there are a wealth of biochemical, physical and molecular descriptions of viruses. A dictionary definition of *virus* might read: infective agent that typically consists of a nucleic acid molecule in a protein coat, is too small to be seen by light microscopy, and is able to multiply only within the living cells of a host. This is an apt description of a virus, but it has some shortcomings. The use of the qualifier "typically" is prescient. Most viruses do adhere to these principles, but there are notable exceptions. Some viruses get along just fine without a protein coat and some have particle sizes larger than some bacteria (refer to Chapter 8). To formulate an understanding of their fundamental nature, it is worth exploring a more refined and inclusive definition of viruses.

The Virosphere and Its Metagenome

The virosphere is the collective of all viruses in all ecosystems, and in all hosts in the biosphere. Notionally, when we think "virus," we think of virus particles and their nucleic acid contents. It is the nucleic acids, ribonucleic acid (RNA) or deoxyribonucleic acid (DNA), which are polymers of either ribonucleotides or deoxyribonucleotides that constitute the essential genetic blueprints of viruses. The genetic code of the nucleic acid that makes up the viral genome contains the information fundamental to its distinct identity. Just as different species of living organisms have different genetic blueprints recorded in their genome sequences, so too do viruses. Today it is possible to visualize different viruses under an electron microscope. This may well reveal particles that are indistinguishable in shape and size, which can belie their differences; their unique identity is in the information encoded in their genomes and it may be distinctly, even radically, different. The true diversity of the virus world can only be realized when their genetic contents—their individual unique bar codes—are cataloged and compared. It is therefore useful to consider the virosphere not simply in terms of the collective of distinct species of viruses but as the collective of their genetic informational content—the *viral metagenome*.

A *metagenome* catalogues the collective genomes of all of the organisms, which can be recovered from an environmental sample. An "environmental sample" may be a gram of soil, a milliliter of seawater, or an organism, each of which represent distinct ecosystems. The most inclusive use of the term collects the genomes in the biosphere, and this includes the genomic information of all living organisms and their viruses. The human metagenome captures the collective of genomes associated with it and therefore includes not only our own genome sequence, but also those of the organisms making up the microbiota that shares our body space. These symbiotic bacterial and archaeal cells constitute the human *microbiome* and occupy our external surfaces—our skin, the mucosal epithelia of the gut, the nasal and oral cavities, and our genital tracts. The human *virome* is the aggregate of viruses that infect both our own body cells and those of our microbial passengers. Their respective gene complements would be considered their metagenomes.

The study of metagenomes has been made possible by major technological advances in molecular biology. It is rooted in our ability to read and interpret the nucleotide sequences of the genetic material of organisms and viruses in a given sample. Prior to this development, the recognition and identification of microorganisms and viruses in a given sample was strictly limited to those that could be grown in culture or directly observed under the microscope. Today the detection of nucleotide sequences in even tiny samples of environmental or biological material can be used as effectively as a fingerprint to identify microbes and viruses.

Over the last decade researchers used these tools to probe for potential links between the composition of the human microbiome and health and disease. It is estimated that this microbial community is made up of 75 to 200 trillion individual microbes—a number comparable to the 100 trillion cells that make up the human body. An equally astonishing fact is that for each of the trillions of microbial cells there may be tenfold more viruses! This population of viruses—largely bacteriophages (phages for short)—is the major contributor to the human virome. The remaining contributors to the human virome are viruses infecting our own cells, human viruses. Although still poorly understood, the three-part interplay between the human body, our microbiome, and our virome is increasingly considered central to our health, and very often to our diseases.

Key tools in the exploding field of metagenomics are new generations of DNA sequencing technologies and sophisticated computational tools. Scientists can determine the nucleotide sequence of trace amounts of DNA from multiple organisms in a single sample. It is no longer necessary to culture the organisms separately and isolate the DNA from each organism. Massively parallel DNA sequencing allows complex mixtures of DNAs to be sequenced simultaneously. Together with sophisticated bioinformatic algorithms, the different DNA sequences and their relative abundance in the sample can be determined. Once it became unnecessary to culture organisms to characterize and catalogue their genome sequences, the principal barrier to researching the biology of our microbiome was overcome. In fact, though the vast majority of microbial species that make up the microbiome cannot currently be cultivated outside the body, today massively parallel sequencing can identify which organisms are present and in what abundance in samples of the gut microbiota. A key factor facilitating this analysis is that, without exception, the chromosomes of cellular life-forms encode genes required to build ribosomes. These are the biological machines responsible for interpreting the messenger RNA templates and manufacturing proteins from amino acids. The 16S ribosomal RNA (rRNA) genes of the small subunit of the prokaryote ribosome have been particularly well conserved throughout evolutionary history. Small differences in the sequence of these highly conserved genes allow accurate deduction of phylogenetic relationships between bacterial species. Comparing these unique "fingerprint" rRNA gene sequences with DNA sequences stored in genomic databases, researchers rapidly identify the bacterial or archaeal species in a given sample. The frequency of the particular rRNA gene sequence in the DNA sequence data indicates its relative abundance in the sample.

Unfortunately, no such tool exists to assist viral metagenomics. Accordingly, its progress lags behind microbial metagenomics. We cannot classify viruses in a given sample using the same approach used for prokaryotes. Virus genomes have no rRNA genes since they do not encode their own protein synthesis apparatus. Furthermore, virus genomes exhibit an unprecedented and quite remarkable diversity of genes and gene sequences. In fact, there is not a single gene or descendant of a single gene that can be found in all virus genomes; no unique viral fingerprint can be used to deduce their presence in a sample and determine their phylogenetic relationships.

Families of related viruses do, however, share similar replicative strategies and consequently have in common certain types of enzymes or structural proteins that are intrinsic to their respective lifestyles. Such genes have nucleotide sequence similarities that allow deduction of viral lineage relationships. Integrase proteins are examples of viral enzymes possessed by many different viruses. Although they can be quite different and highly divergent in amino acid sequence, integrase-related proteins are found in most viruses that integrate their genome into the host chromosome as part of their life cycle. Equally, many virus families employ capsid proteins. Despite the genetic diversity in the viral world, only three different structural templates for capsid proteins have been observed. It appears that only a limited number of viable solutions to the "problem" of virus capsid construction have evolved. Capsidated viruses all have related capsid proteins patterned on one or another of these three different three-dimensional templates. It is these protein amino acid sequence signatures, together with powerful computational tools, that the viral genomics scientist relies on to divine the origin and relatedness of viral sequences in a sample. It is not an exact science, and is complicated by the fact that only a fraction of viral sequences has been catalogued and recorded in genomic databases. It is also confounded by the rapid evolution of viral genes, as well as by the promiscuity of viral genetic information, frequently exchanged, lost, and gained. It is fair to say then that any assessment of the complexity of a viral metagenome is likely to be an extremely conservative estimate. Our computational methods detect similarity between a sample viral nucleotide sequence and those in existing viral databases. Truly novel sequences or those that may have evolved to have no perceptible similarity to known virus genes cannot be definitively assigned to a virus species.

Today, scientists are exploring the viral metagenomes in a variety of ecosystems. It is no great technical challenge for researchers to enumerate viruses in natural bodies of water. Quantitation of nucleic acids recovered from virus particles, isolated by passing samples of ocean water through a 0.5 micron filter, revealed an astonishing fact: each milliliter of seawater teems with 1 million microbial organisms, but there are 10 to 100 million viruses in the same sample (Bergh et al. 1989). The ocean at the seaside is a solution of virus particles. Conservatively then, it can be estimated that the virosphere is composed of 10^{31} individual viruses and they are the most abundant biological entities on earth, outnumbering the *Bacteria* and *Archaea* by a factor of 10 (Brüssow and Hendrix 2002; Suttle 2007; Breitbart and Rohwer 2005). To an alien with the sensorial ability to detect both the microscopic and the macroscopic world, we and the other members of the *Eukarya* would be lost in the earthly crowd—we are that tiny a minority in the planetary community.

Complexity and "Dark Matter"

The diversity of viral genetic information recovered from environmental samples is quite simply, astonishing. The field of marine viral metagenomics emerged when researchers identified viruses in seawater using massively parallel DNA sequencing. Since the first decade of this century the field has advanced rapidly. Professor Forest Rohwer, a marine ecologist at the San Diego State University in California, is one of the pioneers of the field. He and his collaborators were some of the first scientists to exploit technologies for viral metagenomic analysis. In 2006 they reported one of the most comprehensive global studies of the marine viral metagenome (Angly et al. 2006; Suttle 2007). They collected and analyzed samples from more than sixty sites in four oceanic regions, sequencing virus DNA from the waters of the Gulf of Mexico, coastal western Canada, the Arctic Ocean, and the Sargasso Sea. Their studies cracked open the door, allowing a first view into the inscrutable world of marine viral populations and their ecology. With data from numerous other expeditions, a coherent picture has emerged, revealing that viral populations in the oceans are extremely diverse (Suttle 2007). Among the trillions of virus particles found in 100 liters of seawater, there are many thousands of distinct viral species, each with a discrete genetic blueprint, or genotype. More than a million different genotypes can be found in 1 kilogram of sediment taken from the ocean floor. Most phages are widespread and found around the globe—they are everywhere—but their relative abundances in different locations varies a great deal. Different environmental conditions must therefore have a profound influence on the prevalence of each virus and class of viruses found in different locales (Angly et al. 2006; Breitbart and Rohwer 2005).

Although the oceanic virome is the most intensively studied to date, an increasing wealth of research explores the virome in other ecological niches. Metagenomes of halophilic or thermophilic bacteria and archaea are studied in salt lakes and hot springs. The hypolythic microbial communities on the underside of translucent rocks in the hyperarid Namib Desert offer opportunity for study (Adriaenssens et al. 2014). Viruses in these collective ecosystems make up the numerical majority in the virosphere. They predominantly infect bacteria and archaea, but interact with them in many different ways. They exploit wide-ranging strategies, which have one thing in common: the singular goal of replicating and perpetuating their genetic information. Their genomes, encoding all the information that dictates their lifestyles, can be made of RNA or of DNA and may take many different forms: single- or double-stranded, linear or circular, or even segmented. They can in some rare instances be shared (when the genetic information in the viral genome is insufficient and the missing genes are provided by a second helper virus genome). Viruses have the potential for rapid evolution. This is possible due to a number of factors, including the huge complexity of virus populations and their short generation times resulting from fast iterative replicative cycles. Here, the genetic complexity to which I refer is a reflection not only of the large number of individual viruses but also the large diversity in their genetic information content. Another major catalyst of viral evolutionary rates is the promiscuity and the proficiency with which they exchange genetic information, both with each other and with their hosts. Finally, the error-prone nature of viral replication turns out a multitude of inexact copies whose mutations contribute to the genetic diversity of the population. Viruses ride on swift evolutionary currents, propelling their own evolution and the adaptive evolution of their hosts. The viral metagenome is a veritable smorgasbord of useful genetic functionality. It evolves in service to the success of the viral genome but if assimilated by hosts it can provide new competitive advantage for survival in changing and hostile environments.

Scientists estimate that each second, 10^{25} phages initiate infection of a host cell—each of these viruses breaks down into component parts before its genetic blueprint directs the manufacture of replicates. A physicist seeing this phenomenon would likely have in mind the second law of thermodynamics, noting the total mass of the phages and the large scale of thermodynamic energy release and expenditure that these cycles of infection must entail. The breakdown of the organized phage particles must result in a net release of thermodynamic energy and increasing disorder. The rebuilding of more virus particles requires an even more substantial input of energy, which the viruses must capture from the metabolic machinery of their host cells. A biologist, on the other hand is more likely to ponder the enormous biomass involved and how these cycles of infection must have an impact on the various ecosystems, affecting the flux and availability of nutrients in the food chain. A geneticist examining the system will note that when the genomes of the 10^{25} phages replicate, countless mutants will be created. These variants arise when the genetic information in the phage genomes is copied incorrectly and when pieces of genetic information are lost or exchanged with other phages simultaneously infecting the same host cell. Sometimes, host cell genetic information is added to the phage genome and can later evolve as part of the phage identity and serve the purpose of the virus.

This "combinatorial biology" is occurring on a grand scale. It began 3 billion years ago and it is the lifeblood of natural selection, which operates on the unimaginable diversity of genetic information in the viral metagenome. It has created and sustained a world of viruses that exploit a vast variety of replication strategies and relationships with their host cells and host cell populations. The almost infinite number of genetic variants in the virosphere that can be "prototyped" and road tested has allowed viruses to extensively sample and explore "evolutionary space." They can be likened to lotto players with unlimited resources who can purchase every ticket; if there is a winning number, they will have it. All that is needed is one winning number that represents the prototype with an advantage. This underlies the enormous potential of viruses to adapt quickly to changes in their hosts and the enormous success that they have enjoyed preying upon every niche of the domains of life.

The genetic information encoded in the genomes of phages makes up the majority of viral genetic information in the virosphere. If we are to consider all of the genes currently identified in all biological entities in the biosphere, phage genes comprise the vast majority. In 2003 scientists exploited a computational algorithm called Chao1, mining data from all of the identified phage gene DNA sequences in Genbank. This global repository of publicly available nucleotide sequence data is maintained by the National Center for Biotechnology Information, part of the U.S. National Institutes of Health in Bethesda, Maryland. The conclusions were extraordinary. The scientists were able to extrapolate that 2 billion phage genes have yet to be discovered (Rohwer 2003). Taking into account the number of phage genes and their sequences recorded in the Genbank database at the time, this meant that more than 99.9998 percent of the phage metagenome had yet to be sampled (Rohwer 2003). Indeed in the numerous oceanographic surveys of the marine phage metagenome, a large percentage, more than 75 percent, of the sequences collected, do not appear in any existing database and cannot be identified—these DNA sequences have been termed "dark matter" (Breitbart et al. 2002; Pedulla et al. 2003). Who knows what gene treasures will be uncovered in this pool of genetic information: the processes of evolution will certainly use it. These new genes could allow for the emergence of new phage lineages or new virulence traits. Perhaps humans could benefit by using them to fashion new biotechnological tools or medicines.

Viruses are obligate parasites, replicating within a living host cell. Natural selection can act only on the outcome of an infection and the success of the emerging progeny viruses. The virus-host cell relationship is one of reciprocity and natural selection also operates on the host itself. The survival of the host cell is a measure of success for the host genome. This symmetry of natural selection on both the virus and the host genetic information creates the phenomenon by which viruses and their hosts closely coevolve to forge the most mutually satisfactory relationship. The use of the term *mutually* here is gratuitous, as the genomes of both organisms are in conflict, striving for the best outcome for themselves in terms of replicative success. It is commonly referred to as "an arms race" or "Red Queen dynamics" between prey and predator, the host and its invader (Van Valen 1973; Dawkins and Krebs 1979). The allusion to Lewis Carroll's Red Queen is particularly apt, as she tells Alice, "here, you see, it takes all the running you can do, to keep in the same place" (Carroll 1871). Both the virus and host must constantly evolve; each time a new variant of one partner emerges to become more successful to the detriment of the other, the other partner must equal the balance through selection of a countermeasure: punch and counterpunch. In this way extraordinary and mutually complex relationships evolve between the virus and its host. The evolution of viruses is therefore inseparable from the evolution of their hosts. With viruses so abundant and globally pervasive, and with the viral metagenome harboring a diverse reservoir of genetic information to fuel evolution, we can readily conceive the enormous influence that they wield in our ecosystems.

Selfish Information and the Essence of Being Viral

Earlier I called for a definition of viruses based on their essential nature, rather than simply on their composition or size. In 2008 Raoult and Forterre proposed the division of all biological entities into "two groups of organisms: ribosome-encoding organisms, which include eukaryotic, archaeal and bacterial organisms, and capsid-encoding organisms, which include viruses." The three domains of cellular life, *Eukarya*, *Bacteria*, and *Archaea*, all possess the capacity to synthesize proteins and to conduct their own metabolism. As discussed earlier, the molecular machinery of all living cells responsible for assembling proteins from amino acids has at its core the ribosome. Viruses have no such machinery and rely on ribosome-encoding cellular life-forms to make proteins for them. On the other hand, cellular life-forms do not appear to require, nor do their genomes encode, self-assembled capsids. This definition goes a long way toward distinguishing viruses from living organisms based on their distinct gene content. While accurate, I believe it is still inadequate because it fails to capture the quintessential nature of viruses. It also ignores the existence of many biological entities that share the essential features of viruses but lack a capsid gene. In some cases their evolutionary origins can be traced directly back to capsid-encoding ancestral viruses.

These exceptions are viruses that adopt a variety of lifestyles. Some are rather simple genetic replicators that appear to be evolutionary relics—primitive replicons called viroids—which may have originated early on and persisted through evolutionary time (Chapter 8 will discuss conjectures on the origin of viroids). Others may be mobile genetic elements capable of replicating themselves and moving to new host cells by a variety of mechanisms. In some instances, these mobile genetic elements are found in the genome of other viruses and can hitch a ride between cells as part of the chromosome of their ride. In other instances these elements ride between host cells alone, but within a viral capsid provided by a collaborating "helper virus" that infects the same cell. In summary, the absence of a genome-encoded capsid does not preclude it from replicating its genetic information and even from moving between host cells.

The ribosome versus capsid-containing definition is therefore only partially informative as to the true "essence of being viral" as it excludes some biological entities that behave as viruses. A more useful definition is needed, one that recognizes the origins and evolution of viruses and more effectively captures their nature. My opinions are informed by the writings of the eminent bioinformatician Eugene Koonin, who, among others, has written extensively on the evolution and phylogenetic relationships of viruses. Koonin described the virus world as one of "viruses and capsidless selfish elements" (Koonin and Dolja 2014). This does capture the viral world in its entirety. The definition also is more inclusive than the aforementioned ribosome versus capsid-containing definition. For myself, however, I see no value in specifying "capsidless selfish elements" separately. I argue that viruses are simply quintessentially selfish elements, parasitic genetic information, regardless of whether or not they encode a capsid. The virus is the information itself, but this information only constitutes a virus if it can sustain transmission between hosts.

Minimally this genetic information is sufficient to direct its own replication in a host cell and to be mobilized to a second cell. Many virus-like elements found in the chromosomes of prokaryotic and eukaryotic cells are capable of mobilizing their nucleic acids within the cell, amplifying themselves and proliferating in the cellular chromosome. Similarly, plasmids of prokaryotes are genetic elements that replicate in bacterial cells and are vertically transmitted during cell division and transmitted between cells during bacterial conjugation. Neither of these classes of replicating elements achieves transmission between cells by an extracellular infectious entity. Under this definition such elements should not properly be considered viruses. Viruses are certainly selfish genetic elements, but they must also be intrinsically infectious entities: natural selection on a viral lineage acts on the genetic information that undergoes transmission between cells. Genetic variants are tested at infection. Viruses are defined by the information that is transmitted—the viral genotype.

Consider for a moment, a useful definition to describe "selfish" genetic elements. Fundamental to such an element that replicates within a cell is that it undergoes genetic variation and can evolve. Natural selection acting upon it is independent of the powers of natural selection operating on the host cell genome. Each have their own interests at heart, evolving independently, and often in conflict, while under distinct selective pressures. It should not escape our attention that a virus with no available host has no infectious potential, and that a selfish genetic element which kills all of its potential host cells is also lost to evolution—paradoxically the success of these parasitic genetic lineages is inextricably linked to the survival of the host lineage. But evolution has no plan; some lineages of selfish genetic information (including viruses) must have become extinct as a result of this evolutionary conflict. Viruses then are a subset of selfish genetic information, independently evolving and infectious.

Coevolution of viruses arises from the genetic conflict between virus and host and is the process by which a viral lineage and a host cell undergo adaptive evolution to optimize their respective prospects. Often, the host cell adapts to ameliorate the damage caused by the viral parasite, and equally, less virulent viruses may evolve from this relationship. This outcome is not a "desire" for mutual harmony; each has only selfish motives. The virus and the host must tally their evolutionary success independently. It is indeed an arms race, and in many instances, the race in our evolutionary snapshot of time is a standoff. In other instances, notably those involving newly emerging human pathogenic viruses, we have not yet achieved even a quasi-stable equilibrium in our relationship with the virus. As alluded above it is likely that some virus lineages have taken their host species to the grave, leaving viral genetic information whose potential can no longer be realized.

At this point I have used the term *selfish* several times—drawn from Koonin's definition of the virus world (Koonin and Dolja 2014). Informed by the many eloquent writings of the scientist-author Richard Dawkins, most notably his now forty-year-old work, *The Selfish Gene* (Dawkins 1976), today we recognize that the smallest units upon which natural selection acts are inherently selfish. For the sake of discussion, Dawkins chose the gene as this smallest unit, but it may equally be a locus, a polymorphism within a gene, or an extragenic sequence with a function that can be subject to selective pressures. Within Dawkins's framework of argumentation each gene within a virus genome is itself selfish. This is hard to dispute; variants of virus genes are obviously in competition to survive and prevail as the dominant form of that gene locus in the viral population. Purifying selection of deleterious gene variants with the emergence of more successful genes in the virus populations is highly effective. Each new prototype virus variant is road tested independently when each single virus particle initiates infection of a host cell. The fittest virus variants quickly outpace their less fit brethren and come to dominate the population.

Notwithstanding that viral genes competing in a viral gene pool can be equated with Dawkins's selfish genes, I used the term *selfish* in a different context when describing independently evolving elements and a key distinction needs be made. While selfish genes of the host cell evolve exclusively to the benefit of their organism, their "gene survival machine" in Dawkins's parlance, viral genes evolve in conflict and independently of their cellular gene survival machine. The selective pressures acting on viral genes and determining virus fitness and those acting on cellular genes and determining host fitness are quite distinct. Viruses and their genes evolve independently of the host and host cell genome. The success of a cellular gene is immutably linked to the success of its host cell, however, virus genes are under no such constraint. Viruses are obligate parasites of cells, but act egotistically and without regard to the success of their host's genome: they are *egotistical, independently evolving infectious information.* This egotism of viruses fuels the genetic conflict that is central to all evolution of viruses and their hosts. In later chapters you will be introduced to the fate of viral genes that become integrated into the host cell chromosome, losing the capacity for replication and transmission between hosts. From that moment onward they are simply Dawkins's selfish genes, their success inextricably aligned with that of the host gene survival machine.

Most of us understand and readily acknowledge that natural selection is the principal force governing the evolution of life on earth. Equally, although not fulfilling the classic criteria to be living organisms, viruses have evolved under the same laws of natural selection, first posited by Charles Darwin in 1859 in his book *The Origin of Species* and refined in the twentieth century by Huxley, Fisher, Haldane, Dobzhansky, and others in their modern synthesis of evolutionary theory. Why are viruses not considered alive? The key distinction of life and nonlife is that living entities have an ability to autonomously generate energy from sunlight (in the case of photosynthetic organisms) or from complex energy-rich compounds taken in from their environment. The self-sustaining cellular metabolism of life is the discriminating factor. Viruses are obligate parasites of life-forms, dependent on them for energy, infrastructure, and raw materials; a definition of a virus as *egotistical, independently evolving infectious information* does not in and of itself require them to be life-forms. Can natural selection act on nonliving entities if we accept viruses to be such? The answer is fundamentally yes. These natural laws are expected to operate if the following criteria are met: (1) the entity upon which natural selection acts must have the capacity for replication; (2) there must be the capacity for variants of the entity to emerge; (3) there must be an environment in which competition exists and where the best replicators have an advantage and proliferate more than their less successful kin. In the case of viruses, however the lines are blurred and these criteria are only met when the virus hijacks the cell's machinery to direct its own replication. The dynamism of viral replication in a host cell is remarkable; some have argued that the infected cell itself should be considered the independent living entity associated with viruses and have dubbed it the *virocell* (Forterre 2013). It is intriguing to consider viruses as biological entities that fall between inert matter and information on the one hand and autonomous life-form on the other. We can recognize that it is the blueprint, in our case the genome of the virus, that undergoes variation and is subject to the laws of natural selection acting on its capacity for replication and transmission. It is, however, impossible to argue that absent a living host cell, a viral phenotype can be expressed. The knot therefore appears to be woven tight: as entities, viruses are certainly not autonomous and living, but there is remarkable potential bound up in the information content and ordered energy rich virus particle. This potential is released on contact with a living host cell, when the vitality of the virus is revealed.

It is arguable that the notion of a virocell does not usefully advance our concept of the essence of viruses and whether a virocell should be considered a separate biological entity is doubtful. The concept of a virocell has been extended by some to explore the possibility that some viruses (discussed in Chapter 8) might have evolved *from* a now-extinct fourth domain of life. These hypotheses will be laid to rest. It serves best to acknowledge that virus information fulfils its potential and comes closest to being alive when it is exploiting the metabolic functions of its host cell. Nevertheless, it is the transmissible egotistical and independently evolving genetic information that constitutes the virus.

The Emergence of Egotistical Replicators

How will these arguments hold up when we contemplate the earliest steps in the evolution of life and how we believe that the virus world emerged? It is a generally agreed notion that viruses emerged before the evolution of true cellular life-forms (Koonin and Dolja 2014). The criteria for natural selection were presumably first met when simple primordial elements, the precursor genetic elements, probably RNA-based, developed the capacity for self-replication. Presumably, a process including error-prone replication and the linking and exchange of simple elements to form a more complex species became the subject of natural selection. One can envisage that initially there would have been many replicators, composed of distinct genetic elements and replicating primitively in different fashions. Natural selection would come in to play when these replicators began to compete for resources, perhaps the availability of chemical building blocks, or when conditions changed to favor one or another class of replicator. This would reveal differential fitness for replication and favor the predominance of the most successful, and therefore selected, species of replicator. Let us accept for the moment that pre-cellular replicators are the first and precursor life-forms. Then how did viruses emerge? The favored solution is that they segregated from these early precursors of life as they began to develop more complexity and functional modularity. Parasites likely arose as replication-incompetent elements evolved to exploit the chemistry of the replicators and be themselves replicated. After this initial establishment of a parasitic relationship, natural selection could operate independently on both the parasitic, nonautonomous replicators and on the autonomous host replicators. Thus, separately evolving lineages formed. It is perhaps provocative to suggest the possibility that the parasitic nonautonomous replicators, the precursors of viruses, actually started out as members of the replicator population but segregated from them by loss of information, leading to defective replicators which then became obligate parasites exploiting the same chemistry as the autonomous replicators. To follow this line of thought we must conclude that it is possible that the first viruses got their start as early precursor life-forms, but lost the capacity to replicate autonomously as they continued to evolve in parallel with their hosts, which ultimately assumed all the accoutrements of what we today consider to be life.

The Viral Empire

As a final topic of introduction to viruses as egotists and as vehicles of selfish parasitic genetic information, it is worth reiterating that while they are indeed inert in every sense, they are unique in their ability to reinvent themselves in every cell that they infect, starting only from a blueprint of RNA or DNA. Living cells are incapable of this feat. Although we commonly associate the process of natural selection with the evolution of living organisms, these inert biological entities are definitively products of natural selection that evolved together with life to become the most abundant and diverse replicators on earth. As we will see, their capacity for rapid evolutionary adaptation has allowed them to penetrate every domain of life where they have been potent catalysts of the evolution of their hosts. Their profound influences in forming and maintaining the earth's ecosystems today and on our health and disease is reminiscent of the influence of the great empires in history, which exercised pervasive and lasting influences spanning the globe's geography and all of its cultures. The "viral empire" may be inert matter, incapable of animation absent help from a living host cell, but it should not be underestimated. Its potential to continue to evolve in an egotistical fashion, without a thought for humanity, remains today. For viruses, it is still a work in progress.

· 2 ·

VIRUSES, GENES, *and* ECOSYSTEMS

OUR BY-THE-NUMBERS consideration of the virosphere in the previous chapter illuminates the refinement and complexity of these minimalist vehicles of genetic information. It does not, however, capture the varied repertoire of replicative strategies and relationships that they establish with their host cells and host cell populations. Nor does it speak to the evolutionary processes that shaped viruses and how the viral metagenome has influenced the evolution of living organisms in all domains of life, including their ecosystems. Here we will begin to move in these directions.

We begin with the most abundant contributors to the viral metagenome—phages. They have ancient roots and infect some of the most primitive, yet clearly durable and successful genetic lineages: the *Bacteria* and *Archaea*. By virtue of their historic relationships, they exhibit some of the most elegant examples of viral host coevolution. Let it be clear, as we move through these pages, we are simply viewing the viral world and its relationships with living cells as they are today. We may infer perhaps how they *were* but certainly cannot predict how they *will be* in the future. The diversity of the viral metagenome, hidden in genetic "dark matter," is a stark reminder that there are yet many opportunities for evolutionary change catalyzed by the viral world. It will have ramifications for viruses themselves but also most certainly for their hosts and the ecosystems that they inhabit.

Lifestyles and Life Cycles

By virtue of their abundance in natural environments, phages play an important role in shaping our global ecosystems. This is true for the tailed, double-stranded DNA phages that constitute the very large and diverse order of viruses aptly named *Caudovirales* (tailed viruses). Tailed double-stranded DNA phages are the oldest known group of DNA viruses. Infecting both *Bacteria* and *Archaea* (Krupovic et al. 2011), they are prominently represented in the viral metagenomes of all of the ecosystems that have been surveyed to date. They have in common a genome composed of double-stranded DNA contained within a diminutive icosahedral protein capsid, usually less than a tenth of a micrometer across, which is endowed with a tail spike. This similarity in morphology is the basis for classifying this group of phages together as *Caudovirales*. The group is immensely diverse; genes with similar functions have often diverged dramatically in amino acid sequence during evolution from their common ancestor. Moreover, the size of their genome varies from less than 18,000 base pairs, barely sufficient to encode 30 proteins, to almost half a million base pairs, large enough to encode some 675 proteins. The very small phages encode the bare-bones equipment to replicate in a host cell. They are "dragsters," stripped to the chassis for speed, while the phages with large genomes are "luxury sedans," fully loaded with indulgences. These differences are plainly a result of radically divergent trajectories of evolution in different lineages of the double-stranded DNA phages in various hosts. We must conclude that the supplementary complement of genes in the larger-phage genomes has conferred upon them a competitive advantage. The supplementary functionalities that they possess must have improved the replicative success of the respective phage lineages in their individual niches.

Caudovirales, the oldest of phages, infect autotrophic as well as heterotrophic prokaryotes (Hendrix, Hatfull, and Smith 2003). *Cyanobacteriae* (once known as blue-green algae), able to fix carbon by oxygenic photosynthesis, are autotrophs and the oldest bacterial primary producers known. Heterotrophs depend on the primary production of other life-forms and take up organic compounds from the environment to fuel their production of energy. It follows that these phages infected the common prokaryotic ancestor of the bacterial and archaeal domains of life. Furthermore, it is not wild conjecture to suggest that tailed phages in the oceans were, once upon a time, the principal predator of prokaryotic life. Their evolutionary origins certainly predate the emergence of other predators, such as single-celled flagellates and ciliates, which today are major consumers of primary and secondary producing organisms in our oceans. This long-standing predator-prey relationship between phages and their hosts has endured throughout the evolution of cellular life. With their aptitude for rapid adaptation under changing selective pressure and in response to the evolution of their hosts, phages forge a diverse array of intricate and highly evolved relationships.

Some tailed phages, called *lytic* or *virulent* phages, behave simply as predators of prokaryotic hosts. Following invasion of a cell, they quickly direct the expression of their own genes, allowing them to take over the metabolism of the cell. Phage genomes and structural proteins are synthesized, virus particles assembled, and the host cell broken apart (or lysed) to release hundreds of replicate virus particles. In other instances, much subtler relationships unfold. The invading phage may elect to postpone the fatality of its host, eschew its lytic replicative cycle, and become a symbiont of the host cell, inserting its DNA into the cellular chromosome. The resulting *prophage* maintains itself, essentially dormant and harmless to the cell, behaving as part of the cellular chromosome, being replicated only when the host cell divides. In return for this temporary protective custody within the cell, prophages provide their host with immunity to infection by related phages. They may also supplement the host cell genome with new and useful genetic information, providing the cell new tools as a competitive advantage over its uninfected counterparts. Natural selection acts upon phage genomes and their hosts independently, but after a multitude of phage-host encounters, a state of mutualism often evolves benefiting both the cell and the virus. Under these circumstances both have an improved probability of survival and of perpetuating their respective genetic lineages.

To interpret the often-complex interrelationships of viruses with their prokaryotic hosts, we need to consider the process by which phages invade their hosts and commandeer the infrastructure. Their objective is, of course, to accomplish their selfish goal of replication. This is the prime directive of viral genomes, but it equally applies to each of their genes and indeed to all genes of living entities. Phages are viruses that only infect prokaryotes, unicellular organisms that lack a nucleus. Although these are the simplest of life-forms, their phages exhibit a remarkably broad spectrum of lifestyles and infection strategies that recur throughout the viral world. The exquisite coadaptation of host and phage is the product of a long history of coevolution potentiated by the capacity of phages for rapid genetic innovation.

The canonical virus is "an infective agent typically composed of a nucleic acid molecule in a protein coat." The infectious agent itself is the virus particle, the entity that Ivanowski and Twort first observed to pass through the pores of a porcelain filter (Bos 1999; Twort 1915). The particle is the form of the virus released from infected cells, free in the environment; it is simply a nucleoprotein, inert until it chances upon a suitable cell to infect. A phage particle that languishes in the environment for too long will be inactivated or naturally decay as a result of exposure to adverse physical or chemical conditions. Inside the capsid is the nucleic acid genome or chromosome that encodes the heritable information, defining the virus, its structure, and its processes. The protein components of the viral particle have two primary purposes: (1) They must protect the valuable contents from the environment since the longer the virus, along with its genetic payload, survives in the environment, the more chance it will have to infect a new host cell. (2) They must also mediate attachment of the virus to a host cell and the passage of the viral chromosome across the cell wall and cell membrane to the interior cytoplasm. With some exceptions, phages deliver their naked genome into the cell while disposing of their capsid at the exterior. Every cycle of phage replication starts from genetic information alone. Viral progeny are essentially recreations of the original virus built from its genetic specifications. They are identical copies of their progenitors and, save the genetic variation that drives evolution, each behaves in an identical fashion.

The most ancient independently evolving egotists contemplated in Chapter 1 were simply information-encoding elements in the primordial soup that remained separate from (but depended on) other information-encoding elements for their replication. At some point, the precursors of unicellular organisms must have evolved to replicate more successfully within an organized structure (Koonin and Martin 2005; Woese 2002). Presumably, this would be advantageous to the clonal expansion of the replicator by allowing it to maintain the necessities for replication within close proximity, perhaps within a boundary layer. It is only conjecture to speculate on the order or the precise events, but it seems most likely that virus precursors coevolved with their hosts. It is also possible that viruses started out as pieces of accessory replicons within these structures but at some point evolved an "extracellular phase." Having the ability to replicate yet incapable of replicating themselves, they became parasites. Now the parasites needed to bind and gain entry into these enclosed structures.

It is worthwhile examining in a step-by-step fashion the cascade of events that plays out during infection of a host cell by viral parasites. While the molecular details of the processes differ and vary depending on the virus and its host, all viral infections share the same basic necessities: entry, replication, and egress. We will begin with virus entry. The selfish genetic information of our tailed virus encodes capsid proteins and tail assembly proteins. The capsid proteins assemble into a protective shell to house the genome, and the tail assembly accomplishes its introduction into the host cell. Some protein components of the tail, near its tip, have molecular affinity for proteins exposed on the surface of the cell. A chance contact between the phage tail protein and this cellular "receptor" protein results in binding the phage to the cell surface. The physical principles at work here are similar to those that govern the binding of an antibody to its antigen. The molecular interaction results in the formation of an energetically favorable complex. The phage is thus brought into proximity with its prey. The multiprotein tail assembly is a sophisticated molecular machine, evolved to deliver the nucleic acid contents of the phage capsid into the cell cytoplasm. Our phage's tail apparatus can function as a syringe, injecting the phage chromosomal DNA into the cell. This series of linked molecular events, tripped by the first physical interaction of the virus particle with the cellular receptor, can be viewed as a cascade of events with a thermodynamically favorable outcome. The energy captive in the ordered structure of the virus particle is used to drive the process.

After gaining entry into the host cell's cytoplasm, the virus has access to the resources that it needs to replicate its genome and assemble new virus particles. The apparatus of cellular metabolism was originally dedicated to fueling the growth and proliferation of the cell itself—the essence of life—but shortly after the first phage-encoded gene products are made, the same machinery is harnessed to support replication of the virus. The host cell is now destined to die, as its resources are drained by its parasite. Nevertheless, the cell's infrastructure remains vital enough to complete replication of the viral genome and morphogenesis of new virus particles. It serves the best interests of the virus to allow the host cell to live long enough to complete these tasks (indeed, some viruses—filamentous phages—have evolved to replicate and egress without adversely affecting host cell viability at all). Upon completion of the viral replicative cycle, newly assembled phage particles accumulate within the cytoplasm. Typically, the virus encodes specific gene products whose role is to degrade the host cell wall, permitting the release of the newly synthesized viral progeny.

It is hard to grasp the necessary iterative genetic experimentation required to evolve such intricate and refined machinery that permits viruses to exploit the energy and infrastructure of living cells. These evolutionary processes are governed by the same powerful forces of Darwinian selection that forged multicellular life and sentient life-forms, through a process of trial and error and the survival of the fittest variants. It is wondrous to consider sentient multicellular organisms as the pinnacle of achievement of natural selection, far outstripping that of the evolution of these primitive parasites of cells. It is also fair to point out that the diversity of the viral world and the genetic information viruses have created exceeds that of all cellular life. If you search in DNA sequence databases, you will find that the vast majority of unique genetic information in our biosphere is viral and is not shared with, nor derived from, cellular genomes. Viruses created the planet's storehouse of genetic diversity. As parasites of organisms as diverse as cyanobacteria and amoeba to Neanderthals and *Homo sapiens*, viruses have successfully explored a vast swath of evolutionary space, unparalleled by any of the domains of life.

Lysogeny: Exercising Temperance

Some phages only pursue a predatory lytic replicative strategy, resulting in cycles of host cell destruction and the release of new infectious virus particles. These are the bare-bones drag racers of the phage world, built for cruel efficiency, but limited in flexibility. They have a singular purpose.

Others are more flexible and, to extend the automotive metaphor, are *grand tourismos*: fast, yet capable on the open road and acquitted with the necessary comforts for touring and various road conditions. These phages have the luxury of being able to make a lifestyle choice after infecting the host cell. They are called *temperate* phages, after their ability to exercise self-restraint and moderation, denying themselves the immediate gratification of lytic replication. Although they often do proceed to replicate by lysis of the host cell, under some circumstances they can opt for an alternative strategy and take up residence in the host cell's DNA. Now they become heritable genetic information, replicated along with the host chromosome. To accomplish this, the virus encodes the necessary enzymes for incision in the cellular chromosome and insertion of its DNA. The resulting integrated *prophage* expresses a very limited set of genes for repressive proteins that maintain it essentially in a dormant state within the host chromosome, behaving as if it were just another module of cellular genes. This process is known as *lysogeny*. The prophage endures, passed down to daughter cells during cell division.

The evolution of these two alternative phage lifestyles—lytic replication and lysogeny—must offer a survival advantage to the phage lineage. The key to this advantage is making the right decision: to lyse or lysogenize the host cell. Which tactic offers the highest probability of achieving the virus's objective of perpetuating its genes? To make this decision, the virus encodes a genetic switch mechanism that can "sense" the state of the host cell upon infection. If the cell has the metabolic signatures of a rapidly growing healthy population, the phage tends to opt for lytic replication. This strategy allows for rapid amplification of the virus and release of its progeny to the exterior, providing them a good chance of coming upon another healthy cell to attack and perpetuate their growth cycles. On the other hand, if the virus "senses" telltale signs that its new host cell is not rapidly dividing, or if there are many viruses vying to infect the same host cell, it may opt for its temperate strategy and lysogenize the cell, by integrating its genome into the cell chromosome and hunkering down. This is a good strategy under these circumstances because to lytically replicate in a cell that is part of a declining population can lead to release of progeny with too few hosts available to infect. All would be lost, and the virus particles would languish in the environment until they perish. Better to sit tight and pass its genetic information down through the generations of its host. This strategy is not one for rapid amplification of the phage genetic material, but for its preservation.

In any case, the prophage has another survival tactic up its sleeve. As its name suggests, the prophage can be a progenitor of a phage and can reenter its virulent replicative cycle. Should the lysogenized host cell undergo substantial stress, and be at risk of dying, the prophage "senses" the situation and is induced to save its genome. Bacteria commonly respond to environmental stresses with a variety of programmed stress responses. Particularly important for induction of prophages is the bacterial SOS stress response that is triggered by damage to the bacterial genome (Ptashne 2004). It causes the prophage to switch from its repressed dormant state to its virulent mode of replication: it mobilizes by excising its genome from the host chromosome and regaining its replicative form. The prophage then expresses phage gene products for DNA replication and assembly of infectious virus particles. The phage replicates before the host cell perishes. Despite long odds, the phage progeny are released to await an encounter with a new susceptible host cell and another opportunity to replicate.

In a population of lysogenized bacterial cells, *lysogens*, each cell contains a copy of the same prophage in its chromosome. The induction of prophages occurs at high frequency if the population is exposed to stress, leading to massive destruction of the host cell population. However, in a growing healthy population of lysogens, prophages induce spontaneously, at a very low frequency. In the laboratory, reports show a rate of one phage induction in 10,000 individual bacterial cells in each generation. Although such spontaneous and rare induction is fatal to the individual host cell, we deduce that the lysogenized population can benefit from the possession of prophages.

What could the advantage be to outweigh such a poison pill? One explanation lies in the immunity of lysogens to superinfection by related phages. The same repressive functions that maintain the prophage also act to preclude lytic replication of superinfecting phages. Phages released as a result of phage induction therefore cannot prey upon genetically identical lysogens, but may infect non-lysogens or other susceptible host species. Furthermore, phages that lysogenize a cell often bring with them protein encoding genes that are beneficial to the host. This is termed *phage conversion*. It is reasonable to speculate then that being lysogenized can have a net beneficial effect on the cell population. The population of lysogens as a whole is improved in fitness and is more successful.

This advantage provided by the phage more than offsets the disadvantage that is manifested in just a small minority of the population in which phage induction and attendant cell death occur.

Following similar lines of thinking it is also reasonable to speculate that the capacity of prophages to be induced at a low frequency must in itself be advantageous to the phage genome. It is attractive to think of this as a hedging strategy, in which the genetically identical phage population can simultaneously exploit two different phenotypes—in this case, to optimize its probability of genetic success. Lysogeny can be considered as phage conservatism, a strategy suited to survival in adverse conditions. Lytic replication is high-stakes gambling that pays off with confident prediction of outcomes. A phage that never takes advantage of the rewards of the high-stakes game (except under dire and uncertain circumstances) will not be as evolutionarily successful as the generally conservative phage with an occasionally successful flutter that, rewards with a burst of more rapid amplification. It seems likely then that phages have evolved to spontaneously induce, in a stochastic manner, in order to take advantage of lytic replication while not jeopardizing the genetically identical population of prophages still languishing in the chromosomes of their slowly dividing hosts.

Kill the Winner

In aquatic ecosystems our tailed virus friends are the major players; most have small genomes and are virulent lytic phages that infect and directly lyse their hosts. Constructed on a bare-bones design platform with a minimalist genome, they carry only the information essential for entering and hijacking the cell, to duplicate themselves and kill the host in the process. This cycle of carnage takes place on a grand scale in our oceans. Host cell death results in the release of nutrients back into the environment, making them again available for life at or near the bottom of the food chain, including the phage's own host cell population. It has been estimated that lytic phage infections in our oceans account for as much prokaryotic mortality as the ciliate and flagellate protists that graze on them; 20 percent of the microbial biomass of the oceans is destroyed by phages each day (Rohwer and Thurber 2009; Suttle 2007). Phages are thus critically important components in the equilibrium of marine ecosystems, cycling nutrients and diverting organic matter from its default pathway up the food chain to make it available again for lower forms of life.

Phage infections exert a major influence on the populations of prokaryotic species in these environments. High rates of mortality within a host population are often associated with phage infections. Healthy, growing populations of a dominant species of prokaryote provide particularly fertile ground for phage predators. The high density of rapidly dividing host cells favors rapid cycles of lytic phage replication. This phenomenon has been termed "kill the winner" (Short 2012). The epidemic phage-killing of the dominant species of prokaryotic hosts results in a collapse of the population making way for competitor unicellular organisms to proliferate in their place. In turn, this population will succumb to viral predation. Thus, a boom and bust cycle of population expansion followed by viral killing is established. These cycles allow the coexistence of multiple competing prokaryotic species in the same environment and support the maintenance of microbial diversity that is essential for the integrity of global ecosystems.

The ecological impact of epidemic kill the winner phage behavior can be far reaching and an example is found in the Great Rift Valley, a region of extraordinary beauty in Kenya. Volcanoes, some of which are still active, rise from fertile plains, which are home to black rhinoceros, lion, giraffe, kudu, and other exotic wildlife. A series of more than fifty lakes draws wildlife to the valley. Two of the lakes, Bogoria and Nakuru, are home to 75 percent of the world's population of the lesser flamingo, a species listed as threatened on the Red List of the International Union for Conservation of Nature. The hundreds of thousands of pink flamingos that occupy the lakes are one of nature's grand spectacles. Bogoria and Nakuru are called "soda lakes" because of their high salinity and alkalinity caused by a lack of drainage. Their waters are blue-green in color due to thriving populations of cyanobacteria, the most important of which is *Arthrospira fusiformis*. This photosynthetic picoplankton is the main food source of the lesser flamingos. Over the last forty years, visiting flamingo numbers were radically diminished by mysterious die-offs. Suspected culprits such as heavy metal and pesticide pollution or cyanobacterial toxins and infectious disease could be exacerbated by the scarcity of food. An increased frequency of die-offs over the past two decades showed populations of lesser flamingos varying widely between less than a thousand in some seasons and more than half a million in others. Studies of the plankton biomass in three soda lakes, including Bogoria and Nakuru, hinted at a possible cause, revealing a more than fiftyfold variation whose nadir coincided with reduced flamingo numbers. Interestingly, the populations of *A. fusiformis* followed a boom and bust pattern with its dominance in the population interrupted sporadically by the outgrowth of *Anabaenopsis* or the competing picoplanktonic chlorophyte *Piocystis salinarum* (Lothar and Kiplagat 2010). The prevailing notion was that lack of food organisms was rendering the population at higher sensitivity to pollutants or infectious organisms and that habitat quality and environmental changes were at work. The basis for the population collapses in picoplankton, however, remained a subject of conjecture.

In 2013 a team of scientists led by Michael Schagerl at the University of Vienna published the results of their research, approaching the problem from a new angle (Peduzzi et al. 2014). They realized that viruses are likely the most prevalent biological entities in the Rift Valley lakes, and, as in other aquatic ecosystems, may be an important cause of mortality in cyanobacterial populations. They set out to monitor the abundance of cyanobacterial species, the major food organisms of lesser flamingos, as well as the cyanophages that infect them. Remarkably, the scientists recorded virus particle counts higher than any on record from any ecosystem: an utterly ineffable 7×10^9 viruses per milliliter of lake water. When the scientists measured the population density of the cyanophage *A. fusiformis* in the lakes over time, they were able to associate collapses in plankton population density with microscopically visible signs of cyanophage infection. Not surprisingly, this coincided with diminished flamingo populations. It appears most likely then that periodic kills of *A. fusiformis* are caused by lytic cyanophages killing the winner. It may also explain earlier observations where periodically other plankton species temporarily bloomed in the lakes—a signature of kill the winner cycles that maintain the diversity of microbial populations.

The Rift Valley soda lake ecosystem provides the first documented example of a phage infection having such a dominant effect on a complete food chain. The simplicity of this ecosystem, in which a virus infection directly impacts the primary food source of the lesser flamingo at the top of the food chain, makes these effects so dramatic. It remains to be elucidated why the ecology of the lakes changed over the last five decades. Could the changes noted be the simple result of improved record keeping? It is more likely that environmental changes or stochastic factors caused a reduction of microbial diversity in the lake. The simplified food web is now dangerously susceptible to significant fluctuations in biomass due to phage predation. In recent years, the lesser flamingo populations of Lake Nakuru varied between a hundred individuals and more than a million, illustrating the powerful influence that phage-induced mortality can exert on our ecosystems.

Gene Brokers

The phage metagenome treasure trove of unique genes and metabolic functions is a valuable currency of adaptive evolution in the microbial world. Here we will discuss how phages act as gene brokers, facilitating the movement and exchange of genetic currency, fueling the evolutionary economy. The genetic economy is driven by the selfishness of phage genomes levying a commission. The phage genomes acquire a competitive advantage that benefits the replicative success of their own genotype. Phages are powerful catalysts of genetic innovation and evolutionary adaptation of their microbial host species (Casjens 2003; Penadés et al. 2015; Ochman, Lawrence, and Groisman 2000). In a previous section, I introduced you to the alternative lifestyles of phages, including how phages interact with their individual host cells and populations of cells to influence whole ecosystems. We will now explore how these interactions affect their own evolution and the evolution of microbial cells in such ecosystems.

Recombination refers to the exchange of gene information within and between genomes. Recombination can occur between the genomes of different phages infecting the same host cell, resulting in transfer of phage genes between lineages. Importantly, phages often mobilize host cell gene sequences. Host cell DNA sequences, usually those flanking a prophage insertion site, can be mistakenly incorporated into the phage chromosome and packaged into infectious particles. Thus, phages can mediate the transfer and acquisition of genetic information in prokaryotic cells by means other than direct inheritance. This process of *horizontal gene transfer* makes phages instrumental in expanding the microbial gene pool while providing fuel for accelerated microbial adaptive evolution.

In some ecosystems, such as those in the coastal marine environment, ten or more phages exist for each microbial cell. For this reason, coinfection of the same host by two different bacteriophage types must be quite common. The genes in bacteriophage chromosomes that encode proteins working together in functional pathways of gene activity typically cluster together. This mosaic of gene clusters on the chromosome promotes the potential for entire modules of genetic activity to be readily shuttled between bacteriophage species (Weinbauer and Rassoulzadegan 2003). Such recombination-mediated gene exchange creates chimeric phage genomes, potentially the basis for new adaptive capabilities. The newly created phage may have altered host range, allowing it to infect previously naive host cells. Thus, through recombination, together with *generalized transduction* of host genetic information between different microbial hosts, phages efficiently promote the shuffling of genes between species of host cells. When phages complete lytic infection of host cells, free DNA from the dead host cell is released into the environment. Bacteria are quite adept at taking up such DNA and incorporating it into their genomes, a process known as *transformation*. The new bacteriophage genome, equipped with a payload of bacterial genes, may be at an advantage with its new genetic information and form the basis of a new lineage. The new genetic information may transfer into a new host cell and between different bacterial species. The transduced fragments of host cell DNA often encode genes or gene clusters not previously found in the progenitor cell. These changes may lead to a new host genotype with a competitive advantage that allows it to dominate the population or to colonize a new habitat previously inaccessible. Ultimately, this can be the basis for speciation, the creation of a new species and distinct lineage of prokaryote.

DNA transduction is common in the crowded aquatic environment where free phages are so abundant, and each free-living prokaryote plays host to at least two prophages. The role of phages and horizontal gene transfer in microbial evolution and speciation is not, however, restricted to these ecosystems—it is equally significant in other ecologies. For example, researchers at the University of Pittsburgh (Lawrence and

Ochman 1998) studied the enteric bacteria *Escherichia coli* (*E. coli*) and *Salmonella enterica*, which they estimated had diverged into separate species some 100 million years ago. These researchers took advantage of the knowledge that genes of different evolutionary origins are betrayed by their particular DNA sequence content. Different species of bacteria evolve genes of different nucleotide content, that can be measured in terms of percent guanine-cytidine base pairs (GC) and different patterns of amino acid codon usage. Analysis of the complete genome of *E. coli* identified 755 genes that differed significantly in these respects. They had distinct evolutionary origins in other bacterial species. Today these genes make up almost 20 percent of the *E. coli* chromosome and were acquired in more than 200 independent horizontal gene transfer events. Further analysis examined the location of the horizontally transferred genes in the chromosome. The data was telling: they were often situated in the chromosome close to transfer RNA genes. This strongly implicates lysogenic phages in their horizontal transmission, as many are known to integrate preferentially in the proximity of these genes. Their work illustrates the powerful potential of phages and horizontal gene transfer to shape bacterial evolution and speciation. The acquisition of novel genetic information undoubtedly provided *E. coli* with adaptive functionalities that allowed it to exploit environmental niches inaccessible to its ancestor species.

Selfishness Drives Adaptive Evolution

The dense population of microbes and phages in the marine environment offers an extraordinarily fertile breeding ground for genetic exchange and experimentation. The *Cyanobacteriae*, one of the most diverse groups of bacteria, have been successful in both terrestrial and aquatic habitats. They are unique within bacteria as oxygenic photosynthetic organisms, the smallest known to exist. Like higher plants they produce energy and organic building blocks by utilizing light and carbon dioxide and releasing oxygen. At a diameter of only 0.5 micrometers, *Prochlorococcus* is one of the most diminutive organisms with the capacity for photosynthesis. It escaped study until the 1980s when oceangoing research vessels were equipped with flow cytometers sensitive enough to detect them. *Prochlorococcus* is the most abundant photosynthetic organism in the oceans and probably on earth. *Prochlorococcus* and its close relative *Synechococcus* dominate picoplanktonic photosynthesis in the euphotic regions of the oceans, the layers in the water column receiving enough light to support photosynthesis. Together they account for 25 percent of all photosynthesis on our planet (Partensky, Hess, and Vaulot 1999; Field et al. 1998). The photosynthetic machinery of these cyanobacteria is similar to plants in our gardens. Indeed, today's cyanobacteria are likely descended from the precursors of the chloroplasts found in eukaryotic plant cells.

The chloroplast's photosynthetic apparatus is composed of two photosystems, photosystem I and II (PSI and PSII), which are light-harvesting complexes of proteins and light-sensitive pigments connected by an electron transport chain. The energy released by the movement of electrons between these two photosystems is harnessed, moving protons (H+ ions) across a membrane. This proton-motive force generates ATP, the power source of the cell, and feeds the Calvin cycle that synthesizes glyceraldehyde 3-phosphate, the fundamental chemical building block for the cell's materials. The light-sensitive pigments of PSI and PSII are central to photosynthesis, differing in their sensitivity to intensities and wavelengths of light. Clades of *Prochlorococcus* have evolved to have different photosystems, adapted to be most efficient at various levels in the ocean water column (Moore, Rocap, and Chisholm 1998). The algae living at greater depths can better utilize the wavelengths of light penetrating most deeply, while those in surface waters use the wavelengths most readily filtered out by passage through the water. Photosystems are chemically affected by incident light. A danger facing all prochlorococci, particularly those in surface waters is exposure to excess light, which damages their photosystems and results in photoinhibition. The PSII reaction center is composed of a dimer of proteins D1 and D2, which contain the light-sensitive pigment essential for its photochemistry. Protein D1 is particularly susceptible to damage by light, requiring frequent replacement if the cell is to remain healthy and able to conduct photosynthesis. If the cell is unable to replenish PSII with new D1 protein, photoinhibition occurs, and the energy production of the cell is diminished.

Cyanobacteria are host to a plenitude of phages, primarily *cyanophages*. There are many types that infect both *Prochlorococcus* and *Synechococcus*. For all viruses, an essential feature of an infected cell is that it is metabolically viable, producing adequate energy and materials to support the replication of the virus. Phage S-PM2, a member of the *Myoviridae* that infects *Synechococcus*, is exquisitely adapted for infecting its photosynthetic host. Scientists studying the infectious cycle of S-PM2 discovered that its adsorption to *Synechococcus* cells depends markedly on light. It appears that, just as the cyanobacteria themselves grow and divide most actively during daylight hours, their phage predators are triggered to begin a wave of infection at dawn, with progeny being released at dusk (Clokie and Mann 2006). The virus has thus synchronized its infection of the host cell population to the time of day when it is most metabolically active and can best support the energy needs of the predator. When scientists first sequenced the genome of S-PM2, it was clear that it was not a bare-bones lytic phage but a complex one with a 193-kilobase double-stranded DNA chromosome. Moreover, they were astonished to discover that the genome contained photosynthetic genes resembling those of its bacterial host cell (Mann et al. 2003). The coding sequence for the PSII component proteins D1 and D2 were found in a 4-kilobase portion of the phage genome. The genes were highly homologous to their counterparts in *Synechococcus*, confirming their bacterial origin, acquired by horizontal gene transfer. A detailed analysis suggested that the phage had picked up the two genes in independent gene transfer events, a reflection of the prevalence of these mechanisms in genomic evolution of phages and their hosts (Lindell et al. 2004; Sullivan et al. 2006). It is very likely that the phage benefits from incorporation of these host bacterial photosynthetic genes into its genome. It is notable as well that the genes encoding the most photosensitive proteins were acquired. In the later stages of infection, the housekeeping functions of the host cell halt. All cellular resources focus on supporting phage replication and the production of viral proteins. It is, therefore, advantageous to the virus to supplement synthesis of the critical PSII components. These phage genes permit photosynthetic energy production by the infected cell for as long as possible, optimizing the potential for the virus to complete its infectious cycle before the cell dies.

In attempts to determine whether these observations reflected a rare occurrence unique to S-PM2 and its host strain of *Synechococcus*, researchers explored *Myoviridae* and *Podoviridae* that infect its close relative *Prochlorococcus*. Indeed, all three phages of *Prochlorococcus* selected for sequencing had in common the genes encoding bacterial PSII protein D1 and a gene called *hli* coding for high-light-inducible protein (Sullivan et al. 2006). In certain strains, scientists discovered additional bacterial photosynthetic genes. While all were clearly of cyanobacterial origin, based on their sequence similarities, it appeared that the genes were from different but related cyanobacterial species. Notably, the *hli* gene is present in multiple copies in the bacterial chromosome. Scientists speculate that horizontal gene transfer into phage genomes and subsequent reacquisition of the gene from phage DNA may have played a role in the redundancy of the gene at the bacterial *hli* gene locus. A picture emerges of bacterial photosynthetic genes shuffled between phages and bacterial genomes in both directions. It is obviously a benefit to the phage to encode these additional proteins. On the other hand, we can hypothesize that such gene exchange is illustrative of a process that drives microbial diversification, offering benefit to the host organism. It provides a reservoir of genetic variation tapped by microbial cells to facilitate rapid adaptive evolution. Such variation can be advantageous in fluctuating environments, where new selective pressures arise. Adaptation to new circumstances is paramount for the success of the genome. Of course, this is not restricted to genes involved in photosynthesis, and could equally apply to genes that provide other metabolic functions with the capacity to influence the ecological niche occupied by the recipient microbe. One can envisage the phages of a particular host acting as an extended gene pool, a possible source of genes for innovative genetic experimentation. Imagine a particular gene, loaned via generalized transduction, entering the phage gene pool where it will be subject to selective pressures different from those that operated on it in the host. The gene will undergo independent natural selection, as part of the rapidly replicating bacteriophage metagenome later acquired by the same host or a different host. The gene is retained if it provides a selective advantage.

Think of the phage metagenome as a corporate development program designed by a company's Human Resources Department to train genetic talent. In corporate America talented employees work temporarily in different environments, geographies, and business functions, accumulating diverse business experiences and skill sets for success. Such programs are advantageous to the employee (the gene) and benefit the company (the organism). The value of diversity on teams is broadly recognized and sought after in American companies. It promotes team creativity and effective troubleshooting of complex problems. It appears that similar principles operate on a biological level with viruses as their conductor. Of course, in my corporate business example, the employees assigned to development tracks and teams are (hopefully) carefully selected by Human Resources and management. In the natural world no such preselection exists. We must assume that every gene is chosen, with retention of only the genes that increase the fitness of the recipient genome. It is an extremely inefficient process, but one that can operate successfully over time when the number of "tryouts" is adequate to overcome the low probabilities of a beneficial outcome.

Temperate phages make an extended gene pool available to host cells following phage transduction. This is especially important in environments with fluctuating and potentially hostile conditions. With this in mind, researchers sought to characterize the phages in the waters directly above hydrothermal vents deep down on the ocean floor, comparing them to the bacteria in the surrounding seawater (Williamson, Cary, et al. 2008). Scientists recovered free phage particles from the seawater above the vents where warm and cold waters mix. The researchers concluded that in the varying environment prophages were induced at a higher rate than in the stable surrounding seawater. Moreover, a quarter of the DNA sequences in these particles were "dark matter"—completely new to our databases. The bacterial and phage species in these harsh deep-sea environments must be a storehouse of novel adaptive genetic information that can be mobilized by fluctuating environmental conditions.

Phages and the Microbiome

The oceanic ecosystems we have been discussing are home to a millions of microbes per milliliter of seawater and ten times more viruses, but our large intestine, the colon, is an even richer ecosystem. It has a diverse and populous microbiota termed the gut microbiome. A typical one-gram sample of human fecal matter contains up to 10^{13} bacterial cells. After performing the necessary math, one can project that the bacterial cells in our intestinal tract, our microbiota, equal or outnumber the cells in our body. While a milliliter of seawater from many oceanic ecosystems often contains tenfold more virus particles than microbial cells, this is not the case in our gut. Free virus particles counted in feces samples from adults, number "only" between 100 million and a trillion per milliliter. Nevertheless, the basic tenet that where bacterial populations thrive there will be a thriving phage population holds true in the gut. Despite the relative scarcity of free phage particles compared to bacterial cells, temperate phages pursuing a lysogenic lifestyle are abundantly represented in the microbiome and are the dominant forms of virus infecting bacteria in our gut (De Paepe et al. 2014).

Virulent phages that pursue a lytic lifestyle contribute the majority of virus particles that can be filtered from marine samples. Such phages are most suited to relatively stable environments with unfettered access to large, relatively homogeneous populations of healthy and rapidly dividing prey organisms. This is the case in most oceanic ecosystems. The lytic lifestyle is not a winning strategy in the gut, where the microbiota, most of which occupy the large intestine, are in fierce competition for limited nutrients and are not often freely accessible to phage particles. It is therefore of no great surprise that temperate phages have been reported to predominate here (De Paepe et al. 2014). Their innate ability to perceive and respond to the physiological status of their host cell allows them to elect lysogeny over lytic replication. They are so calibrated by natural selection and historical success that this is the prevailing strategy of phages in our microbiome. Genetic lineages that have this capacity appear to be the most genetically successful in the microbiome. They exploit lysogeny because it offers the highest probability for perpetuating their genome lineage in this demanding ecosystem. The interaction of this population of viruses, most of which spend a portion of their existence as lytic viruses and a portion as host chromosomal loci, most certainly has profound effects in shaping the composition of gut bacterial populations. With the wealth of emerging knowledge implicating the gut microbiota in health as well as disease it also follows that they must exist in delicate balance in the healthy individual's gut.

The makeup of the microbial population of the healthy gut includes various classes of symbiont. Some exist in symbiotic mutualism, in which each partner benefits from the relationship, and others commensally, in which the relationship is beneficial to one partner but is of no significant detriment to the other. Others are pathobionts that may start out as mutualists or commensals, but become pathogenic when the balance of the gut ecosystem is disturbed. Over the past decade or so it has become clear that imbalance of the bacterial population of the gut, a phenomenon known as dysbiosis, is frequently associated with disease states (Clemente et al. 2012; Kaser, Zeissig, and Blumberg 2010). It is yet far from clear what specific role phages play maintaining the healthy microbiome or initiating dysbiosis, but it is self-evident by analogy with their profound influence on marine ecosystems, that they have the potential to dramatically affect the abundance and composition of the gut microbiota. Epidemic predation and lysogeny with attendant phage conversion and the potential for phage induction will each play a role. Lysogenic phage can influence host bacterial phenotypes and their evolution, while phage induction (either as a stochastic event or in response to environmental stimuli experienced by lysogenized host cells) will release infectious particles and mobilize genetic material. Phages may also impact our immune system, which plays a very active role in defending this, our largest "international border." They are implicated in preventing invasion by potentially harmful bacteria in the gut, limiting them to residence in the gut lumen (Barr, Youle, and Rohwer 2013). The influence of phages of the microbiota is not restricted to their hosts, but extends also their host's host.

We saw that host—phage population dynamics in marine ecosystems is dominated by virulent phages that display a conventional predator-prey dynamic, referred to as kill the winner. In that ecosystem, phages replicate lytically in robust communities of bacteria until the population collapses and it can no longer sustain the chain of transmission of the phages: the epidemic fizzles and dies. New microbial species bloom in their place and are in turn controlled by phage predation. Such boom and bust cycles sustain balanced diverse microbial ecosystems. The relative paucity of free phage particles in the gut compared to bacteria dictates that kill the winner is not operative there (or at least not as a dominant modality of phage infection). The lower intestine is very densely populated and a highly resource-limited environment; the bacterial colonists are not dividing rapidly and are distributed in a variety of ecological niches that may not be readily accessible for infection. They may even be protected from phage infection by close association with gut wall cells and structures or by the formation of protective biofilms, which phages cannot readily penetrate. In the gut then, lysogenic phages rule the roost, with lysogeny the preferred lifestyle.

Unfriendly Competition

It certainly benefits phages of the gut ecosystem to exist in the chromosome of their host as prophages. Bacterial residents of the gut avail themselves of a variety of mechanisms to remain associated with the gut wall and avoid being expelled in the feces. It follows that this will also benefit the prophages. But what of their lysogen? A survey of bacterial genomes and prophage-related sequences acquired by horizontal gene transfer tells us that unless the functionality of prophages offer a selective advantage to the host, natural selection will favor their inactivation (Ochman, Lawrence, and Groisman 2000; Lawrence and Ochman 1998; Nicholson et al. 2012). Since large populations of lysogens in the gut microbiota have prophages in their genomes, it is reasonable to believe that they gain some advantage from them. It must be adequate to offset the inherent "poison pill" liability associated with phage induction and death of the host cell. Prophages have the potential to be a source of genomic innovation and fuel for bacterial evolution: phage conversion often provides valuable metabolic functions that make the microbial cells more competitive in their environmental niches. Lysogeny results in the introduction of novel genetic information by horizontal gene transfer into the bacterial chromosome, and successful phage lineages often carry genes with functions that have value for the host cell. Others are adept at generalized transduction in which pieces of bacterial chromosomal DNA are incorporated into the phage genome. A subsequent host that becomes lysogenized by this modified phage will receive not only phage genetic information but also information from another bacterium—perhaps of a different species and having novel metabolic capabilities. Prophages, which can make up substantial portions of the bacterial specie's identity (i.e., its uniqueness and distinction from related bacterial strains), are of course potent catalysts of adaptive evolution, and can be anchor points in the genome serving as landing pads for acceptance of new genetic information.

Lysogeny also provides the host cell and all of its daughter cells, which share the same genetic material, with immunity to infection (and potential lysis) by closely related phage species. Bacteria of a different genetic lineage, whether they are different species or simply siblings that are not lysogens, will lack immunity. Such bacteria may be in direct competition to occupy the same ecological niche in the gut. Under these circumstances, spontaneous induction of phages in the lysogenized population can cause an epidemic infection in a competitor susceptible bacterial strain (De Paepe et al. 2014). Here, the phage acts as an effective bioweapon targeting microbes that are competing for the same ecological niche. The lysogen population garners a competitive advantage. "Death-by-induction" of an individual bacterium can be beneficial to the population as a whole and promote the replicative success of its genetic identicals.

Chemical Warfare

The human nasopharynx forms another ecological niche in which commensal bacteria coexist and compete for dominance. It is commonly colonized by *Streptococcus pneumoniae* and by *Staphylococcus aureus* and can be a source for transmission of these bacteria to other anatomical sites and between individuals. Many years ago it was recognized that *S. pneumoniae* and *S. aureus* compete for occupation of the nasopharynx. Epidemiologists who took swabs from the nasopharynx of healthy children found that if they were able to culture *S. pneumoniae* from these samples, they were less likely to also culture staphylococci. On the other hand, staphylococcal colonization was more frequent in children whose pharynges were not colonized by *S. pneumoniae* (Regev-Yochay et al. 2004; Bogaert et al. 2004). Soon it was shown that *S. pneumoniae* uses a chemical weapon to get the upper hand on its competitor (Park, Nizet, and Liu 2008; Regev-Yochay, Trzciński, and Thompson 2006). It releases the chemical compound hydrogen peroxide, a well-known disinfectant, into its surroundings. It remained enigmatic that the levels of hydrogen peroxide produced by *S. pneumoniae* were harmless to itself yet efficiently killed the competitor staphylococci. In 2009, however, scientists from universities in New York State and Spain found the answer to this enigma (Selva et al. 2009). Hydrogen peroxide and its dangerous reaction products cause DNA damage that typically triggers the SOS stress response of the bacterial cell (discussed earlier in this chapter). *S. pneumoniae* appears to have evolved to withstand exposure to the levels of hydrogen peroxide that it produces, and despite DNA damage to its own chromosome, its SOS stress response is not activated. In contrast, staphylococci detect DNA damage caused by the hydrogen peroxide in the environment and strongly induce their SOS response. One consequence is that their DNA repair mechanisms are activated. These changes in the bacterium are sensed by the resident prophages of the staphylococcal cells, which are programmed to undergo induction and enter into their lytic replicative cycle. It is the resulting lysis of the staphylococcal cells by wholesale induction of their prophages that results in their death and clears the playing field, providing a competitive advantage to *S. pneumoniae* (Selva et al. 2009).

Staphylococci are known to harbor a multitude of inducible prophages and their induction in the population at large is a poison pill not only for the individual but also for the whole population. These conclusions were elegantly confirmed in laboratory experiments by Selva and colleagues. They constructed by genetic manipulation *S. aureus* strains that were not lysogens and, therefore, had no viable prophages that could be induced by stress signals. These bacterial cells resisted hydrogen peroxide exposure that was sufficient to kill their lysogenic relatives and were thus immune to the chemical warfare of *S. pneumoniae*. Of course, the vast majority of *S. aureus* in nature harbor multiple prophages, rendering these tactics broadly applicable as a competitive strategy for *S. pneumoniae*. Such "remote control" of prophage induction might be one broadly applicable phenomenon at play in determining the composition of complex bacterial communities such as those that coexist in and on multicellular organisms and in the environment.

These descriptions only begin to catalogue the profound influence that phages exert on living cells, organisms, and whole ecosystems. The reservoir of genetic information in the phage gene pool, or metagenome, is a resource for microbial diversification and adaptive evolution. The facility with which it moves both between phages and between bacteria, in large part via phage transduction, shines a light on a crucial issue in phylogenetic categorization of not just viruses themselves but also their hosts. I grew up with the notion of a tree of life, with diverging branches that independently evolve. This is probably an accurate reflection of the evolution of higher life-forms. For the domains *Bacteria* and *Archaea* and for viruses, the movement of genetic information laterally, between branches of the evolutionary tree by horizontal gene transfer, has been most influential in their evolution and "speciation." Some scientists suggest that the notion of "species" in the microbial world is meaningless when applied to the current crop of bacteria and viruses studied by comparative genomics today. Much of their distinctive identities are attributed to their differential acquisition of independently evolved genetic information. Phages have played a central role in creating this genetic melting pot. Phages facilitate the enormous evolutionary potential of the microbial metagenome and are remarkable agents of evolutionary invention.

· 3 ·

POTENTIATION *of* BACTERIAL DISEASES BY PHAGES

PHAGES PLAY A CENTRAL ROLE in our global ecosystems, influencing the flux of biomass in food chains and promoting the diversity and evolution of their abundant microbial hosts. However, phages also infect and interact with bacteria that call multicellular organisms their homes. In 1930 Félix d'Hérelle, wrote that "the actions and reactions are not solely between these two beings, man and bacterium, for the bacteriophage also intervenes; a third living being and hence, a third variable is introduced" (D'Hérelle and Smith 1930). Here I will introduce you to how the phages that infect microbes are overlooked "backseat drivers" in human health and disease.

If you should wander through an old churchyard in England, examine the inscriptions on the gravestones dating to the early to mid-nineteenth century. It is not unusual to see them engraved with the names of children who died in infancy, often the victims of infectious disease. Infant mortality in those days was not uncommon and epidemics of infectious diseases were undoubtedly primed by poor nutrition and living conditions. Along with the plague, typhus, and cholera, malignant scarlet fever took its toll on whole families of young children. Two of these legendary epidemic killers, cholera and scarlet fever, reflect the ponderings of d'Hérelle, who predicted the triangulation of human, bacterium, and bacteriophage. These diseases are the creation of microbial viruses. D'Hérelle wrote those words in 1930, just a few years after Frobisher and Brown (1927) reported the isolation of a filterable agent from scarlatinal strains of hemolytic *Streptococcus*, the bacteria that cause scarlet fever. The agent they isolated could transfer the ability to make the erythrogenic toxin, the hallmark of the disease-causing bacterium, to erstwhile nonscarlatinal streptococcal strains. This property was inherited in the progeny of the new strain. They were unknowingly describing the phenomenon of gene transduction and phage conversion by a temperate phage. The phage infection and lysogenization of the naive streptococcal strain transformed it into a more pathogenic variant. In other words, the strains of streptococci that caused scarlet fever did so because they were infected and lysogenized by a phage. The prophage had become part of the bacteria's heritable DNA: *phage conversion* had occurred.

When English physician Thomas Sydenham first described *scarlatina*, which we refer to today as scarlet fever, in 1675, he described it as "[attacking] whole families at once and more especially the infant part of them. The patients feel rigors and shivering just as they do in other fevers. The symptoms are however moderate; afterwards, however, the whole skin becomes covered with small red macula, thicker than those of measles, as well as broader, redder, and less uniform. These last for two to three days and then disappear. The cuticle peels off and branny scales remain lying on the surface like meal."

This is a description of a quite moderate illness, an observation belied by the devastating epidemics of scarlet fever that occurred throughout Europe in the nineteenth century. The severity of that disease was aptly compared with typhus and plague, both killers. This transformation of the disease into an epidemic killer, and the periodic outbreaks of milder disease, clouded the epidemiologic understanding of scarlet fever for some time. However, with today's knowledge of the molecular genetics underlying the pathogenicity of streptococcal isolates, scientists can offer a compelling synthesis that explains the variability in the severity of perennial scarlet fever epidemics. We now understand how different strains of this same bacterial species can cause a spectrum of illnesses ranging from tonsillitis or mild skin infections to toxic shock–like syndrome and even necrotizing fasciitis—the much feared flesh-eating disease. The genesis of scarlet fever illustrates how phages serve as critical catalysts of the pathogenicity of bacterial diseases.

For a Charm of Powerful Trouble

The new generation of DNA sequencing technologies that have allowed scientists to explore environmental phage metagenomes are also fundamental to comparative genomics. This, the study of related genome sequences, allows the phylogenetic relationships of organisms to be examined at the level of their genomic DNA sequences. Comparative genomics can be used to study the genomes of closely related pathogen isolates and the relationship of genome sequence with a phenotype, such as drug resistance or the capacity to cause different types of disease. Musser and colleagues (Banks, Beres, and Musser 2002) used comparative genomic analysis to explore the underlying basis for the different diseases caused by different serotypes of group A streptococci. They chose three closely related strains of *Streptococcus pyogenes*: one was associated with epidemics and invasive infection, a second strain had the propensity to cause toxic shock syndrome or necrotizing fasciitis, and the third was associated with outbreaks of acute rheumatic fever. They discovered that the broad spectrum of diseases caused by these different strains of the same bacterial species could be attributed to differences in the prophages in their genomes. More than one-tenth of the genome of some strains was phage DNA and their distinct complements of prophages were the major contributors to their genetic differences. Polylysogeny with different phages drives strain diversity in *Streptococci* and ultimately dictates their pathogenic potential (Banks, Beres, and Musser 2002).

The genome of a pathogenic bacterium is indeed a witches' brew made virulent by a concoction of active ingredients. The many different prophages found in *Streptococci* contribute distinct genes to the broth, changing its potency and its effects. The prophages contribute gene functions to their host cell that promote its successful replication and transmission in the human host population. In turn, the phage parasites benefit from the success of their host, being amplified along with the bacterial genome. The genes in question provoke more severe streptococcal disease; they include genes for exotoxins that wreak havoc with a patient's immune system, causing fever, shock, and other severe manifestations, as well as genes that promote the survival of the bacterium in the diseased patient in the face of the developing immune response. Natural selection is at work on *Streptococci* and on the phage genes that they harbor. It has resulted in many distinct lineages with remarkably diverse phenotypes and capacities to cause more or less severe human disease. It is salient to note that streptococcal lineages of both low and high pathogenicity circulate; sometimes one or another becomes epidemic. In the nineteenth century and today it is evident in the varying nature and severity of group A streptococcal outbreaks. The nature of each epidemic depends on the prevailing strain of bacterium; it is not always the most pathogenic strain that is the most successful. As a species we should be grateful that natural selection operates on pathogen populations without concern for pathogenicity per se. It promotes the survival of genes that provide traits that offer the best probability of being inherited. The genes can be associated with the acquisition of increased virulence, but certainly not always. The viruses and microbes with which we coexist and coevolve are not unerringly evolving to become more virulent, but they have that potential if it offers improved replicative success and more efficient transmission to new hosts. In this instance, the microbe and the virus share the same objectives and reap the same rewards from their relationship.

Toxic Enablers

For centuries, the mention of cholera caused fear. Today, epidemics caused by *Vibrio cholerae* are still one of the world's most persistent ambulance chasers, often emerging in the wake of natural disasters. Cholera is contracted from contaminated food or water, resulting in severe and protracted watery diarrhea, which rapidly causes severe dehydration. If a patient is not adequately cared for, by replacing water and electrolytes, the disease quickly becomes life threatening, particularly in the young or the weak. The severe watery diarrhea associated with pathogenic cholera is central to understanding the success of this enteric pathogen (Faruque, Albert, and Mekalanos 1998). This symptom of the disease is the means by which it is transmitted between hosts and the basis of epidemics when basic sanitation breaks down.

Cholera in one form or another has afflicted mankind for centuries. It originated in Asia in the Ganges Delta, where the bacterium flourished in warm brackish estuarine waters. From there the disease emerged in 1817 to cause the first well-documented cholera pandemic. It spread rapidly throughout continental Asia to the rest of the world, undoubtedly accelerated by British trading practices. The East India Company held sway over trade with India. Having begun as an enterprise to import spices from South Asia, it exercised a British monopoly on commerce with the Indian subcontinent. The trade routes from India were plied by British sailing ships that brought, in their bilges, contaminated water from the Bay of Bengal, which they then expelled into the estuarine water of their home ports. The first epidemic in London in 1832 took thousands of lives, and over the next fifty years the disease made its way via shipping to Montréal and thence to New York. Cholera was soon a global phenomenon.

By the mid-nineteenth century, epidemiological studies established that cholera was acquired from contaminated drinking water. In 1883 Robert Koch, the eminent German Nobel Prize–winning bacteriologist, headed the German Cholera Commission, sent to study an outbreak of the disease in Egypt. There, he was first to identify the etiologic agent of the disease, isolating and culturing the bacterium *V. cholerae*. There are many strains of *V. cholera*e, but only those of two serogroups, 01 and 0139, cause disease. These pathogenic strains express a pilus protein on their surface (termed the toxin-coregulated pilus), which forms small appendages on the cell surface that allow them to colonize our small intestine. They are also armed with genes that encode an extremely potent exotoxin that we know as cholera toxin. It is secreted into the small intestine by pathogenic strains and penetrates cells in the gut wall. Within the cells, its toxic effects destabilize cellular homeostasis, causing a massive efflux of salt ions across the gut wall. This in turn generates a salt gradient that pulls water from the body into the lumen of the gut by osmosis to create the "rice water" diarrhea typical of the disease. Today it is clear that the acquisition of pathogenicity by *V. cholerae*, an otherwise harmless marine microbe at home in estuarine and coastal waters, is the result of horizontal gene transfer by phage conversion.

The first player in our cast of characters is a 41-kilobase-long segment of DNA that contains a gene cluster encoding the proteins for the toxin-coregulated pilus. The second is a temperate, filamentous single-stranded DNA phage called CTXΦ which carries the cholera toxin genes, *Ctx*AB (Waldor and Mekalanos 1996). The pilus on the surface of the bacterium not only helps it to cling to the wall of our small intestine but also acts as the cellular receptor for CTXΦ. *Vibrio* cells possessing the pili can be recognized and infected by CTXΦ. It seems then that the ancestral form of *V. cholerae* acquired the pilus-encoding gene cluster first, opening the way for infection and phage conversion by CTXΦ (Davis and Waldor 2003). The CTXΦ prophage and its genes for toxin synthesis are part of the host genome and are inherited by daughter cells. Since this initial infection, the host and its CTXΦ prophage have undergone extensive adaptive coevolution. A subtle interplay between the prophage and host cell genomes exists and is most evident in the manner of regulation of the prophage toxin genes. Their expression appears to be predominantly governed not by the phage itself, but by proteins encoded by bacterial host genes. This mutualism in the form of cooperative gene regulation must have evolved for optimal bacterial virulence and transmission (Davis and Waldor 2003; McLeod et al. 2005). Filamentous phages are very different from our phage poster boy, the tailed-phage, discussed in Chapter 2. Productive infection by filamentous phages does not result in lysis and death of the host cell. Rather, following replication and assembly of progeny virus particles, they are secreted from the cell via a bacterially encoded pore protein complex. Unlike lytic phages, CTXΦ can actively replicate within *V. cholerae* and generate progeny phages without killing its host cell (Faruque, Albert, and Mekalanos 1998). The process of phage induction for CTXΦ is also different; it occurs without excision of the prophage from the host cell chromosome. Consequently, the CTXΦ lysogen can be induced to enter its replicative cycle and release progeny phage particles while preserving both the host cell and the prophage. CTXΦ can, therefore, pursue lysogeny, being replicated as part of the bacterial genome as well as productive infection and release of progeny phage. It can thus simultaneously be propagated vertically and horizontally between host cells. It can have its cake and eat it too.

There is a supporting actor in this drama. It is also a phage, a close relative of CTXΦ called RS1 that is often found integrated into the host DNA next to the CTXΦ prophage (Faruque et al. 2002; Davis and Waldor 2003). However, RS1, unlike CTXΦ, is a defective phage and does not have the full complement of the genes necessary to direct its own replication and assembly of new virus particles. In fact, RS1 encodes no viral structural protein genes at all. Rather, it exploits the virus particles made by CTXΦ to package its genome. It is an accessory phage and parasite of CTXΦ. RS1 is so frequently found with CTXΦ in toxigenic

Vibrio strains that it is believed its presence offers a competitive advantage to the lysogen. This is in fact the case. A protein produced by RS1, but not by CTXΦ, can upregulate the expression of the cholera toxin genes in the CTXΦ prophage. The RS1 prophage is beneficial to the CTXΦ lysogen since its genes contribute to the increased transmission of the bacterial pathogen to new hosts. Natural selection acting on the CTXΦ genome also favors the maintenance of the parasitic relationship with RS1 since CTXΦ genes replicate more successfully in the context of the more efficient epidemic spread of *V. cholerae*.

The acquisition of lysogenic phages and foreign DNA sequences by *Vibrio cholerae* is essential for its pathogenesis and epidemic potential. Environmental *V. cholerae* do not possess cholera toxin genes and are not toxigenic, suggesting that the toxin offers no advantage to the bacterium in its natural marine environment (Faruque, Albert, and Mekalanos 1998). However, *V. cholerae* lysogens that possess the toxin genes can successfully colonize a new environmental niche, the human intestine. The cholera toxin genes potently cause diarrhea and benefit the bacterium by mediating efficient spread of the infection between individual human hosts. Moreover, since filamentous phages do not lyse their host cell, the cell is not killed during the replication of the phage and production of phage particles. There is no "downside" for the bacterium. The phages themselves benefit in kind from the success of their host as they also benefit from its increased virulence and epidemic spread. Phage genetic information that flows with the epidemic of cholera has more opportunity to be amplified, to explore new hosts and pursue its selfish objectives.

The origin of the cholera toxin genes, *Ctx*AB, remains obscure. Did they come from the phage or bacterial metagenome? We know that ancestral forms of the CTXΦ phage did not have the toxin genes, since some natural nontoxigenic isolates of *V. cholerae* have been found to be lysogenized by a CTXΦ phage that has no toxin gene cassette (Boyd, Heilpern, and Waldor 2000). It thus seems likely that CTXΦ picked up the toxin genes in a generalized transduction event in which unfaithful recombination events incorporated a segment of DNA from a host cell or another phage into its own genome. These observations reveal that phage conversion of *V. cholerae* has occurred on multiple independent occasions and that it is a natural host of the phage. Generalized transduction witnessed here is a core capability in the phage toolbox of evolutionary tricks. The toxin genes are beneficial to the phage allowing it to be spread in environments where it otherwise had no foothold. In this case virulence and disease-causing power come together with selective pressure to create an extremely efficient agent of disease, forged by countless generations of genetic tinkering and accelerated by phage intervention.

Choose Your Poison

In common with natural marine isolates of *V. cholerae*, *Escherichia coli* typically exists as a harmless bacterium living in the mammalian gastrointestinal tract. Pathogenic strains of *E. coli* also emerge as a consequence of phage conversion and are responsible for many human illnesses: often diarrhea, due to intestinal infection, but also urinogenital and respiratory tract infections. Here we will discuss enterohemorrhagic *E. coli* (EHEC), a strain of *E. coli* more often than not traced back to contaminated ground beef. EHEC causes severe disease in humans: specifically, bloody diarrhea, hemorrhagic colitis, and hemolytic uremic syndrome. Our particular provocateur is *E. coli* 0157:H7, first recognized in 1982 as a foodborne zoonotic bacterial infection (Riley et al. 1983). *E. coli* 0157:H7 resides for the most part in another animal species, in this case, cattle or ungulates. It emerges as a pathogen only when passed to humans in contaminated food or water. Pathogenic *E. coli* strains, of which there are many, are equipped with a variable constellation of virulence genes, but the unusually severe illness caused by *E. coli* 0157:H7 is principally attributable to the production of Shiga toxins. *E. coli* 0157:H7 acquired the toxin genes, Stx1 and Stx2, by phage conversion after infection by a tailed double-stranded DNA phage (O'Brien et al. 1984). The disease manifestations are provoked by the release of the toxins into the lumen of the gut. They readily enter the cells lining the gut wall, where they target and disrupt the cellular protein synthesis machinery, resulting in cell death and tissue necrosis. Some of the toxin penetrates into the systemic circulation and can attack susceptible cells and organs, particularly the kidney, resulting in hemolytic uremic syndrome and potential renal failure. The possession of toxin genes by *E. coli* 0157:H7 is an advantageous trait for human infection. As with cholera, the provocation of diarrheal symptoms promotes transmission of the bacterium to new human hosts. Of course, these benefits for the bacterium also benefit its lysogenic phage.

There is one major hitch for the *E. coli* lysogen. The Shiga toxins of *E. coli* 0157:H7 can only be produced during lytic replication of the phage. Toxin production is therefore associated with prophage induction and cell lysis (Wagner et al. 2001; Wagner and Waldor 2002). The diarrhea that is essential for disease transmission therefore occurs at the expense of the pathogenic *E. coli* lysogens. For the individual *E. coli* cell this is unfortunate, but not all of the *E. coli* lysogens in the gut become induced, and transmission and propagation of viable, and genetically identical lysogenized *E. coli* 0157:H7 is assured by the severe diarrheal illness instigated by the toxins. Just as for cholera, the probability of successful transmission of the pathogen to a new human host is directly promoted by the intense diarrheal symptoms of the illness. There are other phenomena at play in the disease caused by pathogenic *E. coli*. It is thought that phage induction and the release of infectious phage particles plays a role in amplifying the disease by initiating lytic infection of bystander bacteria in the gut. These cells in turn make and release the Shiga toxins (Mills et al. 2013; Gamage, Strasser, and Chalk 2003). The possession of a lysogenic phage then is a poison pill for the individual bacterium in which it is induced, but it has a net beneficial effect on the bacterial lineage, increasing its proliferative success and transmission in the human population.

The possession of prophages in the genomes of bacteria can be a double-edged sword. Studies show that spontaneous reactivation of prophages in a population of lysogenized bacterial cells occurs at a low but significant frequency. At any given moment, it is occurring in between 1 in 1,000 and 1 in 10,000 bacteria in a culture. This is a fatal event for this small minority of individuals in the population, but the very fact that many bacterial lineages have a stable association with inducible prophages suggests that there is a net fitness advantage to the host. For EHEC we have readily deduced this. The inducible prophage behaves essentially as a host gene whose phenotype is associated with the suicide of some individuals. However, the phage genes are associated with a second phenotype: toxin secretion. This phenotype of a few individuals is beneficial for the entire population of cells that shares the same inducible prophage in its genome.

The severity of the disease caused by EHEC correlates with the degree of prophage induction. More toxins are released, and more damage to the gut lining ensues. Although induction of prophages occurs spontaneously at a low frequency, it is promoted under conditions of stress for the bacterial cell. This response is beneficial to phage genes that are at risk of demise within a dead bacterium. Indeed, this is likely the case during an episode of disease. Among the conditions that induce prophages are reactive oxygen species, a form of biocide released by immune cells, as part of our response to infection (Wagner, Acheson, and Waldor 2001). Antibiotics are also agents that mediate prophage induction. Treatment with fluoroquinolone antibiotics inhibits bacterial DNA replication and causes chromosomal DNA damage, inducing the bacterial SOS response, phage induction, and attendant toxin production (Zhang et al. 2000; Ubeda et al. 2005; De Paepe et al. 2014; Maiques et al. 2006). There is abundant evidence that antibiotics stimulate toxin production by EHEC and therefore, not surprisingly, antibiotic therapy can exacerbate EHEC disease (Wong et al. 2000; Zhang et al. 2000). For this reason, treatment of EHEC-infected patients eschews the use of antibiotics and is limited to the provision of supportive care.

Today, cases of EHEC and sizeable outbreaks of the disease caused by 0157:H7 are rare occurrences. Most of the time, the bacterium (together with its phage passengers) replicates in its natural host, most likely in the gut of ruminant cattle, where it causes no disease. Why, if the bacterium reproduces here without causing disease, does it naturally retain functional toxin-encoding prophages whose reactivation is a poison pill? It would seem that if the vast majority of the 0157:H7 population replicates in cattle and their disease-causing Shiga toxins play no role in their replicative success, then they would derive no benefit from the prophage. It follows that there must be positive selective pressure for *E. coli* 0157:H7 to retain the prophage while it lives in its major reservoir species. Comparative genomics of many bacterial genomes has documented that they most frequently have a substantial complement of prophages, but that most are defective or fragmentary remnants of the genomes of erstwhile functional prophages (Kuo and Ochman 2010; Lawrence, Hendrix, and Casjens 2001). From this observation, we can infer that natural selection often favors bacterial genomes in which many of their prophages have been neutered and are unable to reactivate. Bacteria often evolve to benefit from phage conversion without retaining the functional prophage and its associated liability of induction, causing cell death. Phage induction, while usually limited to a few cells in the population, can occur at profoundly higher rates in conditions that stress the host cell. Since each bacterial cell may have multiple prophages, it is easy to see that there should be strong selective pressure for their inactivation unless there is counterselection for their functionality. In the case of 0157:H7, scientists hypothesize that Shiga toxins provide a survival advantage to the bacterium living in the gut of cattle (Steinberg and Levin 2007). It is thought that possession of Shiga toxins protects the bacteria from predation by grazing zooplankton that also inhabit the gut of cattle. If this is indeed the case, then the pathogenic phenotype of these toxigenic *E. coli* in humans is just an unfortunate coincidence and a legacy of their struggle for existence and their evolution to survive in their natural environment, the ruminant gut.

Treasure Islands

Staphylococcus aureus is a notorious bacterial pathogen that, in the decades since the advent of antibiotic therapy, has succeeded in acquiring a repertoire of drug-resistance genes which render obsolete whole classes of antibiotics in our arsenal of treatments for these acute invasive microbial infections. Methicillin-resistant *Staphylococcus aureus* (MRSA) has become an increasingly worrisome group of Gram-positive bacterial pathogens, responsible for severe and life-threatening infections that can be acquired in community and hospital settings. In addition to their resistance to β-lactam antibiotics such as methicillin, they have often acquired a multidrug resistance, along with a suite of virulence factors associated with increased pathogenicity that confers improved replicative success in patients (Gordon and Lowy 2008; Otto 2010). These virulent MRSA strains are now global and can dominate outbreaks of *Staphylococcus* in both communities and hospitals, where severely compromised patients are at higher risk of their devastating potential. Today the emergence of drug-resistant and highly virulent bacterial pathogens has reached a crisis point. The challenge to public health resides in the ability of bacteria to undergo rapid adaptive evolutionary change in the face of antimicrobial therapies. *S. aureus* has this ability in spades. At the root of the problem is the selection of drug-resistant variants and variants with increased or altered disease-causing properties. These evolutionary adaptations emerge following mutation and change in the bacterial genome, but most importantly they result after incorporation of new genetic information acquired by horizontal gene transfer from both related and unrelated bacterial genomes.

As for *V. cholerae* and *E. coli*, which become more virulent as a result of the acquisition of new genetic information imported by phages, *S. aureus* strains evolve through acquisition of a vast and various complement of mobile genetic elements. These vehicles, each common in the bacterial metagenome, range from prophages, plasmids, transposons, and pathogenicity islands (Baba et al. 2002; Kuroda et al. 2001). They can be thought of as comprising a "mobilome," a library of mobile genetic material that can, with remarkable facility, be exchanged between both related and unrelated bacterial species. The mobilome is of particular value to communicable pathogenic bacteria that are continually under varying selective pressures in different hosts, and subject to different antibacterial therapies. Phages, plasmids, and pathogenicity islands are all formidable vehicles of genetic variation, as they can mobilize whole genes and gene clusters between bacteria. Complex phenotypic traits can therefore be acquired in one momentary event. Plasmids are alone among these mediators of horizontal gene transfer in that they have no infectious extracellular form, relying on host cell conjugation for their mobilization. As such, plasmids are not classifiable as viruses. Putting them aside in this discussion is not intended to diminish their importance in genetic exchange in microbial systems.

Phages are responsible for fueling microbial diversity and can be directly incriminated in aiding and abetting bacterial pathogenesis by phage conversion. They facilitate adaptation to new hosts and environments by providing phage-associated genetic information that becomes integrated into the host cell chromosome. This can be the basis of a selective advantage, in terms of replicative success, for the bacterium's genome as well as for the phage DNA itself, which selfishly reaps the collateral benefit of its host's improved replicative success. Phages go along for the ride, but also have the option to occasionally "play the field" and "go lytic," hoping to win big, explore new hosts, and replicate their genome even more rapidly.

Pathogenicity islands are mobile genetic elements found in the genomes of a broad range of both Gram-positive and Gram-negative bacteria (Novick, Christie, and Penadés 2010). They comprise large clusters of genes and often many kilobases of DNA. The emergence of so many pathogenic lineages of *S. aureus* in such a short period of time showcases the influence of pathogenicity islands on the evolution of pathogenesis in bacteria. These gene cassettes are all flanked by directly repeated DNA sequences and encode integrase-related genes that are reminiscent of those encoded by temperate phages, that is, those that can lysogenize their hosts. Their different payloads of many genes encode proteins with a wide variety of functions including drug resistance genes and virulence genes such as toxins, superantigens, and other novel gene products. These are the distinctive and valuable genetic currency of pathogenicity islands. Their acquisition by a bacterium can permit the immediate expression of multiple new genes, providing the recipient cells with a radically new phenotype upon which natural selection can exert its effects. Their propagation as parasites of bacterial genomes depends on the competitive advantage provided by their genetic cargo, and it is this cargo that makes them such important catalysts of adaptive evolution in bacteria. Central to their success is their mobility and, of course, you may have already guessed: central to their mobility are phages!

Pathogenicity islands are independently evolving egotistical genetic elements that parasitize phages to carry out their life cycle. In this regard, they are qualified to be included in my broad definition of viruses. They might be considered the ultimate lysogen. They lack the genes to mobilize themselves and rely on the functions of an associate, a "helper virus," to complete their independent replicative life cycle (Ram et al. 2012; Tormo-Más et al. 2010). Phage infection or prophage induction in their host cell activates pathogenicity island–encoded genes that initiate their replicative life cycle and mobilization. The pathogenicity island DNA is excised from the host cell chromosome, forming a circular genome that is replicated by the host cell DNA synthesis machinery. More pathogenicity island gene products are now produced to permit it to hijack the structural proteins of its helper phage to generate infectious virus particles containing its genome. These virus particles are made in large quantities and are the vehicles of efficient horizontal gene transfer of pathogenicity islands between bacteria. Pathogenicity island replication is detrimental to the helper phage that cannot efficiently make its own virions in the face of the activated parasite. This interference with helper phage reproduction has been described as a "paradigm of molecular parasitism" and characterized in elegant studies on *S. aureus* pathogenicity islands and the helper phage 80α (Ram et al. 2012; Novick, Christie, and Penadés 2010). A triangulated relationship between the host cell, the pathogenicity island, and phage 80α has coevolved; while it is clear that the bacterial host benefits from the pathogenicity island, which in turn benefits from receiving "goods and services" from its helper phage 80α, it is less clear why 80α plays along in this game. It is an egotist and selective pressures are expected to optimize its own replicative success. Providing help to the pathogenicity island parasite appears to be detrimental to that objective since 80α replication is diminished as it can make fewer of its own infectious particles. Nevertheless, multiple instances of similar relationships have been documented, involving different pathogenicity islands which exploit different species of phages. Since all of these relationships appear to be evolutionarily stable, we must conclude that the helper phage genome is extracting some replicative benefit from this three-way relationship. One can posit that the reduced lytic fecundity of the phage is perhaps more than offset by the increased and robust bacterial cell populations available as hosts.

Prophage Induction and Antibiotic Drug Resistance

Mutations resulting from errors in chromosomal DNA replication occur apace in bacteria due to their relatively short generation times and are a significant source of genetic variation driving adaptive evolution. Mutations are most often discrete changes in the nucleotide sequence of chromosomal DNA that affect the function of a single gene or protein. Only infrequently do they offer improved fitness to the bacterium and become subject to positive selection. It can be argued that the most significant contributor to evolution in bacteria is the acquisition (or loss) of DNA sequences by horizontal gene transfer. Whole genes and collections of genes are acquired, instantaneously providing new potentially adaptive traits to the bacterial cell. Pathogenicity islands found in virulent strains of staphylococci are particularly potent vehicles of horizontal gene transfer. Their payloads of genes introduce valuable functions to the recipient cell by producing proteins that mediate drug resistance or enhanced pathogenicity. This currency buys them the benefits associated with positive natural selection operating on their host's genome. The improved replicative fitness of the host cell benefits both the host and its pathogenicity islands. Pathogenicity islands have evolved to resemble defective lysogenic phages, dependent on the replication of a helper phage to complete their replicative cycle and infectious transmission. This dictates that either infection of the cell by a helper phage or induction of a prophage can trigger pathogenicity island mobilization. It is a cruel irony that the very treatments that we use to combat bacterial infections promote phage-mediated DNA transduction, the movement of pathogenicity islands between cells and phage conversion (Modi et al. 2013; Penades et al. 2015; Ubeda et al. 2005). Our treatments amplify horizontal gene transfer and lay the groundwork for quantum leaps of adaptive evolution of the very pathogens we are battling.

This is an issue of urgent global concern today. Our health systems increasingly find that the therapeutic options for dangerous and life-threatening bacterial infections are rendered ineffective by antibiotic resistance. The discovery of antibiotics and their profound benefit to humankind globally was one of the great achievements of medical science, but today we are losing ground. The infectious cause of death in hospitalized patients in Western Europe or North America in the 1990s was dominated by those succumbing to acquired immunodeficiency syndrome (AIDS/HIV-1 infection) and its associated opportunistic infections. Today those same statistics are almost devoid of AIDS deaths but are dominated by deaths caused by pathogenic bacterial infections. Many of these infections used to be readily curable. Now we are faced with new strains of bacteria equipped with the ability to withstand our most advanced antibiotic therapies. Such is the emergent crisis of virulent, multidrug-resistant bacterial pathogens that international societies representing infectious disease physicians and researchers have issued desperate pleas and a call to arms for governments to take steps to reinvigorate antibacterial research and development (Stagnates 2004).

This worrying turn of events is attributable to horizontal gene transfer between bacterial species in our microbiome, the environment, and pathogenic species of bacteria (Wright 2007; Martinez 2009; Penades et al. 2015). The overprescription of antibiotics and their profligate use in feedlot cattle are significant contributors to the emergence of antibiotic and multidrug resistance in our commensal microbiota (Martinez 2009). However, even the metagenomes of pristine environments are rife with genetic information ready to be tapped in the evolution of pathogenicity and drug resistance (Colomer-Lluch, Jofre, and Muniesa 2011; Wright 2007). Mobile genetic elements, particularly plasmids, and certainly in some instances phage-mediated transduction, all have a role to play in juggling drug resistance and virulence genes among commensal and pathogenic bacterial species. We live cheek by jowl with our gut microbiota and their repository of mobile pathogenicity and antibiotic resistance genes are of particular concern. Phage induction within our non-pathogenic commensal microbiota is a phenomenon that is likely central to the mobilization of such genes. The concept is highlighted by recently published work from laboratories led by Dr. James J. Collins at Boston University and Harvard Medical School (Modi et al. 2013) that provides tremendous insight into the mechanisms underlying the development and transmission of antibiotic resistance among bacterial species resident in our gut.

It is known that environmental stressors such as antibiotics, even at subtherapeutic levels far below those used in the treatment of infections, can increase prophage induction in bacteria in the gut (Maiques et al. 2006; Ubeda et al. 2005). Normal but rare stochastic events become frequent and deterministic events for the bacterial population. The Collins team of investigators sought to recreate the conditions of antibiotic treatment in an animal model using mice. They investigated the perturbation of the *phageome* (the collective sequences of all of the phage genomes in the sample) in the feces of mice that were treated with antimicrobial drugs. Employing antibiotic doses equivalent to those that would be used to treat a human subject suffering from a bacterial infection, they chose two commonly used antibiotics with different mechanisms of action. The first, ciprofloxacin, is a fluoroquinolone that exerts its antimicrobial activity at the level of DNA replication by inhibiting the bacterial topoisomerase II enzyme ligase activity. As discussed above, fluoroquinolone treatment is known to induce the SOS stress response and phage induction in EHEC patients. The second antibiotic was ampicillin, a β-lactam whose antibacterial effects result from inhibition of bacterial cell wall synthesis. Following treatment, phage DNA was isolated from feces of the mice, and the phageome was determined by deep sequencing technology. The DNA sequences were computationally compared to known DNA sequences in databases, allowing their origin (bacterial or phage) and the function of the corresponding gene to be determined.

The observations were striking, and the researchers' conclusions have far-reaching implications for how we understand the pivotal role phages play in genetic diversification and as catalysts of genetic adaptation in the gut microbiota. In the mice treated with antibiotics as if they were patients under treatment, there was an increase in the recovery of phage DNA encoding a broad range of drug resistance gene sequences. To the researchers' surprise, however, these phage-mobilized genes included genes that confer drug resistance to the respective administered antibiotic and also genes associated with resistance to many unrelated antibiotics. This generalized transduction of DNA significantly enriched the genetic complexity of the phageome. It arose from mobilization of host bacterial DNA from the most abundant bacterial phyla populating the mouse gut. It was also enriched in bacterial genes known to be functionally beneficial under stress-related conditions. For example, the phageome of fluoroquinolone-treated mice was enriched in DNA replication and repair-related genes, while that of ampicillin-treated mice had an overrepresentation of genes related to carbohydrate and cell wall synthesis the pathways that ampicillin disrupts. In each case, the genes mobilized are linked to the pathways perturbed by the antibiotic. The induction of phages and the generalized transduction of host DNA sequences seemed to favor genes that provide bacteria with adaptive capacity for stress situations. The researchers also tracked down fragments of DNA containing phage and bacterial sequences linked together in a single contiguous molecule. These fragments of DNA were evidence of infection and recombination between the genome of a particular phage and host species. Remarkably, they found host-phage recombinant DNA fragments indicating that unexpected novel phage-host interactions were taking place. It seems that in the stressed gut ecosystem, phages not only act as an expanded adaptive gene bank, they also increase access to this expanded genetic repertoire by increasing the promiscuity of phage-host interactions.

The influence of phages in the evolution and dissemination of virulence and multidrug resistance is therefore greater under antibiotic selective pressure on the gut microflora. The destabilizing effect of antibiotic treatment on the gut microbiota is well known to increase the potential for new niche colonization and expansion of novel pathobionts in the gut ecosystem. Such destabilization of our normal gut flora is the principal underlying cause of the pathologic expansion of the opportunistic pathobionts *Clostridium difficile*, which otherwise causes us little or no distress (Young and Schmidt 2004; Owens et al. 2008). Collins's work illustrates that antibiotics also provoke unforeseen changes in our gut ecosystem. Promiscuous phage-mediated mobilization of resistance and virulence genes among our gut microbiota is a salutary observation. It has the potential to accelerate microbial evolution and the emergence of adaptive pathogen variants with new virulence and drug resistance phenotypes.

The passive collaborator of pathogenic bacteria is, of course, the vast repository of genetic information accessible in the viral and the bacterial metagenomes. Collectively, they act as a reservoir for genetic adaptation, aided and abetted by phage-mediated replication. This hoard of genetic information can be mobilized and tapped for adaptive evolution. So significant is the role of horizontal gene transfer in bacterial evolution, that it is the primary driver of intraspecies divergence and often the major contributor to microbial speciation. Doubtless, bacterial species must have a collection of genes that make up the core species genomic identity, but the boundary can be blurred by an almost limitless supply of accessory mobile genetic information. Collectively this has been dubbed the microbial pan-genome.

Whether it be scarlet fever, cholera, diphtheria, toxic shock syndrome, necrotizing fasciitis, or newly virulent forms of MRSA, all have backseat drivers in mobile genetic information and phage-mediated biological events. Clinical management of bacterial diseases and public health strategies must also account for the role of phages and horizontal gene transmission in the rapid emergence and spread of antibiotic drug-resistant and newly virulent organisms. We will do well to be most vigilant of viruses and develop therapeutic strategies that avoid the promotion of adaptive evolution in the pathogens that we are battling.

· 4 ·

VIRUSES *and* HIGHER ORGANISMS

We will now shift focus to our own viruses, those that make up the human virome. We have a diversity of relationships with these viruses each forged by various durations of coevolution. They can be transient and inconsequential, life-threatening or fatal. They may be long-term persistent infections, in which the disease is chronic and progressive or latent and relapsing. Many are clinically unapparent. It is estimated that each of us is chronically infected with eight to twelve different viruses (Virgin 2014). This number will vary with our age, our ecosystem (where we live), our economic status, and above all, our genetic makeup, none more influential than the genes that govern our immune system. It is also a significant underestimate, as it takes no account of the viruses of our microbiome (Chapters 2 and 3) and cannot account for viruses that are yet to be recognized.

Viruses, Cells, Organisms, and Populations

The viruses of primitive prokaryotes display remarkably evolved and nuanced relationships with their single-celled microbial hosts. Our perception of them as "primitive" is somewhat misleading: they have coevolved with their hosts since before the third domain of life, the eukaryotes, emerged. We may believe that our own success is measured in our biological and cultural sophistication—our sentience, the development of language, or our complex social structures, but our genome (and any of our genes for that matter) could care less. If it could care at all, it would consider success simply its genetic perpetuity. In this regard phages and their genomes are hugely successful, having shepherded their genetic information from the very earliest roots of evolutionary history to the present day. Nevertheless, we are now moving into the territory of viruses that infect us—vertebrate animals, metazoan organisms, and members of the *Eukarya*. Regardless of how one qualifies success on a genetic level, these nanomachines must navigate much more complex territory if they are to be successful and perpetuate their genes.

While *Homo sapiens* as a species is thought to have emerged no more than 200,000 years ago, many of the viruses that infect us have likely been constant companions of the ancestors from which *H. sapiens* descended. They coevolved and codiverged faithfully as successive speciation led to the great apes and ultimately to man. Others are likely just as ancient in their origins, but relatively new diseases for humankind. Just as we think of ourselves as the pinnacle of evolutionary accomplishment, they, too, have evolved sophisticated lifestyles that ensure the propagation of their genetic lineage and identity. Our current snapshot of the human virome records only the successful lineages of viral information—those that have evolved the necessary informational content in their genomes to allow them to persist as species today. It is a singular phenomenon (and one we will return to later) that unlike living organisms, viruses leave no tangible fossil record from which evolutionary relationships can be referenced to geological time. All of our knowledge of viral evolution is a backward projection, gleaned from the genetic composition of the lineages of viruses that prevail today; extinct viruses and their genetic information are utterly lost to research. As selfish genetic parasites, these viruses, if they were to have motives, have none different to the phages that infect bacteria: self-replication. The parameters that influence the success of a eukaryotic virus lineage are no different either. We will see that, at a high level, the factors that drive the evolution of human viruses parallel those that operate on the simplest phages.

A significant distinction, however, needs to be addressed at the outset. It resides in the complexity of the host: it is an organism, a cooperative community of cells. When we discussed lytic phages, the viruses that infect the bacteria and archaea, we considered only the individual infected host cell. A successful phage infection begins with the infection of a single cell and is ultimately realized by the release of progeny phage particles into the external milieu, free to infect other susceptible individuals in the population. These particles are the vehicles of the phage's genetic information, responsible for transmitting it to new host cells. Individual and genetically identical host cells within the population are infected in successive amplifying cycles of viral replication for as long as the host cell population remains. Successful transmission is vital for the replication and propagation of viruses in a host population. The phages of bacterial cells adopt a variety of strategies that ensure the success of their lineages. Virulent lytic infection results in immediate genome amplification while lysogeny postpones productive replication pending reactivation of the prophage at a later time. In each case, I adhere to my definition of a virus, which requires it to be an independently evolving selfish genetic entity. Transmission to and infection of a new host cell are central to the essence of being viral. The viruses that infect metazoan organisms such as ourselves have the same fundamental challenge to their survival: transmission. They must infect the organism, replicate, and be released in such a way that they can access a new host. This is a singular challenge: viruses must gain entry into and navigate the vast collective of cells, almost 50 trillion in total, that make up its tissues and organs. They must find the particular cells in which they can replicate and also survive multiple layers of defenses. Vertebrates, for example, have highly evolved innate and adaptive immune systems designed to protect against foreign invaders. Finally, they must engineer the release of their progeny virus particles for transmission to a new host.

Even for animals such as ourselves, a virus infection and the disease it causes can be broken down into its fundamental units, the infection of individual single cells. Nevertheless, animal viruses must go beyond replicating and spreading within the heterogeneous population of cells that make up the tissues and organs of the body. They must also reproduce and spread among the population of organisms, which are more than likely separated spatially. The distinct pathologies of different virus infections are the best window through which to scrutinize the machinery and processes that they have evolved to meet these challenges and emerge by natural selection as successful viral lineages today. Animal viromes are composed of an array of more or less pathogenic viruses, which pursue different and often surprising strategies in their relationship with their host. As with phages and their prokaryotic hosts, the relationships of animal viruses and their hosts are equally diverse, representing a continuum extending from parasitic to mutualistic symbioses. These virus-host relationships are also products of the reciprocal evolutionary trajectories of the host and pathogen, each pursuing the arms race to equip itself for survival. The evolutionary adaptation of respiratory viruses to the human host provides us with a vivid illustration of evolved pathogenic processes. Here the viruses have evolved to ensure their successful replication and efficient transmission, fueling their epidemic spread among human hosts.

"Just a Virus"

One early fall afternoon a three-year-old child is listless and unusually needy. Later in the evening she has a mild fever and her cheeks are flushed; the parents bring down the fever with acetaminophen. The following day the fever persists, she is congested and has a profuse nasal discharge that defeats any attempt to keep her upper lip dry; a rasping cough is starting to develop, and she wheezes. Quite rightly concerned, the parents take her to their pediatrician. The diagnosis is reassuring but certainly not definitive: "Not the flu, probably just a virus." He advises, "Take her home, keep a close eye on the cough, and keep her comfortable and well hydrated, but if it takes a turn for the worse, come back in." He will prescribe no medicine. None exist; there are no specific treatments, and this constellation of quite generic symptoms might be caused by any one of scores of different viruses that infect our respiratory tract. Chicken soup works as well as anything; members of my generation were subjected to a variety of home remedies. One involved the ordeal of sitting in a bath of steaming yellow mustard water to "draw out the toxins." It was as successful as any other, and the astringency of the mustard was not, in fact, unpleasant at all. The toddler's symptoms will likely persist for a day or so, after which the familiar cheery child will return, no worse for wear.

This is the course of almost all acute respiratory infections caused by viruses. They are usually transient and self-resolving infections that rarely lead to serious complications. The doctor's diagnosis and his inability to pinpoint the particular viral culprit are satisfactory only in that a more definitive diagnosis would not be helpful. There are no specific treatments he can offer. There are a bevy of different viruses that cause colds (not to mention bacteria that cause a small proportion of cold-like illnesses) and in everyday life you will rarely receive a definitive diagnosis of the etiology of your child's upper respiratory infection.

We are concerned with viruses and the first human virus we will explore is that seemingly most ubiquitous of pathogens, the "common cold" virus. The viruses that infect us and cause the symptoms of the common cold can include influenza (if it is an unusually mild case), parainfluenza viruses, respiratory syncytial virus, metapneumoviruses, Coxsackie viruses, enteroviruses, adenoviruses, coronaviruses, and rhinoviruses. Nevertheless, human rhinovirus is the poster child for acute respiratory virus infections. It is the cause of up to 70 percent of viral respiratory infections, particularly those in the spring and the fall (Jacobs, Lamson, St. George, et al. 2013). The success of this human pathogen is self-evident. But why is it so staggeringly successful? It causes only mild and transient respiratory illness but our collective experience teaches us that colds spread through households and classrooms with ruthless efficiency. How has its particular infection strategy and its relationship with humans (its exclusive host) evolved such that it is our perennial and irksome companion today?

Human Rhinoviruses

The common cold virus is the human rhinovirus (HRV). *Rhino* derives from the Greek, meaning "of the nose or nasal." HRV was recognized in the 1950s as a virus that causes colds and today more than 100 different serotypes are characterized (Jacobs, Lamson, St. George, et al. 2013). All are closely related genetically and were isolated from patients with the same illness. The virus replicates in the epithelial cells lining our nasal passageways. For decades, the disease pathology caused by the virus was attributed to localized viral replication in the nose and the nasopharynx. Scientists knew that HRV replicates best in cultured cells kept at temperatures between 33 and 35 Celsius. This is far below body temperature, so they assumed the virus would infect only the cooler upper nasal passageways. This was wrong. In a minority of patients rhinoviruses cause lower respiratory tract symptoms as well as the typical constellation of cold symptoms restricted to the upper airways. Replication in the lower respiratory tract has now been definitively demonstrated in many studies by the detection and recovery of viral genome RNA in the lower airways (Gern et al. 1997; Kaiser et al. 2006; Mosser et al. 2005). Scientists at Case Western Reserve in Cleveland conducted precise studies on the prevailing temperature in the human airways. They directly measured the actual temperatures in the lower airways and discovered that temperatures approaching 37 Celsius prevail only after the airways have branched at least four times and have burrowed deeply into the lung (McFadden et al. 1985). We now accept that rhinovirus infection can extend to the lower respiratory tract and that virus replication can be detected there. This aspect of rhinovirus pathogenesis is particularly important for more susceptible individuals. Patients with other respiratory diseases such as asthma or chronic obstructive pulmonary disease (COPD) are susceptible to severe exacerbations of their preexisting illnesses. It appears then that when your mother pronounced you to have a "chest cold," she might well have been medically accurate in her diagnosis.

There exists an apocryphal notion that colds are associated with cold temperatures; "you will catch your death of cold" was a frequent admonition to children venturing out underdressed for the weather. Cold weather was thought to benefit the transmission of the virus. What better place then to carry out studies on the transmission of rhinovirus colds than at McMurdo Station, a U.S. research outpost in Antarctica? A team from the University of Wisconsin did just that and reported their results in 1989. One of their conclusions was that long-term exposure to intense cold had no significant influence on the transmission of the common cold among the scientists overwintering at the station (Warshauer et al. 1989). These studies debunked yet another myth about the common cold.

So what of the "way of life" of rhinoviruses? What is the secret of this successful genetic lineage of human disease pathogens that visits us with such regularity? Let us begin at the beginning as we did in our consideration of tailed DNA phages. Rhinoviruses are members of the large and diverse *Picornaviridae* family; they are similar in size to the prokaryote-infecting tailed phages and also have an icosahedral-shaped protein capsid in which the genome is packaged. Instead of a double-stranded DNA genome, the human rhinovirus's genetic blueprint is composed of a single-stranded polymer of RNA, approximately 8 kilobases in length. RNA is closely related to DNA and is generally accepted as its evolutionary precursor as the hereditary material of life. This positive-sense strand of genomic RNA directly presents the protein coding information to the host cell's ribosomes, where viral protein synthesis is immediately undertaken. The viral RNA is translated into a large protein that is processed into eleven smaller proteins, together composing the structural components of the virus particle (viral proteins 1–4) and the nonstructural proteins responsible for conducting the replication of the genome. The rhinovirus particle is an effective vehicle for the virus: infection is achieved when a virus comes into contact with the nasal epithelia or the conjunctiva of the eye. The virus can remain viable in that environment for up to four days and for two hours on undisturbed skin (Jacobs, Lamson, St. George, et al. 2013). Surveillance of volunteers reveals that 40 percent of people with colds have detectable virus on their hands (Gwaltney, Moskalski, and Hendley 1978). Hand-to-eye, hand-to-nose, and direct contact with nasal secretions or aerosols are the most common modes of infection. The virus that reaches the nasal passageways must gain entry into host cells, those of the nasal epithelium, in which it will replicate. This is secured by recognition and binding to receptor proteins displayed on the surface of the nasal epithelial cells. The specificity of this physical interaction is exquisite. The exterior shape and structure of the viral capsid are precisely complementary to the three-dimensional structure displayed in the extracellular domain of the receptor (a "lock and key"). In this way the vast majority of rhinoviruses recognize and bind the intercellular adhesion molecule 1 (ICAM-1), a protein on the surface of epithelial cells. Under normal circumstances, it plays a central role in the trafficking of leukocytes, cells of our immune system, to sites of infection or inflammation. A small minority of rhinovirus types, about 10 percent, exploit binding to a different cell-surface protein, the low-density lipoprotein (LDL) receptor. This binding event between the virus particle and a cell surface protein fulfills the same purpose; it results in the attachment of the virus to the cell. Regardless of which receptor the rhinovirus recognizes and binds, the resulting complexes are taken into the cell by endocytosis. The virus hijacks the same process that the cell uses to retrieve and recycle its own proteins from the cell surface. The complex is internalized in a series of membranous vesicles created by the involution of the cell's bounding membrane. The virus and receptor now dissociate, and the virus particle disassembles to release the genome into the cytoplasm—just as with phages, the genome enters the world of its host naked. The virus genome is immediately recognized as a template for protein synthesis by cellular ribosomes. These viral proteins quickly arrest the normal functioning of the cell and assemble into replication complexes that duplicate the RNA genome, ready for packaging and release. Unlike phages, the rhinovirus has no lifestyle choice to make after it enters the cell. It has a lifestyle equivalent to a virulent lytic phage; it has no option for temperance (the lysogeny exhibited by temperate phages), and it has no mechanism for persistence in the cell. It is a bare-bones racer, encoding only essential gene products; it must pursue hit-and-run lytic replication in its host cell and in our bodies.

Luckily rhinovirus replication rarely, if ever, escapes the respiratory epithelial tissues, and it is ultimately overcome by concerted immune responses that are mobilized as soon as the first cells become infected by the virus. Immediately upon infection, the nasal epithelial cells detect the invading virus and trigger the activation of defensive networks of gene expression. These responses comprise the innate antiviral response of the cell and the adaptive immune response, which is subsequently triggered in a cascade of events (Jacobs, Lamson, St. George, et al. 2013). These reactions immediately cause the infected cells to secrete inflammatory mediators, kinins, and prostaglandins, which lead to vascular permeability and exocrine secretion. The runny and itchy nose that first signals your oncoming cold is the result. This first nasal discharge is plasma leaking from the vasculature of your nasal passageways; only later is it composed predominantly of virus-laden mucus resulting from hypersecretion. Other lines of communication between cells are also opened in the form of interleukins, signaling molecules such as interleukin 8 (IL-8) and chemokines that are released from the infected tissue. IL-8 is a potent chemoattractant, which summons immune cells (neutrophils and monocytes) to the location of an infection to take up the battle against the invader. The amount of IL-8 measured in nasal discharges seems to be closely related to the severity of cold symptoms.

It is our immune system itself that is responsible for aggravating our cold symptoms as it goes about the process of containing and clearing the virus-infected cells from our body. The manifestation of symptoms is, of course, essential for the transmission of the virus, which is released from nasal epithelial cells and carried out of the host with the flow of nasal secretions. A rhinovirus infection that failed to provoke the immune-inflammatory response of the host would not achieve its end: successful communication to susceptible new hosts. It is interesting to note that one response of rhinovirus-infected cells is to increase the expression of the ICAM-1 molecule. This is the very receptor of some HRV serotypes but not of others, and ICAM-1 is upregulated by all rhinoviruses indicating it is not a direct result of the engagement of the receptor by the virus. Papi and colleagues have speculated that noncytopathically rhinovirus-infected respiratory epithelial cells may become more susceptible to further infection as a result of receptor upregulation (Johnston et al. 1998; Papi and Johnston 1999), but this remains speculative. It would in any case be a phenomenon restricted to rhinoviruses that use ICAM-1 for cell entry. Why then would all rhinoviruses, irrespective of receptor selectivity, have evolved to have this effect on the infected cell? It seems most likely that increased ICAM-1 expression has no influence on the outcome of a productive virus infection in any one particular cell but that it has evolved to increase the efficiency with which leukocytes migrate to the infected tissue. This is certainly the most important biological response as it is in this manner that rhinovirus provokes a more efficient mobilization of immune cells to the nose, increasing the severity of the symptoms responsible for its transmission (Papi and Johnston 1999).

A commonly held belief of evolutionary biologists is that long-term coadaptation of a virus and its host will result in the evolution of a relationship that minimizes the negative consequences of the disease in the host. The logical conclusion of such a premise is that infections will over time naturally evolve to cause minimal or no disease in a host. This outcome is often evident in viruses that have coevolved and codiverged with their host species over long periods of evolutionary time. As the result of an extended arms race between host and virus, such evolutionary détente can only be achieved if the virus can be successfully transmitted in the absence of disease. In the case of rhinoviruses that rely on disease symptoms for their transmission, natural selection has by necessity maintained their capacity to cause clinically apparent disease in the host. During their transient relationship with the host organism, the viruses must create the conditions that allow them to move from host to host. It is not unreasonable to speculate, however, that evolution has modulated the severity of the disease that rhinoviruses cause, optimizing the potential for transmission. These evolutionary pressures are likely responsible for creating what we recognize as the common cold, a vexing infection that causes marked upper respiratory symptoms. The relatively benign nature of the disease and its mild symptomology do not undermine the success of the virus: it is the highly effective vehicle of its spread among us. Colds cause only minor constitutional symptoms that do not restrict our daily life. We continue to have social interactions with other potential victims, despite being infectious centers for further disease spread. It is the perfect crime.

Uncommon Diversity

The disease that rhinoviruses cause is a "storm in a teacup": no massive tissue destruction occurs, and only a small minority of cells in the nasal epithelium become infected and die. After resolution of the infection, we acquire immunity to the rhinovirus that was responsible. Reinfection with the same strain of rhinovirus is precluded by immunological memory that quickly responds, nipping the infection in the bud. Reinfection by the same virus strain may only occur, if it occurs at all, after immunological memory has dimmed sufficiently, which may for some virus infections occur over more extended periods of time. The success of rhinoviruses as the cause of multiple colds per year in our children is not due to reinfection by the same virus (and here I use the term "same virus" to refer to a particular genetic lineage of rhinovirus), but to infections of genetically distinct viruses. It is a reflection of the large number and diversity of rhinovirus types that circulate simultaneously in our population. This is illustrated convincingly in observational studies conducted at the University of Wisconsin. Thirty-four children with asthma were studied over the course of September, the month during which the children returned to school (Olenec et al. 2010). In the third week of the month, sixteen of thirty-four children had a rhinovirus cold; during the entire month the team detected seventeen different rhinovirus strains in the cohort. Although these were studies of asthmatic children, for whom susceptibility to rhinovirus and the duration of their colds may be greater than for their healthier counterparts, it serves to illustrate that, at any one time, multiple different strains of rhinoviruses circulate among us.

To explore the evolution of rhinoviruses, comparative genomics researchers have analyzed the genome sequences of various distinct isolates to elucidate the underlying basis of their genetic variation and the selective pressures that have created their genetic diversity. Errors in RNA replication are central to the capacity of rhinoviruses to adapt and evolve. Surveillance of the variation in the nucleotide sequence of different rhinovirus strains has provided remarkable insights into the forces that have driven their evolution. This will need some explanation: our genetic code (and that employed by viruses) is "degenerate" in nature. It needs to be able to specify each of the twenty-one different amino acids that in various combinations and sequence are coupled together to make up proteins; it also must encode a stop signal. A nondegenerate code would have a lexicon of twenty-two words: one per amino acid and an additional one to tell the ribosome to stop making protein. In fact, our alphabet of ribonucleotides has four different bases (think of them as letters): guanidine, adenine, cytidine, and uracil. These are combined to make words, or codons, of three letters in length that specify which amino acid to incorporate into the protein. Thus, there are $4^3 = 64$ available words while only twenty-one amino acids and a stop signal are needed. In fact most amino acids can be encoded by more than one different codon, hence the degeneracy, or in other words redundancy, of the code. This feature of the genetic code provides a valuable tool for the study of evolution. Since words (that are codons) with different spellings (various triplet sequences of the four bases) can denote the same amino acid, some spelling errors in the genetic code may result in no change in meaning, and the same amino acid will be incorporated. Such a change would be considered a *synonymous* mutation. Should the change in the codon result in a new codon corresponding to a different amino acid, it is regarded as a *nonsynonymous* mutation.

Below are examples of synonymous and nonsynonymous mutations in a codon for the amino acid leucine:

CUU (leucine) -> CUC (leucine) = synonymous mutation

CUU (leucine) -> CGU (arginine) = nonsynonymous mutation

Examination of the entire genomes of multiple rhinovirus isolates and an assessment of the number of synonymous changes compared to nonsynonymous changes in the genome allow comparative genomics researchers to infer the type of natural selection and nature of the selective pressure that different regions of the genome have been subject to. If purifying selection is the prevailing selective pressure, it will favor the maintenance of the preexisting protein amino acid sequences. Synonymous changes may be tolerated because they result in no changes to the viral protein products. Nonsynonymous nucleotide changes will, however, be selected against because they are deleterious to the fitness of the viral genome. On the other hand, if conditions prevail in which changes in the viral proteins are favored by selective pressures, a greater proportion of nonsynonymous changes to the viral genome sequence will be observed. This is termed positive selection.

The observations made by Kistler and colleagues at the University of California, San Francisco are worthy of particular note (Kistler et al. 2007). They studied the complete genome sequences of a representative set of thirty-four rhinovirus strains. They concluded that most of the rhinovirus genome evolves under purifying selective pressure. Most of the genetic variation caused by errors in HRV genome replication, therefore, result in changes that are deleterious to the fitness of the virus. Natural selection purges the genomes of mutations that alter the protein amino acid sequence. The hallmark of this purifying selection is that synonymous mutations are tolerated and thus more frequently observed than nonsynonymous mutations. This is consistent with what one might expect for a virus that has become highly, perhaps optimally, adapted to its host. There is no selective pressure for genetic change. Given the prevalence of the common cold, the conclusion that rhinoviruses are well adapted to their hosts is inescapable.

Other observations provide some clue as to the driver behind the origin of multiple serotypes of rhinoviruses. In some discrete regions of the genome, they saw a higher than expected frequency of nonsynonymous changes compared with synonymous changes: positive selection had been at work on these portions of the genome. The regions affected were restricted to those that encode the proteins which form the exterior capsid of the virus and to discreet regions of some nonstructural proteins.

The portions of the nonstructural proteins that were under positive selection appeared to be restricted to those that the researchers considered were not central to their functions in the cell. It is most likely that these selective pressures for change result from the rhinovirus being under attack by our immune system. The *virion* proteins, those associated with the virus particle, are recognized by circulating protective antibodies that evolve pursuant to infection. After recovery from the cold, antibodies with this specificity render a person immune to reinfection by the same serotype of rhinovirus. These are the antibodies that recognize different rhinovirus serotypes and were the initial basis for classifying the more than 100 serotypes that have evolved. There is constant selective pressure on viruses to circumvent these neutralizing antibodies; they do so by disguising themselves within capsids that are altered and thus cannot be recognized by an immune system which may have experienced other strains of the cold virus before. It is reasonable to speculate that the evolution of multiple serotypes of rhinoviruses permits them to persist and flourish in our populations because they have access to a conveyor belt of susceptible hosts.

The reason for positive selective pressure (for a change) in the amino acid sequences of the virus nonstructural proteins (to which no neutralizing antibodies develop) has not been explained in the literature. I speculate, however, that this is also driven by selective pressure for immune escape. The immune-inflammatory cascade initiated by the infected cell is also responsible for recruiting cytotoxic killer T cells to the infected tissue. This arm of our immune system—termed the cell-mediated immune response—has the unique ability to recognize and target infected cells for elimination. Infected cells betray themselves to the immune system because, on their surface, they display small segments of proteins—peptides—that are derived by degradation of the proteins made within each cell. The many peptides usually displayed by the cell are derived from cellular proteins and are recognized by the immune system as "self-antigens." The fingerprints of a virus infection are detected when fragments of viral proteins find their way onto the cell surface. These are "seen" and recognized as foreign by cytotoxic T cells, which then hone in on the cell and destroy it. The particular viral peptides that are displayed and recognized on the surface of infected cells thus educate our immune response and render the virus-infected cell susceptible to cell-mediated immune clearance. Cells that are infected by variant viruses with changes in these peptides will not be recognized by the cytotoxic T cells. These viruses will have a selective advantage in that host and evade this mechanism of immune clearance. They will be positively selected.

It is, therefore, a reasonable supposition that the principal driver of rhinovirus diversification is our (and our prehuman ancestors') immune responses. The facility of rhinoviruses to evolve diversity has been critical to their success as ubiquitous pathogens that rely on the availability of naive hosts to infect. Each of us has experienced many colds, all likely attributable to infection by different rhinoviruses; our children are particularly susceptible to colds for just this reason—they are naive to the majority of rhinovirus serotypes. They are fertile ground for infection because they are equipped with antibodies to repel only those rhinoviruses that they have already experienced. Colds are a moving target for our immune response; not only are they adept at causing mild disease with symptoms designed for optimal communicability between persons, they are also genetically diverse, allowing their genetic information to be propagated in the greatest available number of hosts. Rhinoviruses cannot rest on their laurels; the perpetuation of each rhinovirus lineage depends on continuous cycles of replication. Do some rhinovirus serotypes run out of hosts and become extinct, and are new serotypes emerging? The latter is highly likely, but the former less so. Numerous rhinovirus serotypes are known to circulate simultaneously in any particular population and no individual serotype is successful enough to exhaust its available hosts and burn itself out. Geographical migration and the birth of susceptible infants also provide viruses with susceptible new hosts for each serotype to infect.

Accidents of Pathogenesis

I have alluded to the fact that some rhinovirus infections are more serious than others. Most are mild and restricted to the upper airways, but some take the form of more deep-seated infections that affect the lower airways. It is reasonable to ask if this is based on differences in the virus or of the host. The prevailing wisdom has been that the severity of a cold is more dependent on the individual than the serotype of virus with which he or she is infected. Higher symptom scores have been recorded in patients who mount a less powerful antiviral response. This is evident when the severity of cold symptoms are compared with the amounts of the antiviral mediator interferon-α that are secreted (Message and Johnston 2001; Copenhaver et al. 2004; Sykes et al. 2012). Patients who mobilize less interferon have more severe disease. Individuals with pulmonary diseases and immunocompromised patients are also more susceptible to serious outcomes of rhinovirus infection. Nevertheless, a new clade of rhinoviruses, HRV-C, discovered recently (Lee et al. 2007) does appear to be associated with more severe lower respiratory tract effects, particularly in asthmatic individuals. This segment of rhinovirus disease is understandably of great concern to both patients and doctors, but in terms of rhinovirus evolution it may be of little consequence since it is so rare. It is unlikely that these infrequent disease sequelae promote the selfish objectives of rhinovirus genomes. This is particularly likely if they result from differences in the host and are not the result of genetic variation in the virus that can be subject to natural selection. If the severe sequelae of colds in some individuals were to be determined at the level of the HRV genome sequence and if it provided a substantial benefit to disease transmission, one would expect that more severe forms of the virus infection would gain in frequency. Since this does not appear to be the case, it seems most plausible that the evolution of the pathogenicity of rhinovirus lineages has already achieved the optimal balance between disease transmission and pathogenicity. The fittest rhinovirus genomes appear to be the product of the most frequent encounters of rhinoviruses with their hosts in which typical mild and limited cold symptoms are the hallmarks of pathogenicity. The minority events resulting in more severe disease sequelae do not appear to be influential drivers of rhinovirus evolutionary change. Rhinoviruses will not evolve increased pathogenicity. The prevailing purifying selective pressure on the genome supports the precept that during the long-term relationship of rhinoviruses with humans, evolution has already honed their interactions to an optimum level.

It is still possible that rhinovirus evolutionary change has been (and could be) potentiated by the occasional case of more serious disease. One may speculate that such infections will result in a greater number of replicative cycles occurring within the single individual. The increased number of replicative cycles in the patient could enrich the genetic diversity of the virus population to a greater extent than that created during a short-term mild infection (Tapparel et al. 2011). These variants, if available for transmission, are likely subject to the prevailing purifying selection and purged from the population (in the status quo). Nevertheless, under conditions where the host population might be undergoing evolutionary change, such increased genetic variation could be fertile ground for altered (positive) selective pressures. There is, however, currently no data to substantiate the idea that this phenomenon has played a central role in the perpetuation of rhinovirus lineages or their diversification.

Allow me to introduce you to another virus of the *Picornaviridae*, this one an enterovirus, the poliovirus. Its genetic makeup is remarkably similar to the rhinovirus: its genome is of the same size, it is organized similarly and has a very similar capsid. Poliovirus, however, does not cause respiratory illnesses; it is an enteric pathogen that infects the human gut. The virus in the feces of infected individuals is the primary source of transmission of the infection. In the collective consciousness, poliovirus is first and foremost the cause of the devastating disease, poliomyelitis, which until a vaccine became available in the 1950s, returned as seasonal epidemics. Its primary victims were children and poliovirus was the cause of infantile paralysis. This was a truly horrific disease. However, after the introduction of Jonas Salk's effective vaccine and a successful vaccination campaign, the image of the iron lung, a massive contraption fashioned to assist the breathing of those unable to do so on their own, has faded from memory. The reason I am introducing polio into my discussion now is not simply because it is a devastating disease caused by a virus, but that it illustrates most effectively the point that I have just made regarding the selective pressures that exert the most influence on rhinovirus evolution. Poliovirus transmission is by the oral-fecal route; the virus is taken in orally and ultimately shed in the feces. It first infects and multiplies in the oropharyngeal and intestinal mucosa, but it can also cross the gut wall and transit into gut-associated and cervical lymph nodes. Replication here causes a viremia, in which free virus particles can be found in the blood. Nevertheless, for 98–99 percent of infected individuals that is where it ends. It is a minor disease and results in sore throat and constitutional symptoms such as fever and malaise. Indeed, serological studies of the period in which polio was endemic in our society revealed that large numbers of individuals had been infected by the virus but could not report ever having had the illness. In many instances then poliovirus infection occurs without causing noticeable symptoms, and in most others the disease is benign. On the other hand, poliovirus can invade the nervous system in a small minority of infected individuals and can result in the terrible neurological sequelae that are the hallmark of poliomyelitis. This is, however, an accidental disease, incidental to the natural course of the infection, which is essentially one of the alimentary tract.

Since only a small minority of infected individuals succumbs to neuronal invasion, it is logical to speculate that this minority may be predisposed to the more severe disease. Work with animal models of poliovirus infection has implicated the antiviral response to infection and the interferon response. It appears that the production of lower levels of interferon may preclude the effective control of viral replication in tissues that serve as the staging point for poliovirus invasion of neuronal tissue (Racaniello 2006). It has therefore been postulated that poliomyelitis may be a consequence of the failure of some individuals to mount an adequate antiviral response. A failure to effectively control viral replication will result in more virus particles and an increased risk of seeding infection of neurons and invasion of the central nervous system. The interferon-based antiviral response has a strong influence on the severity of rhinovirus cold symptoms, so too might it govern the susceptibility of young children to paralysis caused by poliovirus.

The selfish objectives of the poliovirus genome, replication, and transmission are entirely satisfied by replication in the gut: it is an enteric virus and has evolved with this as its principal mission. Nervous system infection is incidental. Time and time again virologists must parse the significance of the sometimes-severe disease outcomes in a small minority of infected individuals and the major burden of mild disease. It is the mild or asymptomatic disease that contributes the overwhelming majority of infection-transmission cycles of the virus, and this is where the evolutionary music plays.

Mutation, Diversity, and Quasispecies

The unique ability of viruses to rapidly evolve and influence the evolution of their hosts will be a theme that runs through this book. Rhinoviruses and poliovirus share with all other RNA viruses a feature that distinguishes them from viruses with DNA genomes: their error-prone replication machinery. The RNA-dependent RNA polymerases of RNA viruses are distinguished from DNA-dependent DNA polymerases by their lack of a proofreading mechanism. With rare exceptions, DNA polymerases can edit their text to ensure its accuracy, while RNA polymerases must be satisfied with their first draft. Consequently, RNA polymerases introduce mutations, and hence genetic variation, in progeny genomes at a magnitude greater than their DNA polymerizing counterparts. Such is the error-prone nature of RNA virus replication, that for a virus with a genome of 10,000 bases, each progeny genome can be expected to differ from the parental genome by at least one nucleotide change and from its siblings by two (Malpica et al. 2002). This remarkable generation of genetic diversity is the lifeblood of evolutionary adaptation and over the next chapters we will see how this trick up the sleeve of RNA viruses is central to their success as genetic parasites and pathogens. It is counterintuitive but true that this sloppy replication of the RNA genome is actually an asset to RNA viruses.

We typically discuss a virus species as if it is a genetic lineage with a unique "wild type" genome sequence that defines its identity. In natural infections, however, the virus population is a veritable swarm or cloud of closely related yet genetically distinct individuals that have been termed *quasispecies*. In each round of infection and transmission, viral variants are discarded or retained as purifying or positive selective pressure comes to bear on them, and more variants are created as the quasispecies propagates itself. The constituent population of unique genomes that make up quasispecies has been the focus of much experimentation, theoretical modeling, and debate since it was first described in 1978 by scientists working in the laboratory of the renowned molecular biologist Charles Weissmann in Zurich (Domingo et al. 1978). Weissmann's pioneering work with the small RNA bacteriophage Qβ, a tiny virus with a 4,500-base RNA genome, included the elucidation of its replication scheme within the bacterial cell. His laboratory was also the first to apply the techniques of molecular cloning to the Qβ genome and the first exponent of site-directed mutagenesis, in which mutations could be introduced into genes in the laboratory. Weissmann states in his autobiographical paper that it was during the performance of control experiments for their site-directed mutagenesis work in the mid-1970s that they discovered the genome sequence of a Qβ phage population quickly became heterogeneous during growth in successive bacterial cultures: "We concluded that the population was in dynamic equilibrium, with viable mutants arising at a high rate . . . on the one hand, and being strongly selected against on the other" (2012).

These observations had far-reaching implications for how we view RNA virus evolution. A viral species is actually not a unique genetic entity: it is composed of a complex population of diverse but related genetic entities that act as a whole and perpetuates itself, hence its name, a quasispecies. Natural selection and evolution of a viral species act not upon a single genotype but on the ensemble of genotypes that are represented in the quasispecies. The individual genotypes are the products of random mutational changes caused by erroneous ribonucleotide incorporation during replication. They are subject to selective pressures that discern the relative fitness of the individual components and hence shape the quasispecies's composition. Our concept of "wild type" is one of a singular species, usually having a genome of "consensus sequence" and a phenotype that is the fittest, or best adapted, to the prevailing environment. Wild type remains a useful concept for the virologist, but it might better be considered as the center of gravity of the quasispecies, the consensus wild-type sequence may exist fleetingly in a quasispecies, or indeed sometimes be absent. It should be stressed that "quasispecies theory" in no way undermines the principles of Darwinian evolution; it extends and relies on them.

As we will see, RNA viruses draw on quasispecies to fuel their rapid evolutionary adaptation, to hedge their bets, and to facilitate pathogenesis in the complex physiological milieu of cells, tissues, and organs that make up multicellular organisms. At the heart of the quasispecies, the engine is the error-prone replication machinery of RNA viruses, which is the generator of genetic diversity. The rate of genetic diversification, the rate at which errors are introduced into new genomes, must have itself been finely tuned by evolution. In his article in *Scientific American*, published in 1993, Manfred Eigen, a chemistry Nobelist who later turned his attention to the information concept and molecular evolution, wrote a beautifully lucid and erudite exposition of the properties of quasispecies. He talked of sequence space as a concept to map diverse nucleotide sequences into a multidimensional matrix called a Hamming sequence space. Should the error rate in the replication of the nucleotide sequence be so high that none of the offspring are similar to the parent sequence then the sequence population would expand to occupy uniformly sequence space. He likened this to the molecules of a gas, dissipating to form a uniform cloud within a container. Strictly speaking this would not really be replication, since the integrity of the species is not maintained at all, even in the presence of selective pressure for the fittest genotype. On the other hand a reduction in the error rate would result in reduced dispersal of genome sequences, and to quote Eigen: "At some critical error rate, the effect of selection on the population would change radically: the expansive force of mutation would strike a balance with the compressive force of selection. The diffuse gas of related sequences would suddenly condense into a finite but extended region" (Eigen 1993). This cloud of sequences with its center of gravity at the founder sequence will be a self-sustaining population, reproducing themselves imperfectly, but maintaining their integrity over time as a whole.

Such then is the nature of a quasispecies: the density of the sequence cloud at any one point in sequence space is determined by the relative fitness of the sequence; regions of the cloud representing sequences of lesser fitness will be less densely populated and those with higher fitness, most populated. Here lies the most powerful quality of viral quasispecies: the density distribution of fitness variants dictates that sequences are represented at frequencies in relation to their relative fitness. Genomes with lower fitness will replicate poorly, or not at all, and the fittest genomes will replicate most efficiently. It therefore follows that there is a large bias toward the production of well-adapted genotypes: there are more of them, and they undergo most replicative cycles. This can permit viruses to experience evolutionary adaptation at rates that are orders of magnitude higher than those that could be achieved by truly random unbiased mutation. Sequences rapidly condense around the fittest area of the sequence space. Should the environment change, and, therefore, selective pressures change, a quasispecies can opportunistically exploit its inherent adaptive potential. Genotypes rapidly and ever-faster gravitate toward the cloud's new notional center of gravity. Changes in the fitness landscape of the sequence space that is occupied by a quasispecies are the natural consequence of altered selective pressures operating on the virus population. Such alterations may be the consequence of changed immunologic pressures exerted by the host, the application of antiviral drug therapy, or even cross-species transmission requiring the virus to adapt to a new host. Genotypes that once occupied the "central" space, reserved for the fittest genotypes, are reduced in frequency and now occupy the more sparsely populated fringes of the fitness landscape; the very edge of the sequence cloud if you will. Here too lies an advantage for a quasispecies: it has a memory. The once best-adapted genotypes, now at a fitness disadvantage, can persist in the quasispecies as minor sequence variants. Under circumstances of fluctuating selective pressures, the ability of the population to recall an "old" genome variant is a great asset. The quasispecies can rapidly respond and adapt by plucking out a preexisting variant and quickly coalescing around it to recreate an optimal fitness landscape.

There is now ample empirical evidence from elegant scientific experimentation to definitively demonstrate that RNA viruses have evolved to use error-prone replication to their advantage. It is self-evident that there must be an upper limit to the error-prone nature of RNA virus polymerases. There can be too much of a good thing. If a certain critical threshold of mutations in each genome is exceeded, then the creation of diversity would not be beneficial; the majority of progeny genomes would be inviable. The larger the genome, the more errors per genome will be accrued during its synthesis until a critical threshold is exceeded. At this point *error catastrophe* occurs. Extending this line of thinking, one can deduce why most RNA viruses have short genomes, under 15 kilobases. Longer genomes cannot be replicated with adequate fidelity to avoid error catastrophe (although I must acknowledge that other limitations may also be at play). Coronaviruses typically with genomes approximating 30 kilobases are notable exceptions to the rule. However, our underlying assumption remains valid: the maximum length of an RNA virus genome is dictated by its mutation rate. Recent work has deduced that coronaviruses are unique among RNA viruses in that they have evolved to encode proteins that increase the fidelity of genome replication. The nonstructural protein 14 has been shown to be a 3'–5' exoribonuclease, which contributes to the fidelity of coronavirus genome replication (Smith et al. 2015). The increased replication fidelity of the genome has in turn allowed for genome expansion and increased information capacity and potential for adaptive evolution.

One of the first clues that RNA viruses are at an advantage to have a certain level of error-prone replication came with the isolation of a novel mutant poliovirus by a team of scientists at Stanford University who published their work in 2003. Pfeiffer and Kirkegaard isolated 3D-G64S, a mutant virus, which encoded an RNA polymerase with increased fidelity. It made fewer mistakes during RNA synthesis (Pfeiffer and Kirkegaard 2003). The very fact that this single amino acid substitution of a glycine residue by a serine has not emerged during poliovirus evolution is a strong indicator that, in natural poliovirus populations, this simple change must place its genome at a disadvantage. Under normal circumstances, it seems that a poliovirus RNA polymerase with lower fidelity is favored by natural selection. It could be hypothesized that the generation of genomic diversity, manifested in quasispecies, provides an advantage to a virus population under selective pressures.

The Stanford team and another team at the University of California, San Francisco (UCSF) pursued this line of research further. They were able to illustrate convincingly how replicative infidelity and the resulting swarm of quasispecies is beneficial to RNA viruses, particularly in such complex environments as during pathogenic infection of a host animal (Pfeiffer and Kirkegaard 2005; Vignuzzi et al. 2006). Both teams explored the pathogenesis of poliovirus in infected animals. They each used genetically engineered mice that were susceptible to poliovirus infection. They compared pathogenesis of the mutant virus with a high-fidelity polymerase containing the G64S mutation to a virus containing a wild-type RNA polymerase molecule. The mutant virus, which was expected to exist as a quasispecies of reduced complexity had lower pathogenicity, and while the wild-type virus caused disease that readily spread to the mouse brain, the mutant virus did so with much-reduced efficiency. The investigators concluded that a complex virus population possessing many variants was at an advantage in causing disease in the mouse. The evidence points to the benefit of quasispecies per se and to complementation or collaboration between variants in the quasispecies to be maximally pathogenic in the mouse. Collaborators of Dr. Raul Andino at UCSF (Vignuzzi 2006) demonstrated that only in wild-type infections was the virus that had been isolated from the brains of the infected animals fully pathogenic. It could be reinoculated peripherally into another animal and initiate a vigorous infection that spread to the brain. In mutant virus-infected mice, fewer animals succumbed to brain infection, and the virus that could be recovered from the brains of those mice could not infect the brain when reinoculated peripherally into another mouse. The mutant virus could not generate an adequately diverse quasispecies to be fully pathogenic upon reinoculation.

A picture emerges in which the creation of genomic diversity in the poliovirus population is essential for its pathogenesis within the animal. Perhaps different subpopulations of quasispecies mediate different aspects of pathogenesis, such as replication in the gut, transport between tissues, or entry into the nervous system. Only viruses capable of sustaining adequate genomic diversity can express the full spectrum of viral pathogenicity. If this is indeed the case, then complex or variant disease outcomes caused by RNA viruses must depend on the ability of viruses to create adaptive populations and evolve adaptively within each infected host. Complex ecologies, such as a host organism, offer a variety of challenges to a virus population, and the negotiation of these biological obstacles might only be navigated successfully by a diverse population of viruses. The transmission of the virus to a new host is typically associated with the virus population passing through what is termed a "bottleneck," in which the genetic diversity of the population is highly but transiently restricted. This is due to the limited number of viral particles that pass from one host to another during disease transmission. Nevertheless, unless changed and specific selective pressures are at work, it is likely that genotypes occupying the fittest and most populous region of sequence space will pass through the bottleneck. These will undergo reexpansion (and sequence divergence) in the new host, and reconstitute its center of gravity, from which the cloud of sequences making up the quasispecies radiates. If less adapted genomes are founders of a new infection, the natural properties of quasispecies will allow the diversity of the population to reform and condense around the preferred area of the quasispecies sequence space.

The central importance of quasispecies in the pathogenesis and evolution of RNA viruses has been fully realized only in the last ten years or so. Quasispecies are a constant influence on the evolution of all RNA viruses. It is the powers of natural selection on these viral populations that determine the stability or flux of viral genome sequences over time. Strong purifying selection can ensure that viral lineages are relatively stable over very long periods of time (occupying highly restricted regions of sequence space), despite their inherently high mutation rates. I have used poliovirus to exemplify a quasispecies in action, but remarkably natural poliovirus isolates are highly conserved in nucleotide sequence. On the other hand, viruses for which positive selective pressures prevail and favor change (such as those that are under intense selective pressure to adapt to a new host species) can experience high rates of genetic flux fueled by the complex composition of their quasispecies. The quasispecies thus provides the virus with a reservoir of genetic diversity that can be drawn upon based on the prevailing circumstances. As we examine different viruses and their evolutionary modus operandi, we will discover other powerful tricks that viruses leverage as they out-evolve their hosts and achieve their own survival and replication.

· 5 ·

THE FLU: NO COMMON COLD

It is a great mistake to equate influenza with the common cold. In Arnold Bennett's novel *The Card* penned in 1911, Mrs. Machin strongly asserted, "There was no influenza in my day. Call a cold a cold." Bennett would have considered these lines less plausible if he penned the novel ten years later. In 1918, the pandemic Spanish flu ravaged the world's population, infecting four in ten and taking as many as 40 million lives in a single year. The common cold virus is an extremely effective pathogen but coevolution with humans has ensured that it remains a prevalent yet mild disease, provoking disease symptoms adequate only to support its efficient transmission between hosts. It has not evolved to become a commensal, incurring no significant detriment to its host; it remains a parasite and induces enough damage to its host to ensure its transmission. Its genetic diversity is broad but remarkably stable, at least to our eyes, and it has evolved into multiple genetically and serologically distinct strains. These related viruses co-circulate so that no single serotype is the exclusive epidemic agent of colds at any one time. The biology of the influenza virus, on the other hand, is quite distinct. Human influenza A virus is among the most successful and dangerous of human viruses and a poster child for viral evolution. The bewildering pace with which influenza viruses continuously reinvent themselves genetically can make them one of our most devastating diseases. Our relationship with influenza virus has little similarity to the long-standing and stable equipoise that exists with the common cold virus. The influenza virus is a moving target: although we can be confident it will cause many illnesses and deaths every year, we can never be sure of the nature and severity of the next epidemic.

In February each year, the World Health Organization (WHO) convenes a meeting of international influenza experts. Their purpose is to review the available epidemiological and laboratory data collected internationally in 141 clinical laboratories across 111 countries. They must then divine the particular strains of the influenza virus that should be included in the annual flu shot. This is the vaccine that needs to be made available to clinics and hospitals in the Northern Hemisphere before the winter months and the clocklike onset of our annual flu season. If they get it wrong, hundreds of thousands of lives will hang in the balance. The recommendations are handed down to the appropriate national agencies, who then license the manufacture of the necessary vaccine in their respective jurisdictions. The record of success of these deliberations is splendid, but certainly not perfect. In some years, an antigenically novel and potentially pandemic flu strain emerges to confound their predictions, or the epidemic virus may have changed just enough to be a partial mismatch with the selected vaccine strains. Today medical scientists have a sophisticated understanding of the evolution of human influenza viruses based on the evolution of past isolates and epidemics. In a typically pithy statement, the great Yankees catcher Yogi Berra said, "It's tough to make predictions, especially about the future." Predicting the future evolutionary trajectory of viruses, especially influenza virus, is most certainly fraught with uncertainty.

In February 2014, the Centers for Disease Control in Atlanta, Georgia, announced that the 2014–2015 flu vaccine would be targeted against the following strains:

- Influenza A/California/7/2009 (H1N1) pandemic 2009-like virus
- Influenza A/Texas/50/2012 (H3N2)-like virus
- Influenza B/Massachusetts/2/2012-like virus

The nomenclature, although informative to the flu expert, is opaque in the least and not of undue concern to us here. Casual observation, however, reveals that there are two isolates of influenza A; one representing

"H1N1" strains that emerged as a pandemic virus in 2009 and one representing recent 2012 vintage "H3N2" strains, and a third strain, a recent isolate of influenza B. The viruses are antigenically representative of the flu strains dominating contemporary seasonal epidemics and are anticipated to be a close match to the dominant influenza strains that will begin to circulate in the United States in the early winter months. It is calculated guesswork and sometimes it is wrong. In the ensuing 2014–2015 flu season a vaccine mismatch was discovered. Most infections were caused by influenza A(H3N2) viruses but they were genetically and antigenically distinct from the A/Texas/50/2012(H3N2)-like influenza strain used in the vaccine. The mismatch was uncovered in March, too late for the 2014–2015 influenza season. This was reflected in the severity of the seasonal flu epidemic that year.

To clarify what makes influenza viruses such talented genetic innovators it is important to build a picture of them as viruses and their particularities as human infectious agents. There are three kinds of influenza viruses (A, B, and C), but we will focus exclusively on influenza A viruses. They are by far the most important in terms of the burden of human disease that they cause each year, and they exhibit the most complete repertoire of evolutionary tricks that make influenza viruses such successful pathogens.

Like the human rhinovirus, influenza A virus carries its genetic information coded in RNA; unlike rhinovirus, however, its genome is composed of negative strand RNA. After infection its genes must be transcribed from the negative-sense strand into its complement, which serves as the messenger RNA recognized by ribosomes to direct the synthesis of viral proteins. Furthermore, the genome is composed of eight distinct segments of RNA packaged as ribonucleoprotein complexes and collectively contained within a virus envelope. The envelope, a lipid membrane, is penetrated by three viral proteins, displayed as projections on the exterior surface of the viral particle. These proteins are the viral envelope glycoproteins, hemagglutinin (HA) and neuraminidase (NA) and the Matrix protein (M2). The HA and NA proteins are the major antigenic determinants of the virus and the chief target for antibodies that can provide protective immunity to virus infection. HA is the protein responsible for binding to receptors on the surface of epithelial cells in the upper respiratory tract and mediates the penetration of the virus into the cell (see Chapter 11 for HA proteins with different receptor preferences and their influence on the epidemic potential of influenza viruses). The flu virus has its own distinctive mechanisms for usurping the cellular machinery to replicate its genome and assemble virus particles, which bud from the cell through the lipid bilayer making up the boundary wall of the cell. At this point the NA protein takes on its singular function to mediate the release of the virus particles from the cell surface. The host cell membrane is populated by the viral receptors that mediated virus entry into the cell. Virus binding to cell receptors was essential to infection, but is now a liability. The progeny virus particles are bound at the cell surface and must be released. Here the NA protein intervenes to cleave the virus from the cell surface, allowing for their transmission to new host cells and organisms. Notably, virus particles bud only from the superficial surface of the respiratory epithelial cells facing into the lumen of the respiratory tract. The viruses produced by infected cells are thus directed to their conduit of transmission, the respiratory tract secretions. The human disease is thus typically restricted to the airways.

The onset of the flu is not unlike a cold, although it can be remarkably swift and the symptoms far more severe. As with a cold, a stuffed or runny nose and sore throat signal infection, which most commonly occurs via the nasal mucosa, the paranasal sinuses, or pharynx. Soon the virus is replicating in the trachea and in more severe cases in the bronchioles of the lung where it can cause viral pneumonia. The fever, cough, malaise, and body aches that accompany the flu are as much a symptom of the body's immune defenses taking the fight to the virus infection, as they are a result of direct damage to infected cells by the virus. The upper and lower respiratory tract symptoms causing virus-loaded exudate to be expelled from the airways are of course the principal route of viral transmission between humans. The virus is efficiently spread by airborne respiratory droplets or contact with surfaces contaminated with infectious virus. Its relentless epidemic transmission is evident in the fact that in a typical seasonal epidemic of human influenza, it can infect more than one in ten of the world's population (Nelson and Holmes 2007). The infection is typically acute and self-resolving, but complications can occur, particularly in the most susceptible patients, such as the elderly, and in those with other underlying conditions. Pneumonia, bronchitis, or sinusitis are commonplace. It is easy to underestimate the severity of flu as a human disease. Each year between a quarter and half a million persons worldwide die as a result of influenza infection despite the availability of a vaccine (WHO 2014).

Antigenic Escape Artists

At the core of influenza A's success as an epidemic pathogen is its capacity for creating genetic variants to evade the immunity of the host population—they are veritable immune escape artists. On some occasions, they are so successful that they cause epidemics that spread globally: *pandemics*. We experienced four such pandemics in the last century, most recently in 2009 (Salomon and Webster 2009; Kilbourne 2006). The influenza virus that circulates in seasonal flu epidemics typically has subtle antigenic differences from the virus strain of the previous year. Pandemic strains, on the other hand, are quite different antigenically, and the majority of the population has no useful immunological memory that can neutralize them. The opportunity for this genetic variation is provided by mutation and genetic exchange events. The rapid cycles of viral replication that occur within the host and the sustained chain of transmission between multiple hosts afford mutational variation. Like other RNA viruses, the flu virus RNA polymerase makes approximately one error per 10^3–10^4 ribonucleotides incorporated (Nelson and Holmes 2007). We must then envisage that flu virus also exists as a quasispecies. The quasispecies complexity will be transiently restricted by the bottlenecks that occur during transmission between hosts, but it will rapidly reconstitute itself after infection of each new host. This is an important and constant source of genetic variation, which can be subject to purifying or positive selection, depending on the nature and the strength of the operative immune selective pressures. A second source of variation of influenza A virus is the wholesale exchange of genetic material between influenza viruses that infect the same cell. The segmented nature of the influenza virus RNA genome allows for *reassortment* of entire genome segments to create new chimeric viruses with the potential to possess radically changed pathological and antigenic properties. The repertoire of influenza gene segments is not unlike the dealer's deck in a game of poker. Cards can be turned in, and the player's hand is changed for the better or for the worse. Just as in poker, natural selection for viral fitness ensures that failed hands fold with only viable hands remaining in the game. This phenomenon of genetic reassortment is a form of horizontal gene transfer and is not unlike the acquisition by bacteria of new traits by phage gene conversion. Here, the newly created virus possesses one or more new viral genes that offer benefits. Reassortants that receive a new HA or NA gene segment from a different subtype virus are the primary source of new and potentially pandemic influenza strains with antigenic novelty. It is one of the most powerful and unpredictable weapons in flu's arsenal. Until the advent of nucleotide sequencing, influenza diversity was described solely on the basis of serologic criteria and was limited to description of antigenic diversity among virus strains with different HA and NA subtypes on their surface. Since the antibody response to influenza virus is principally directed toward the exterior viral envelope glycoproteins HA and NA, the description of influenza viruses relied on their antigenic properties. This remains the case today since these are the most influential sources of antigenic diversity that determine the pathogenic potential of flu viruses. Nevertheless, today we can supplement our analysis of viral isolates with knowledge of the nucleotide sequence of the respective viral HA and NA genes as well as that of the other six viral gene segments.

There are eighteen HA (H1—H18) and ten NA (NA1—NA10) subtypes recognized today (Webster and Govorkova 2014); functionally and antigenically distinct, they are used to "type" influenza A viruses. Hence, the subtypes anticipated to be epidemic in the winter of 2014–2015 in the Northern Hemisphere were predicted to possess H3N2 (HA subtype 3 and NA subtype 2) and H1N1. Reassortment of HA and NA subtypes (resulting in *antigenic shift*) and the diversification of HA subtypes via mutation and selection (*antigenic drift*) are the principal determinants of the epidemic potential and pathogenic properties of influenza viruses. Today it is recognized that the genomic context of these variant subtype genes is equally important for successful pathogenic flu viruses to emerge. During the last decade, with high throughput sequencing technologies at their disposal, scientists have obtained the complete genome nucleotide sequences of an unprecedented number of flu isolates. The phylogenetic, epidemiological, and evolutionary dynamics of influenza virus during and between epidemics can thus be examined. Such phylodynamic analyses have served to document how the influenza virus reinvents itself with such aplomb.

Human Influenza A Virus

Influenza A viruses are not restricted to human hosts; they infect a wide range of mammals such as pigs, horses, sea mammals, birds, and bats (Webster et al. 1992). Those that infect humans and can be transmitted among us are limited to viruses of only three of the eighteen known subtypes of HA (H1, H2, and H3) and two of the ten NA subtypes (N1 and N2). In fact, first and foremost, influenza is a virus of aquatic birds, and phylogenetic analysis indicates that all influenza A viruses have evolved from avian influenza viruses (Webster et al. 1992). Aquatic birds are therefore their natural and most historic hosts and their principal reservoir species. Influenza circulates freely in wild bird populations, mainly in ducks, geese, and other wading birds, where it is seasonally epidemic during the late summer and early autumn. Influenza viruses circulate among abundant wild waterfowl not as a respiratory disease, but as an enteric infection of the gut epithelium. The virus is transmitted between birds in fecal matter deposited directly in the environment and in the water that other fowl live in and eat from. It has been estimated that there can be almost a billionfold more virus per gram of infected bird fecal matter than it takes to experimentally infect cultures in the laboratory (Webster 2002). Despite these high viral burdens, the infection of birds is usually benign and causes no disease. This is believed to be a reflection of extensive coevolution during the long-standing virus-host relationship that has existed between influenza A virus and wild birds.

It appears that avian influenza and its hosts have evolved to an adaptive equilibrium (Webster 2002). Comparison of bird endemic influenza viruses over a sixty-year period revealed little signs of evolutionary change in their genomes, compared with influenza viruses from mammals that accumulated substantial amino acid changes in all of their eight genome segments. The evolutionary status quo in birds exists despite the continuous creation of genetic variants during the replication of the flu RNA genome. It appears that there is no adequate positive selective pressure to support the emergence of new variant lineages. Presumably the extant viral lineages have achieved a "fitness optimum" in their natural species, and almost any genetic change is detrimental to the fitness of the genome and subject to purifying selection. Supporting this conclusion is the observation that synonymous nucleotide changes with no consequence on the protein sequence (and hence phenotype) of the virus far outnumber nonsynonymous changes. It appears then that the evolutionary arms race between birds and influenza virus, which must have taken place, is now ancient history, and it was fought to a resolution; a stable entente is in effect. Wild birds exhibit a relatively mild immune response to influenza infection. Presumably this is one reason that no strong selective pressure drives continuing evolution of the virus population in bird species; the need for immune escape or adaptive change is small. This is not the case for influenza viruses in humans; here the genetic conflict continues to rage.

The influenza virus of wild birds frequently infects domestic poultry. In this new and relatively maladapted host species positive selection for genetic change takes over. The change in species environment creates new and diversifying selective pressures on the genome which are evident in the increase of nonsynonymous over synonymous nucleotide substitutions in the genome (Nelson and Holmes 2007). Mutations associated with changes to protein amino acid sequences can be beneficial to the replication of the virus in its new host. It appears that this occurs despite the poor immune response that domestic birds mount against the virus. The positive selection, however, is not focused only on the usual suspects, the antigenic determinant HA and NA genes; it is rather more evident in the other gene segments. It appears that other virulence determinants or targets of cell-mediated immunity are the most dominant drivers of this new evolutionary trajectory. Like all new relationships, competition between partners and the necessity for adjustment to each other's behavior is greatest at the outset.

As the oldest influenza virus lineages, those in the wild bird population possess a vast repository of genetic diversity. It serves as a veritable melting pot in which multiple influenza virus lineages undergo multiple cycles of relatively unhindered transmission and replication, commingling to allow for the exchange of genetic information by reassortment. This reservoir perpetuates influenza A and is the creative force behind the genetic diversity that can be exploited by the virus.

Human influenza A is well established to have its origins in bird influenza virus genetic information. While avian influenza virus does infect new hosts, it rarely establishes lineages within them. It meets a dead end, due either to excessive virulence or failure of virulence within the new host or a failure to establish a chain of transmission. Of the lineages that have become established—seal, horse, and swine among them—human influenza viruses are most akin to those of swine. Human influenza viruses do not circulate in wild bird populations; presumably they have lost this capacity as a result of their evolutionary adaptation to humans. Human influenza virus lineages rarely directly acquire avian influenza genetic information. Swine appear to be important intermediate hosts between birds, probably ducks, and humans. The human influenza viruses that currently circulate have derived their genes from multiple ancestral viruses but the reassortment and acquisition of these influenza genes most probably occurred in pigs from whence transmissions to humans are well documented (Webster 2002; Webster and Govorkova 2014). One of the currently circulating seasonal human influenza viruses, an H3N2 strain, has been circulating seasonally since 1968 when it first emerged as the pandemic Hong Kong flu. As we will discuss later, the extensive time period in which this virus has been a prevailing epidemic in humans presents us with a unique tool to study the evolution of epidemic human influenza viruses over several decades. The comparative genomics of H3N2 strains arrayed both in time and geography provide profound insights into flu evolution. The reassortment event that created the Hong Kong H3N2 pandemic strain resulted in the replacement of the previously prevailing H2N2 influenza virus. Examination of the gene segments of the new 1968 H3N2 virus revealed it to be a result of reassortment between circulating H2N2 viruses and avian H3 viruses. It retained six gene segments from H2N2 but acquired a new HA (H3) and a new gene segment encoding the RNA polymerase.

The previously circulating epidemic H2N2 virus had, in fact, over time evolved into two genetically distinguishable H2N2 lineages (clades). It seems that these viruses continued to circulate at some level after the emergence of the pandemic H3N2 virus. The phylodynamics of the emergent and rapidly evolving new pandemic was complex. While the first isolates of H3N2 received all of their H2N2 genes from one of the two circulating clades of H2N2 viruses, later isolates had gene segments derived from both clades. Scientists conclude that not one, but multiple reassortment events between circulating strains must therefore have played a part in the rapid evolution of the new pandemic strain (Lindstrom, Cox, and Klimov 2004).

Epidemic Influenza: Dress for the Season

Influenza occurs in seasonal epidemics, in the winter months of the temperate regions of both the Southern and Northern Hemispheres. The factors that govern this seasonality are obscure. It was suggested that these weather conditions favor the transmission of the virus between hosts—perhaps assisted by increases in interpersonal contact, as there is a greater tendency for living in close quarters during these months of the year. Others implicated a seasonal diminution of human immune robustness or vitamin D deficiency due to reduced daylight hours as contributors to the seasonality. Some or all of these factors may play a greater or lesser role. It does appear that avian influenza epidemics in birds are correlated with times of premigration congregation after the birth of new and susceptible chicks, so it is quite likely that behavioral factors influencing transmission rates are one of the major contributory factors to seasonality (Nelson and Holmes 2007). It is established that the crucible of human influenza is in subtropical regions of East and Southeast Asia. In these regions epidemics follow no fixed seasonal cycle and the virus can be isolated year-round. It seems likely that human influenza lineages are perpetuated in this region and seed influenza epidemics that spread into temperate regions.

The evolution of influenza genotypes and phenotypes occurring during epidemics is at the crux of their limitless potential to be epidemic in humans. We have come to accept as dogma that the sustainability of annual epidemics is made possible by the capacity for immune escape of influenza strains. A circulating epidemic influenza virus genome is under selective pressures from the host. These pressures are the weapons of the arms race: if the host has immunologic memory of an infection by the same or a related influenza virus, there will be circulating antibodies ready to neutralize it upon infection. There is, therefore, strong diversifying selective pressure for viruses that are antigenically changed. Such viruses will not be recognized by the host and will be the subject of positive selection. Only viruses and their associated genetic information that escape neutralization will sustain a chain of transmission. The virus creates the necessary genetic diversity in the form of mutations in surface envelope glycoproteins HA and NA. The dominant epitopes of the HA protein are found in the most outward projecting portion of the HA protein, known as the HA1. This part of the virus particle is most critical in determining antigenicity and is recognized and bound by protective antibodies. Historically, HA1 has been the most scrutinized influenza protein, both phylogenetically and immunologically. Nucleotide sequences have been collected and analyzed to evaluate the genetic evolution of circulating influenza strains each year. Protective antibodies recognizing HA1 have been characterized extensively and identify five major epitopes, portions of the protein preferentially targeted by antibodies, on the HA1 protein. This is the primary arena in which the arms race is played out.

It is accepted that these selective pressures cause antigenic drift due to gradual mutational change in HA1 epitopes. The result is a gradual loss of the immunity of hosts, which is the primary source of competitive advantage among flu variants that drives the evolution of epidemic influenza. Antigenic drift together with declining immunological memory of the host population and the availability of new hosts are fundamental necessities for the perpetuation of epidemic lineages of influenza. Absent antigenic evolution and immune escape, the virus will face a shrinking susceptible host population. Under these conditions, herd immunity can develop and the epidemic will not be sustained. Herd immunity arises when a critical proportion of the available host population is immune to infection. It may be counterintuitive, but if a population is to survive the predation of an infectious agent, it is not necessary that all individuals in the population are immune to infection. If a substantial proportion is resistant, the infection cannot sustain a chain of transmission within the population due to the scarcity of susceptible hosts and the epidemic burns itself out. A pathogen may persist under these circumstances, occasionally emerging as an epidemic if it finds itself with access to a susceptible host population that can support epidemic spread. On the other hand, some pathogens—and influenza A virus is one of them—have the capacity to promulgate epidemic spread by changing their properties with such alacrity that there are always adequate susceptible hosts to support their spread. How else can influenza find susceptible hosts, since it so successfully infects (and thus renders immune) a substantial portion of the world population annually? We suffer from several "colds" each year, resulting from infections by genetically distinct viruses to which we are immunologically naive. Each year, however, the seasonal influenza epidemics that we experience are dominated by only one or two highly prevalent strains to the exclusion of all others. Flu epidemics therefore represent the propagation and evolution of an individual genetic lineage.

Influenza A virus is uniquely equipped to sustain its potential for epidemic spread. To do so it must be able to access susceptible hosts. To make this possible, it exploits two strategies. The first involves the exploitation of rapid mutation and selection under the selective pressure of the immune responses of the host population. Incremental yet progressive evolution by antigenic drift is associated with the seasonal epidemics of the influenza strain. The second evolutionary mechanism, antigenic shift, involves dramatic changes to the pathogenic and antigenic properties of the virus through the wholesale exchange of genetic information between different viral lineages. Genetic drift contributes primarily to the generation of antigenically distinct flu strains that fuel the annual epidemics of influenza. They are different enough from their ancestral epidemic strain to escape immune control, yet it is likely that the host population benefits somewhat by having "seen" a related strain in a previous year. The antigenic shift in influenza strains represents a radical departure in which the influenza strain is antigenically unrelated to previously circulating strains. These are the viruses that cause pandemic influenza and are the source of major concern for public health planning.

Quasispecies, Sequence Clusters, and Codon Bias

In recent years, the genetic structure of circulating influenza strains has come under intense study as the availability of rapid nucleotide sequencing technologies has allowed multiple virus isolates to be comprehensively sequenced and subjected to computational analysis. The most heavily studied epidemic influenza is H3N2, which emerged in 1968 and continues to circulate today. In two studies published in the *Proceedings of the National Academy of Sciences* (Plotkin, Dushoff, and Levin 2002; Plotkin and Dushoff 2003), collaborating scientists at Princeton University studied the sequence evolution of the H3N2 HA gene in a database of more than 500 viral sequences collected over two decades by the WHO in their surveillance of influenza epidemics. The research developed two concepts that underpin epidemic influenza virus evolution. In the first study, rather than restrict their analysis simply to the phylogenetic relationships among HA proteins, the researchers assigned them to a chronology. They made pairwise comparisons of amino acid codons in all HA gene sequences, assigned a "distance" between each sequence, and placed them in time. Their analysis revealed that HA1 sequences fell into closely related clusters that change over time, with one cluster of sequences being replaced by another each two–five years. The existence and circulation of viral sequences as clusters of related sequences echoes the creation of viral quasispecies or swarms of related sequences by the viral RNA polymerase. In fact, each database entry is the record of one member of the prevailing quasispecies in a particular host. An epidemic influenza virus therefore exists as an ensemble of interrelated sequence clusters (each themselves quasispecies) that are metastable, being periodically replaced by competing swarms. These observations were consistent with theoretical mathematical modeling applied to antigenically diverse infectious agents (Gupta, Ferguson, and Anderson 1998), positing that under strong selective pressure populations would segregate into discrete strains, but that under intermediate levels of selection these distinct strains may vary in a manner either cyclical or chaotic. Is this the basis for such variation in influenza clusters? In any case, the creation and perpetuation of quasispecies representative of clusters must be an evolutionary asset to the influenza virus population, providing it the basis for rapid genetic and antigenic evolution in response to environmental pressures.

The second paper (Plotkin and Dushoff 2003) represents a minor landmark in understanding the evolution and antigenic plasticity of epidemic influenza HA protein. It established that the influenza HA gene has evolved for *evolvability*. On the face of it this is a heretical claim: How can Darwinian selection act on a phenotype that does not yet exist? These studies, too, used the large database of HA sequences, but in this case the collaborators focused on the relationship of HA gene sequence changes to HA amino acid sequence changes. Recall that mutations in a protein-coding gene can be synonymous or nonsynonymous and genes under positive selective pressure accumulate a disproportionate number of nonsynonymous mutations that change the amino acid sequence of the protein. These researchers extended their observations to the H3N2 HA1 coding sequence sampled over a twenty-year period. The study revealed that five discrete portions of the HA1 protein, constituting the antibody combining regions, exhibit significantly more nonsynonymous nucleotide substitutions than other parts of the gene. This is consistent with the expectation that the epitopes of HA1 are under strong diversifying positive selection. The other regions of the protein (and indeed the other influenza proteins) all exhibit a larger proportion of synonymous nucleotide changes. These regions of the genome experience lower levels of immune selective pressure and are potentially subject to more rigorous purifying selective pressure because of their more complex functions or fragility to mutational change. They can remain unchanged over time without jeopardizing the viability of the flu lineage.

The researchers looked deeper into these observations, in hopes of gaining insight into the mechanisms underlying the high evolutionary rate and extraordinary immunologic plasticity of influenza HA. They probed in more detail the precise codons that are used by the virus to encode the influenza HA1 protein. They discriminated between codons on the basis of *volatility*. Each three-nucleotide codon is related by a single nucleotide change to nine "mutational neighbors." Of those nine mutations, some proportion change the codon to a synonymous codon and some change it to a nonsynonymous one, which directs the incorporation of a different amino acid into the protein. More volatile codons are those for which a larger proportion of those nine mutational neighbors encodes an amino acid change. The use of particular codons in a gene at a frequency that is disproportionate to their random selection for encoding a chosen amino acid is termed *codon bias*. Such bias is common and is influenced by many factors, but here the collaborators found strong evidence for codon bias that was particular for and restricted to the amino acids making up the HA1 epitopes. Remarkably, they observed that influenza employs a disproportionate number of volatile codons in its epitope-coding sequences. There was a bias for the use of codons that had the fewest synonymous mutational neighbors. In other words, influenza HA1 appears to have optimized the speed with which it can change amino acids in its epitopes. Amino acid changes can arise from fewer mutational events. The antibody combining regions are optimized to use codons that have a greater likelihood to undergo nonsynonymous single nucleotide substitutions: they are optimized for rapid evolution.

The authors were well aware that this assertion was not readily reconciled with well-accepted views of Darwinian evolution because it appears to violate the "law of causality." Evolution does not anticipate the future advantage of a mutation at a given amino acid position in a protein. The volatile codon is only beneficial after it has been removed from the sequence by mutation and, therefore, cannot be the subject of natural selection. Only the phenotypic benefit of the new amino acid can be subject to selection. Two arguments were advanced to explain how such codon bias is maintained for HA1 epitope-encoding sequences. The first is a retrospective explanation: codon bias is a footprint left behind by previous frequency-dependent selection. The virus infects a substantial portion of its host population per year, and the relatively long-lasting immunity that develops is an active driver of frequency-dependent selection for antigenic variation. Such selection dictates that the selective pressure on an allele is influenced by its frequency in the population: alleles at low frequency are at an advantage and positively selected, while those that prevail in the population are negatively selected. In the context of antigenic variation this means that low-frequency epitopes will be poorly recognized by prevailing host immunity (and represent immune escape variants), while prevailing epitopes are targeted by extant immunity and at a disadvantage. The selective pressure on a particular epitope can therefore be expected to fluctuate over time and in the context of successive different epidemics. The investigators make the assumption that the mutational process at the biased codons is symmetric in time, in that any mutation can also be subject to reversal. Allowing that volatile codons have a high proportion of single nucleotide substitutional neighbors, it is reasonable to conclude statistically, that an encoded amino acid is likely to have itself arisen from a single nucleotide substitution. Moreover, it can also be argued that there is a high probability that the next mutation in the codon will also result in an amino acid substitution. Hence, if frequency-dependent selection is at work, it will imprint codon bias at this nucleotide position.

The first explanation does not assume that codon bias is selected based on its adaptive utility, but a second prospective argument supports the possibility that this may be at play. Simply put, this argument is based on the observation that a correlation exists between the volatility of a codon and the average volatility of codons which are related to it by a single nucleotide change. Since volatility on average begets volatility, it can be argued that along an evolutionary lineage volatility can be a heritable trait. Each of these is a nice argument—rather in its older sense of "uncertain and delicately balanced." In any case, these studies of HA1 are empiric observations and describe the system "as is" regardless of its specific etiology. Codon bias appears to be one exquisitely useful tool that the influenza virus wields to speed its adaptive evolution and compete successfully in the arms race against the human immune system. Despite the lack of such comprehensive data for other viruses, one dare speculate that this will also be observed in other viruses that rely on antigenic plasticity and immune escape to perpetuate their genetic lineages in epidemic or chronic viral infections.

Correlating Genetic and Antigenic Evolution

In 2004 Derek Smith of the University of Cambridge and collaborators in the Netherlands and United States published another landmark paper closing a significant gap in our understanding of epidemic influenza antigenic evolution (Smith et al. 2004). It was generally understood that amino acid sequence changes in the HA1 epitope regions translate into the antigenic change necessary for immune escape of the virus. Nevertheless, the precise relationship of amino acid sequence with viral fitness and thus the phenotype conferred by a particular amino acid sequences was missing. Clusters of HA were defined by amino acid sequence but their relationship to each other in terms of how they interact with the human immune system was not. It is, after all, the viral phenotype resulting from the interaction of the virus and host that is critical. It is the antigenic *structures* of HA1, that are *seen* by the human immune system, not simply the sequence of amino acids, on which the powers of selection operate. The host immune system *sees* the antigenic structures of HA through the eyes of antibodies, which are directed to particular epitopes. While we may reasonably assume different amino acid sequences will be seen differently, the degree to which an individual amino acid change alters the ability of host antibodies to bind HA1 is mostly conjecture. The change in the structural and physicochemical *shape* of an antigenic site manifested by a particular amino acid substitution may have a more or less profound effect on its recognition by circulating antibodies. It is the relative capacity of the antibody to recognize its cognate epitope that determines the phenotype of the genotype and hence its fitness in the face of prevailing antibody responses.

The work of Smith and his colleagues critically bridged this gap. His team studied antigenic data from thirty-five years of H3N2 circulation between 1968 and 2003. They established quantitative methods allowing antigenic evolution (the true measure of phenotype) to be compared directly with genetic evolution. These quantitative methods were essential in order to measure and computationally analyze HA1 evolution. The methods measured cross-reactivity of virus isolate antigens with reactive antibodies. Unlike the virus isolates and sequences that were readily available, reactive antibodies from human patients that had been infected by the respective flu strains were not. In an experimental tour de force the investigators generated the antibodies by experimentally infecting ferrets with selected flu strains. Seventy-nine ferret sera were generated and evaluated for cross-reactivity with 273 influenza isolates, forming a map plot of H3N2 antigenicity over time. The antigenic map revealed that, rather than experiencing gradual antigenic drift over thirty-five years, the antigenicity changed in fits and starts and could be grouped into noncontiguous antigenic clusters. The influenza lineages periodically experienced jumps as successive clusters became dominant then receded with a periodicity of approximately three years. This is not dissimilar to the genetic sequence clusters that Plotkin and colleagues described earlier.

Now the researchers could compare the antigenic and genetic evolution of influenza H3N2 directly. The result was enlightening. There is an overall correspondence of antigenic clusters and genetic maps, but genetic change, that is, nucleotide sequence evolution, occurs rather gradually while antigenic evolution is punctuated. Within each antigenic cluster, substantial sequence evolution can occur, much of which is relatively silent antigenically. Thus, sequence evolution continues at a steady pace but is associated with little, perhaps subtle, phenotypic change. On the other hand, some antigenic clusters were separated by single ground-shifting mutations with profound effects. The rate of evolution between clusters was shown to be higher than within clusters; it is during the transition from one cluster to another that antigenic evolution occurs most quickly.

Since this seminal work was published, further studies have interrogated H3N2 evolution within and between epidemic years, but also in the context of geographical and epidemiological data (Holmes et al. 2005; Nelson et al. 2006). The analysis of H3N2 circulating in local epidemics revealed that multiple sequence clades of the virus commonly circulate in each epidemic season. A further layer of complexity is evident in that these co-circulating clades commonly reassort with each other, providing yet another basis for the creation of genetic diversity within the circulating influenza virus population. This phenomenon, when it involves reassortment between different viral subtypes, is responsible for antigenic shift and the creation of pandemic influenza viruses. Now it was recognized as a common phenomenon contributing to evolution of circulating seasonal influenza viruses within an epidemic. Such intratypic reassortment is associated with increased rates of amino acid substitutions in viral genes (Holmes et al. 2005; Neverov et al. 2014). The precise genome background of each influenza virus gene segment influences its fitness. Gene segments within the same virus are under selective pressure to undergo evolutionary coadaptation for optimal virus fitness. Disruption of a well-adapted genome complement by reassortment therefore creates strong selective pressures for coadaptation and accelerated evolutionary change. This is clearly evident in intersubtype reassortants and a key driver of the rapid evolution witnessed in the genomes of emerging pandemic influenza virus strains.

Seeding of Seasonal Epidemics

East and Southeast (E-SE) Asia are believed to be "central casting" for human influenza virus and the epicenter of emerging flu epidemics. The precise events that lead to the seeding of annual influenza epidemics in the temperate Northern and Southern Hemispheres have remained somewhat uncertain. Do seasonal epidemic strains reemerge locally to fuel the next epidemic or does each annual epidemic emerge in subtropical Asia and spread geographically? The question was recently answered, at least for H3N2. Researchers took up the antigenic and genetic analysis of 13,000 influenza A (H3N2) virus isolates from six continents over the five-year period between 2002 and 2007 (Russell et al. 2008). The data revealed that influenza circulates in E-SE Asia in a regional network of year-round, temporally overlapping epidemics. These witches' brews seed annual influenza epidemics in the rest of the world. This has significant consequences for how we view influenza evolution. If each year's epidemic is always seeded in E-SE Asia, then it follows that the evolution of influenza viruses within the epidemic outside the crucible of E-SE Asia will have no impact on the long-term evolution of the virus. All of the genetic innovation of long-standing consequence to influenza lineages is limited to epidemics cycling in that region. It remains uncertain to what degree both genetic and antigenic evolution of the influenza genome within a regional, seasonal epidemic of the virus might play in the pathogenesis of the disease. Is it possible that under some circumstances seasonal epidemics can be reseeded locally? Typically, however, it appears that the mechanism for seeding the annual epidemics of H3N2 is to go back to the well of virus in E-SE Asia. This is a reminder of the importance of continuous surveillance of influenza strains circulating in that region because we can predict with some confidence that they will be visiting our hometown soon.

These insights into influenza evolution and antigenic drift have a significant impact on our ability to interpret annual epidemiological data on outbreaks and anticipate the likely phylodynamic trajectory of circulating influenza subtypes. Our confidence in the capacity of WHO experts to predict the annual epidemic strains of influenza has never been greater; the vaccine prepared for the upcoming flu season is usually an adequate match for the circulating strains (notwithstanding the noted mismatch in 2014). The risk of antigenic shift, however, is always present and can create a profound public health crisis until an appropriate vaccine can be prepared. Not only is influenza genetically innovative, it also exploits the technical and societal innovations of its host species. New pandemic strains do not require boarding cards and today can travel global airway routes with alacrity, spreading globally in days to weeks to seed regional epidemics via our densely populated global travel and business hubs (Lemey et al. 2014).

Pandemic Influenza: The Emperor with No Clothes

No accounting of human influenza A virus evolution can be complete without due consideration of pandemic viruses, which emerge periodically and reinvigorate the human influenza gene pool. Their very nature is one of unpredictability; they do not arise from incremental evolutionary processes, but from stochastic events, the wholesale shuffling of genes between viruses of different lineages. Influenza viruses are consummate experimentalists combining and modifying genetic resources, in this case whole gene segments in continuous cycles of beta testing. In poker parlance, we might compare it to discarding multiple cards with the hope of then picking up a winning hand; a rarely successful tactic. We are most concerned with pandemic viruses that possess novel combinations of HA and NA subtypes. It is these viruses that have pandemic potential because the host immune system has never seen them before, and there is no preexisting immunity in the population. They are the influenza gene "dream team," a rare combination of talents that can, with very little training, become a truly globally dominant epidemic virus. Reassortment between viruses may occur more frequently than we perceive. Intratypic reassortant viruses are commonly detected circulating in the seasonal epidemic swarm. A successful viral genome, however, must be a well-oiled machine to emerge in epidemic form. Even the genetic reassortment between closely related strains circulating in seasonal epidemics results in the creation of increased diversifying selective pressures. These pressures extend across the entire genome and are caused by the relative incompatibility of the new ensemble of viral genes (Neverov et al. 2014). By definition, viruses with pandemic potential will be the products of reassortment between radically different viruses. It is likely then that the creation of a viable and successful team of influenza gene segments by intertypic reassortment happens very rarely. The epidemic potential of new pandemic strains is not dictated only by recombination of antigenic HA and NA segments; it is also influenced by the origin of the other gene segments. The determinants of flu virulence and pathogenicity are complex and are difficult to attribute to particular gene segments in new pandemic isolates.

For purposes of illustration, let's concern ourselves with the 2009 pandemic H1N1 influenza virus. When this virus emerged, viral geneticists and epidemiologists had at their disposal a wealth of knowledge and data never before available in real time. The most recent prior pandemic virus to emerge was H3N2 in 1968, a time predating most of our current molecular technologies. Understanding the virus behind the 2009 pandemic is intriguing in and of itself, and it provides some remarkable insights into the evolution of other epidemic and pandemic influenza strains over the last 100 years.

The 2009 H1N1 pandemic started in March: a rather perfunctory paragraph in the CDC's *Morbidity and Mortality Weekly Report* noted "reports of an increase of influenza-like illness in Veracruz, Mexico" (CDC 2009). On April 17th, a case of atypical pneumonia was reported in Oaxaca State and two children in two different counties of Southern California in the United States came down with a serious flu. By the 23rd, the authorities reported that the Mexican cases were caused by swine-origin influenza A (H1N1): S-OIV. Swine influenza infection of humans, particularly those who work in the swine industry, is not uncommon. One or two cases are reported each year by the CDC (and others probably often go unreported), but they rarely spread to other humans. The Californian children had no known exposure to pigs, so they must have contracted the infection through human contact. Human-to-human transmission of the virus was taking place; it was soon reported to be a reassortant with a unique combination of gene segments—all hallmarks of an emerging pandemic virus. On April 29th the *New York Times* reported that the WHO had moved its global pandemic influenza alert level to Phase 5, one step from its highest warning. A global pandemic was highly likely. Mexico had already seen 166 deaths, and such was the concern that 176 professional soccer games that week were played in empty stadiums. By now other counties in California and Texas had cases, and the flu had sickened students and staff at a New York City high school and caused illnesses in Canada. The stage was set for a pandemic, even as health authorities scrambled to put in place a plan to rapidly develop the necessary viral strains that could be used in a vaccine to contain the disease. Hospitals in New York City were overwhelmed by patients concerned that they had the swine flu (although most of them did not). By early May, twenty-one countries in North and South America, Europe, the Middle East, and Asia were affected. More than 80 percent of the cases were traceable to a traveler returning from Mexico to their home country; the virus had literally flown around the world in the course of a single month. In-country epidemics were beginning to take hold. On July 6, 2009, the WHO declared a pandemic. By mid-October, H1N1 had sickened 22 million Americans, sent 100,000 to the hospital, and caused nearly 4,000 deaths. The pandemic would last until September 10, 2010, when the post-pandemic phase was announced by the WHO. The global death toll was 200,000 with another 200,000 deaths attributable to consequential illnesses.

This chronology illustrates the rapid and devastating consequences that an extremely transmissible pandemic influenza virus can wreak, even on a society as socially and medically advanced as our own. Pandemic viruses have such a selective advantage through their unique combination of gene segments that they quickly emerge as the dominant globally circulating influenza strain. Their advantage is in the new combination of antigenic proteins HA and NA, to which the human population has never been exposed. Why, you might ask, were we not protected by our prior exposure to the human H1N1 influenza A that had been one of the circulating influenza strains for the past few decades? The answer is complex and to get to it we need first to understand the genetic origins of the 2009 S-OIV (Smith et al. 2009; Garten et al. 2009). It is a story that begins in 1918 with the Spanish influenza epidemic. The 1918 pandemic of H1N1 influenza A was an epidemic perfect storm, infecting as many as 40 percent of the world's population and causing 20–50 million deaths. The prevailing conditions were fertile for the epidemic spread of disease: the world's armies were demobilizing after World War I, and the population was ripe for the picking by an aggressive contagious disease. What is more, the "collaboration" of influenza with the bacteria that cause pneumonia secondary to viral infection was a major contributor to the high mortality rate of the 1918 pandemic, which predated the antibiotic era.

The genetic origins of the 1918 pandemic H1N1 virus have now been extensively researched (Shanta and Donald 2009; Taubenberger and Morens 2006). Its genes originated in avian influenza viruses, and it is widely believed that it entered the human population directly from birds. Shope, an eminent virologist of the day, concluded that the virus was then introduced from human into swine (1936). Swine influenza or "hog flu" as it was referred to on the farms of the Midwest, was recognized after the first wave of the human pandemic in the spring of 1918. Most pandemics proceed with a first epidemic wave, seeded directly from birds (or pigs) during which the virus rapidly adapts to the new host. It reemerges months or sometimes years later with increased fitness, perhaps more pathogenic or better adapted to be transmitted efficiently between human hosts (Miller et al. 2009). This is presumably a result of evolutionary adaptation through mutation and possibly even intratypic reassortment between emergent clades. Today we recognize that a vast reservoir of avian influenza genetic material reassorts freely in wild bird populations. These are the natural hosts of the virus, and they are often infected with multiple subtypes of the virus at the same time. Secondary to this, swine are thought of as a mixing vessel of influenza viruses. They are highly susceptible to infection, having cellular receptor molecules that support infection by both avian as well as human influenza viruses. Swine are thus a common conduit through which influenza viruses enter human populations. It is therefore somewhat ironic that this singular pandemic 1918 H1N1 virus of avian origin was the founder of what we today term "classical" swine influenza. The H1N1 virus continued to circulate in human populations for almost forty years, after which it was replaced by H2N2, a reassortant virus that caused a pandemic in 1957 (Kilbourne 2006; Morens, Taubenberger, and Fauci 2009). In turn, H2N2 was succeeded by the Hong Kong flu, an H3N2 reassortant that caused the last pandemic of the twentieth century in 1968 (Kilbourne 2006; Morens, Taubenberger, and Fauci 2009). As I discussed earlier, the H3N2 virus has been the subject of intense research since its emergence more than forty-five years ago, and it is still one of the circulating seasonal influenza strains.

In 2009 when S-OIV H1N1 made its debut, the world was caught unaware. The strains selected for the seasonal influenza vaccine targeted the strains of influenza A that had prevailed in 2008 and were expected to be in circulation the following year: an H3N2 and an H1N1 strain. Here there is a small twist in our story. After being displaced by pandemic H2N2 viruses in 1957, the H1N1 strain disappeared entirely from seasonal epidemics. Remarkably it only remerged in 1977 (Kilbourne 2006). To the surprise of virologists, the reemergent H1N1 virus appeared almost genetically indistinguishable from the virus that faded from view in 1957 (Nakajima, Desselberger, and Palese 1978; Scholtissek, von Hoyningen, and Rott 1978). If the virus had been circulating in a reservoir species such as pigs, it would have been bound to accumulate mutations. The lack of genetic divergence led to only one hypothesis: for a large part of the intervening twenty years the virus had been "on ice." Today it is believed that the virus may have been accidentally reintroduced to the human population just prior to 1977, probably as a result of a vaccine clinical trial gone wrong (Rozo and Gronvall 2015). Its reemergence in 1977 led to an epidemic restricted mainly to children and young adults. Older people had, more often than not, been previously exposed to the H1N1 flu, while those born after its disappearance were immunologically naive to the virus.

This H1N1 virus, a direct descendant of the 1918 Spanish flu virus, continued to circulate thereafter along with H3N2, with one or the other virus dominating in successive flu seasons until 2008. Of course, this contemporary H1N1 virus had been in continuous circulation for seventy years (between 1918–1957 and 1977–2008). Given the selective pressures exerted by the human immune system, its surface proteins had evolved very substantially by antigenic drift. These were the very same molecular mechanisms that we discussed in detail for the antigenic evolution of H3N2 during forty-five years of circulation as a seasonal influenza virus.

Recall that in 1918, swine became infected by the Spanish flu H1N1 virus. If indeed the virus passed directly from birds to humans, who then infected pigs, this presaged an important phenomenon (I say "if" because it is hard to preclude the possibility that the virus passed from birds to pigs and then to humans). The same phenomenon was reported again in 1977 when strains of human Hong Kong H3N2 virus were found to be circulating in herds of swine in Asia (Shortridge et al. 1977). The strains had undoubtedly been circulating in swine for some time since they were most similar to H3N2 viruses that were no longer in human circulation. It was evident not only that pigs can act as mixing vessels and transmit viruses to humans but also that the virus can move from human to pigs. I observed earlier that the measured evolutionary rate of avian influenza viruses in bird populations is very slow, a phenomenon ascribed to the high level of evolutionary virus-host coadaptation that is extant in the avian viruses. The evolutionary rate of influenza in pigs is slower than in humans. Consequently, the virus circulating in pigs can act as an archive of differentially evolved genetic material that can be tapped by reassortment by either human or avian viruses which simultaneously infect a pig. This may create what are known as "triple reassortant viruses": a fantasy draft pick of a gene complement in which viruses can be assembled in pigs from circulating swine, avian, and human viruses. Such viruses have indeed circulated in American and Chinese swine populations and appear to be frequently transmitted to humans who have contact with infected swine (Smith et al. 2009; Yin, Yin, Rao, Xie, Zhang, and Qi 2014).

Soon after the S-OIV virus began to circulate in 2009, molecular virologists set to work to examine the new virus: What were its origins? How had it arisen? The answers revealed a virus with a complex pedigree (Smith et al. 2009; Zimmer and Burke 2009). Of its eight gene segments, six were inherited from a triple reassortant virus, which itself descended from a human H3N2 virus, an avian virus and the classical swine H1N1 virus. The remaining two segments were derived from a swine H1N1 virus of avian origin that emerged in Eurasia in 1979. Interestingly, the Spanish flu lives on in this pandemic lineage, in the form of three gene segments, the hemagglutinin (HA), the nonstructural (NS), and nucleocapsid (NP), that were each transmitted through the triple reassortant. These two swine H1N1 viruses appear to have reassorted in swine to form S-OIV some time before they were transmitted to humans to cause the pandemic. This is evident because the genetic distance between their genes and those of their parental virus sequences indicate a degree of nucleotide sequence divergence consistent only with circulation for some time in swine. It is quite likely that the reassortant virus that was to emerge as the 2009 H1N1 pandemic strain was simmering for some time, undetected in Mexican swine populations, before emerging to infect patient zero (Smith et al. 2009). These reassortment events succeeded in bringing together HA subtype 1 and NA subtype 1—the same subtypes that were circulating in 2008. Both were directly descended from the 1918 Spanish flu virus, but are genetically divergent: the NA subtype 1 is of avian origin and from a Eurasian swine flu virus, and the HA subtype 1 is derived from classical swine H1N1 flu strains. Although the HA gene is derived from the ancestral 1918 H1N1 influenza gene, it has evolved over the intervening ninety years in a different host species. The HA subtype found in the human H1N1 virus that was circulating contemporaneously (and was included in the 2009 vaccine) had therefore diverged a great deal from that of the swine virus. By 2009, the human population was devoid of meaningful immunological memory of either of the S-OIV viral surface antigens.

The independent evolutionary trajectories of antigen genes in different reservoir hosts are powerful sources of genetic diversity in influenza. The HA or NA subtypes that come together to form a new pandemic virus are not created equal; they can have very different evolutionary histories and play an influential role in the emergence of more successful influenza lineages. Together with the intertypic reassortment of gene segments, these evolutionary processes can create antigenic anonymity of the emerging virus and are certainly a major contributor to pandemic strain evolution. On the other hand, Holmes and colleagues (2005) studied circulating H3N2 virus isolates and concluded that intratypic reassortment between co-circulating divergent clades also makes a substantial contribution to viral genetic diversity. In a related analysis researchers studied seventy-one historical whole genome sequences of human H1N1 virus, the direct descendants of the 1918 Spanish flu, from between 1918 and 2006 to elaborate in fine detail the independent evolution (and thus the phylogenetic history) of each gene segment (Nelson et al. 2008). They constructed phylogenetic trees based on nucleotide and amino acid sequence divergence of each of the viral genes. As observed for H3N2, there was evidence of reassortment events between genetically divergent clades of H1N1 virus. This was evident because the phylogenetic trees constructed for each of the different influenza gene segments differed: they had different branching topology. Rather than being congruent, they were tightly networked. During sixty-eight years of geographically and temporally spatial seasonal epidemics, the H1N1 virus strains repeatedly explored reassortment in continuing cycles of evolutionary optimization, rather like a coach making small line changes in his dream team based on subtle differences in the opposition team's tactics.

This work highlights the potential of inter-pandemic genetic change, over and above antigenic shift, to create unusually pathogenic reassortant viruses. One such intratypic reassortant of H1N1 emerged in 1947: that year the vaccine was ineffective, but the viral subtype was unchanged; a severe epidemic H1N1 ensued (Kilbourne 2006). Analysis of the new strain revealed that the HA gene segment was substantially changed. The consequent antigenic change was enough to give the influenza virus an advantage that year. Other severe interpandemic epidemics, such as that in 1950–1951, might also have been associated with such intratypic reassortment events. An influenza genome is the sum of its eight gene segments, not just its envelope glycoproteins. This team of genes must play optimally together for the best outcome of the genome. This is illustrated by experiments that have been carried in the laboratory to reconstruct the virulent 1918 Spanish flu virus (see Chapter 11). Only viruses reconstructed from *all* eight original gene segments of the virus combine to form a virus with the expected virulence of the ancestor virus; no other combination of eight gene segments could do the job. It is thus the unique constellation of genes that make up the dream team virus; the basis for such optimal combinations remains opaque to us. Influenza viruses do not knowingly change their gene lineups to optimize the performance of the team. Accidental events, mutational changes, and promiscuous exchange of gene segments between viral lineages create the conditions in which influenza can, with unparalleled efficiency, empirically interrogate all combinations of team members. The most successful genetic lineages emerge from the remarkable evolutionary search engine of influenza viruses.

· 6 ·

ALTERNATIVE VIRUS LIFESTYLES

So far we have been preoccupied with viruses that cause short acute and usually self-resolving illnesses. I chose to focus on the human rhinovirus and human influenza virus. Both have lifestyles analogous to virulent lytic phages and are, generally speaking, "hit and run." The transient symptomatic illnesses that they cause are the central plank in their policy platform for epidemic transmission between hosts. Typically, the infected host recovers after the virus has been cleared by the immune system. Thereafter, they are rendered immune and cannot be reinfected successfully by the same virus. Other viruses that fall under this rubric include parainfluenza virus, measles virus, rubella virus, mumps virus, poliovirus, and even variola, the smallpox virus. These viruses can successfully maintain themselves in a host population only when there is an adequate supply of naive and susceptible hosts to support their chain of transmission. It has therefore generally been held that they are modern viral diseases that have infected humans in the recent history of our species. As early as the 1950s it was asserted, based on theoretical calculations, that measles virus would be unable to sustain uninterrupted transmission in any population that was smaller than 250–300 thousand people. This theory was later upheld by the studies of Francis Black at Yale University. His studies of measles in insular populations documented that breaks in the transmission of measles inevitably occurred in communities smaller than 500,000 people (Black 1966). He proposed that populations sufficient to support the endemic measles virus in humans would not have existed in primitive society.

In his discussion of diseases in antiquity, Hare observed that an infectious agent would not persist in these smaller populations unless it could multiply in a nonhuman host (i.e., a reservoir species) or persist in the infected person beyond the acute phase of disease (Hare 1967). Measles is such a virus. It is antigenically stable over many years and causes lifelong robust immunity after the infection is cleared. It is also a virus that can only infect humans (there is no known animal reservoir) and does not persist within individuals beyond the normal course of the disease. So Hare's premises applied. His review of historical accounts of human diseases uncovered no evidence for measles before the sixth century CE, or indeed of smallpox before the first century CE, consistent with his thesis. Any conclusions made from such historical investigations are of course speculative, but it seems most plausible that measles emerged following human's domestication of animals and consequent human infections by the related rinderpest virus of cattle. This may have represented a historic watershed opportunity for many of our modern diseases to emerge as zoonoses, when they jumped from their newly domesticated animal hosts into the human population. It was the dawn of civilized societies and the advent of concentrated population centers that supported their successful emergence as endemic human viral diseases. Mesopotamia may be the cradle of both civilization and modern epidemic diseases. Its population may have been the first metropolis large enough to support the persistence of new endemic viral diseases in the human population.

We will witness other examples of societal change priming the pumps for the emergence of viral epidemic (and subsequently endemic) diseases, but first we will turn to some examples of viruses that are associated, not with modern diseases, but with ancient ones. These are viruses that have plagued hominids since their emergence and whose coevolutionary roots we can project backward to before the emergence of mammals and even to invertebrate life perhaps 400 million years ago. There are, moreover, compelling arguments implicating tailed DNA phages of prokaryotes in their evolutionary past. With a great degree of certainty, one forebear of mammalian herpesviruses has been traced back to a hypothetical virus that infected a mammal-like reptile walking the earth well over 200 million years ago (McGeoch et al. 1995; McGeoch, Rixon, and Davison 2006). The herpesviruses belong to the family *Herpesviridae*. Over 300 species of *Herpesviridae* have been identified to date; they infect invertebrates, bivalve mollusks and abalone to be specific, and vertebrates—fish, reptiles, and mammals (Davison 2002; McGeoch, Rixon, and Davison 2006). Their mammalian hosts range from mouse to elephant, whale to human, bats to gorillas, and ungulates to birds. Eight species infect humans alone, suggesting that many remain as undiscovered infections of our evolutionary cousins, or perhaps have become extinct. This enormous family of successful viral parasites has a long and colorful coevolutionary history with its many host species. Those herpesviruses that have been researched intensively reveal an extraordinary degree of sophistication in their relationships with their host cells and organisms. I will have much more to say about the biology and lifestyle of these fascinating viruses and the lessons they can teach us about viral evolution in the context of their host species.

It goes without saying that our *Homo* genus ancestors in Africa must have suffered infections of herpesviruses. To pursue our train of thought regarding the population sizes necessary to maintain uninterrupted virus lineages, it follows that these viruses must have been able to survive among the relatively sparse, primitive, and distributed populations of early hominids. They are the antithesis of the hit-and-run viruses, and pursue a very different lifestyle. After a primary infection that can be associated with a relatively mild and transient symptomatic illness, they set up shop in the host organism, forming a lifelong persistent infection, termed *latency*. Often unknowingly, the host carries the infection until death (a mentor of mine was fond of telling his students that, for the most part, these were "viruses that you died *with*, not *of*"). Rather like the prophages of bacterial lysogens, these viruses reactivate periodically to produce infectious viruses. Regardless of the distinct and varied modes of transmission that the herpesviruses employ to get from host to host, their capacity for latency offers a well-designed strategy for propagating the viral lineage. The initial infection renders the host infectious and therefore a source of viral transmission to new hosts. Following apparent recovery, the host is periodically infectious, offering the virus additional opportunities to sustain the necessary chain of transmission for continued propagation. "Design" is, of course, not the right operative term; the preservation of the persistent latent-infection strategy across evolution (certainly in the most studied members of the herpesvirus family) indicates that it has offered the genetic information of the virus a selective advantage in the past. It is an advantage that must have been manifest from accidental genetic change and reinforced through favored inheritance under positive selection. We can assert that latency and reactivation of herpesviruses is a primary contributor to their evolutionary success as transmissible infectious agents. The mechanisms at play in this part of their life cycle are of such complexity that it is impossible to construe latency as an accidental product of evolution. It is a highly evolved process that serves herpesviruses well.

Before moving on to examine some of the intricacies of herpesvirus infections, we should review some of the later work of Francis Black that serves as a compelling illustration of the distinct advantage of this viral lifestyle to the preservation of viral lineages in human populations. Black's article in the journal *Science* (1975) concerned studies of primitive cultures on the periphery of the Amazon Basin. He judged that these very isolated land-bound communities, often of only 100 to 300 individuals, were representative of the conditions prevailing in early and primitive human hunter-gatherer societies. Black collected demographic data and serological samples in several such populations to look for antibody reactivity to common human viruses. For some of the most common diseases of today's cosmopolitan society—measles, mumps, rubella, parainfluenza, and poliomyelitis—he found no evidence of serological reactivity in many or all of the tribes. If there was some evidence of the disease, it was found in individuals of a particular age or older, but was entirely absent in the younger members of the tribes. This was evidence of an epidemic of the viral disease sweeping through the tribes at a particular point in time. It then quickly burned out after recovery or death of the infected individuals, leaving no susceptible hosts available for the virus to infect. Children born after the epidemic ended remained disease free. A contrary situation was evident for herpesviruses and hepatitis B virus. A consistently high prevalence of high-level antibody titers, regardless of age, were found in all of the tribes. These infections caused little overt illness and were not a threat to the survival of the tribes. Each of the viruses—herpes simplex, varicella zoster, and Epstein-Barr virus (all herpesviruses) and hepatitis B virus—establish persistent infection in the host after the initial infection and are therefore well adapted to endemic survival in the small community groups. While these viruses spread rapidly through the tribes at an early age and persist across generations despite the small population, the hit-and-run viruses are transiently epidemic and then burn out. They must be reintroduced to the population after sufficient time has elapsed for new immunologically naive hosts to be available to sustain an epidemic chain of transmission. Such diseases, therefore, must have posed little threat to ancient humans, but are the products of modern society.

Latency: Till Death Do Us Part

I became inclined toward pursuing research on eukaryotic viruses during my undergraduate studies at the University of Cambridge in the late 1970s. I vividly recall being captivated to learn about the splicing of messenger RNA transcripts from the hexon gene of adenovirus, a DNA virus that replicates in the nucleus of eukaryotic cells. The recent discovery of split genes at Cold Spring Harbor Laboratories would earn Phillip Sharp and Richard Roberts the Nobel Prize in Physiology or Medicine in 1993. RNA splicing comprises the processing of the primary transcript of a gene in the nucleus, before export to the cytoplasm. It excises whole segments of the noncoding sequence called introns that are included in the transcript. The resulting mature messenger RNA retains the contiguous protein-coding exons. It is transported to the cytoplasm where it is recognized and translated by the ribosome. It was becoming apparent that this was a process fundamental to all cellular gene expression. Just as bacteriophages had provided much of our instruction on the genetics and molecular biology of the bacterial world, these viruses were going to teach us about eukaryotic cells. So my first fascination with viruses was not in the viruses themselves and the diseases that they caused, but in the anticipation of what they would teach us about our own cellular processes, which they utilize so expediently. Nuclear DNA viruses, like their cytoplasmic RNA virus counterparts, must utilize the same cytoplasmic machinery of protein translation, but they must also manipulate the cellular machinery for gene transcription in the nucleus. Their messenger RNAs are transcribed from the viral genome DNA by the human RNA polymerase II. Not only must it encode the necessary signals and codon sequences of amino acids to direct the synthesis of its proteins by cellular ribosomes, it must also contain the information required to orchestrate the coherent expression of its genes by the cellular transcriptional machinery.

Herpesviruses are among the most complex DNA viruses and have genomes that vary in size from 125 to 230 kilobases of double-stranded DNA, encoding between 70 and 200 genes. The complexity of their genomes is a compelling indicator of their potential for highly adapted and nuanced relationships with their host organisms. The herpesviruses are a spectacular evolutionary success story.

A more formal introduction to the herpesviruses is warranted. Just as the *Caudovirales* became recognized as a family of related bacteriophages because of their similar morphologies, so too did the order *Herpesvirales* enter into the viral nomenclature. Examined under an electron microscope they all share virus particles of similar architecture. They are composed of a lipid envelope studded with viral glycoproteins, surrounding an amorphous protein matrix, within which is nested an icosahedral capsid containing the DNA genome. The *Herpesvirales* is composed of three families: the *Herpesviridae*, comprising all the herpesviruses that infect mammals, birds, and reptiles; the *Alloherpesviridae* of fish and amphibians; and the *Malacoherpesviridae* of invertebrates. Most of our knowledge of herpesviruses is derived from the *Herpesviridae*, which has three subfamilies: the *Alpha-*, *Beta-*, and *Gammaherpesvirinae*. Humans are host to three alphaherpesviruses, three betaherpesviruses, and a single gammaherpesvirus. The human herpesviruses will serve as our window on the lifestyles and pathogenesis of herpes infections. All of them share the singular capacity for two modes of infection: lytic and latent. After initial infection of the host, the virus undergoes lytic infection, typically associated with a mild symptomatic disease and the shedding of infectious virus. The establishment of the lifelong latent phase of infection in quiescent cell populations follows the initial lytic phase of infection. Latency is operationally defined by the lack of production of infectious virus, and it occurs in the face of an active immune response (Wilson and Mohr 2012; Roizman and Whitley 2013). Much of the success of herpesviruses can be attributed their capacity for immune evasion. They have evolved mechanisms to counteract host immune defenses and during latency they remain under the radar of host immune surveillance. The mechanisms at play that maintain the viral latent state remain the subject of intensive research, but it seems clear that it is a metastable state to which both cellular repressive mechanisms and active viral mechanisms contribute. Reactivation from latency occurs periodically, and apparently stochastically, with the production of infectious virus that can be transmitted to a new host.

Herpesviruses are very fastidious in their choice of natural host, and herpesvirus diseases are rarely associated with severe or life-threatening illness in healthy hosts. This is indicative of a high level of evolutionary adaptation to their particular hosts. It is, of course, advantageous to the virus if the host remains relatively healthy over an extended period of infection so that it can transmit the virus to a new host during its normal behavior. This is particularly true for sexually transmitted pathogens or those that require close personal contact for transmission. Severe and often fatal herpes infections do, however, result when epizootic infections (of a different animal species) occur. These are most often accidental and transient *dead-end* incursions of the virus into a foreign host species that do not result in the establishment of endemic disease. The macaque herpes B virus provides such an example (Huff and Barry 2003). It causes asymptomatic disease in its natural host, but is lethal to other monkey species and most often fatal in humans who are unlucky enough to contract the infection from an infected macaque. Herpes B virus has a prevalence of about 75 percent in wild-caught macaques used for animal research but today captive breeding colonies are the norm, significantly reducing this zoonotic hazard for laboratory workers. Many other cross-species introductions of herpesviruses have similar outcomes and are provoked by human activities such as intensive farming and curation of zoological collections, which can bring different species into unnatural proximity.

In common parlance, having herpes is associated with an infection of herpes simplex virus (HSV). Genital herpes, caused by HSV, first achieved notoriety in the early 1980s. In 1982, *Newsweek* branded it "The VD [venereal disease] of the 80s," while *Time* called it "The New Scarlet Letter," Twenty million Americans were afflicted by an epidemic that swept through our sexually liberated society. There are actually two species of herpes simplex virus, HSV-1 and HSV-2, and they are associated with herpes *labialis*, a common condition in which herpetic lesions erupt in the mouth and on the lips, and herpes *genitalis*, which involves infection at the genital mucosa (Whitley and Roizman 2001). This infection is sexually transmitted when the virus is shed from genital mucosa during either primary infection or reactivated latent infection. Testaments to success as human pathogens, HSV-1 and HSV-2 are both extremely common in humans, and rates of infection increase with age. The less prevalent HSV-2 is carried by 30 percent of our adult population, while HSV-1 can be detected in 50 percent or more of children aged ten, and in late adulthood rates of seroprevalence can exceed 80 percent. The route of infection of herpes simplex virus is via direct contact of the virus with epithelial cells. As with the influenza virus, attachment to the cell is mediated by molecular interactions between viral envelope glycoproteins and cellular membrane proteins. A fusion of the viral and cellular membranes allows the entry of the nucleocapsid and associated matrix proteins into the cell cytoplasm, where the nucleocapsid makes its way to the membrane that surrounds the cell nucleus. It introduces the naked viral genome DNA through nuclear pores into the nucleus itself, the control room of the cell. Here, the virus undergoes a cascade of gene expression in which waves of successive proteins are synthesized, the cellular defense mechanisms neutralized, and the metabolism harnessed to the virus's own ends. The DNA genome is replicated and virus particles assemble in the nucleus before budding through the nuclear membrane and entering the golgi where the viral envelope is acquired. After budding from the golgi, the particles transit the cytoplasm and fuse with the outer membrane of the cell, releasing mature virions into the surrounding milieu. This lytic replicative cycle in epithelial cells results in the outward symptoms of the infection, appearing as watery blisters on the mouth, lips, or mucous membranes. These lesions are loaded with infectious virus that can be readily transmitted to others in saliva or in genital secretions. HSV can be a particularly efficient and stealthy contagion; often, transmission of the infection to a sexual partner can occur despite the absence of visible symptomatic lesions of the genital mucosa.

During a primary infection, virus particles released from epithelial cells infect the sensory neurons that innervate the infected epithelial tissue. The nucleocapsid containing the DNA genome enters the axon termini and thereafter it and its genetic cargo are transported along the considerable length of the axon to the nucleus in the nerve cell body located in the dorsal root ganglion. The genome is injected into the neuronal cell nucleus but, more often than not, the lytic replicative program does not proceed. The balance of power in this nuclear environment is different: viral gene expression is highly restricted by the neuronal cell and the virus makes only one transcript (termed the *latency associated transcript*), which plays a role in reinforcing the dormant, latent state of the virus in the neuronal cell. The genome persists in the cell nucleus as a circular DNA episome packaged into chromatin in a manner similar to cellular chromosomal DNA. Its expression is highly restricted and tightly regulated. It is from these neuronal ganglia that infectious virus is periodically produced in a reactivation event, commonly called an "outbreak" by herpes sufferers. Such outbreaks typically recur at or very close to the site of the original infection, whether as cold sores on the lips or genital blistering. Typically, overt reactivation may be associated with stresses to the host, perhaps allowing the virus transient escape from the continuous immune surveillance that operates in the host to minimize the pathologic consequences of stochastic reactivation events (Roizman and Whitley 2013).

Human herpes virus 3, or varicella-zoster virus, is similarly a neurotropic virus, which also establishes latent infection in neurons. Its initial infection is the cause of chickenpox, a common disease of childhood, but unlike HSV it causes generalized constitutional symptoms, replicates in mucoepithelial cells, and is shed from an itchy cutaneous vesicular rash on the skin and the oral mucosa. The latent virus in neuronal cells reactivates to cause shingles, or *zona*, a disease that is remarkably quite distinct in nature to that occurring after the primary infection. Cutaneous viral eruptions occur only on a limited area of the skin innervated by the affected neurons. It seems that while the primary infection cannot be locally contained by the immune system, reactivations can be more efficiently subdued. The only human betaherpesvirus, cytomegalovirus (CMV), and the gammaherpesvirus, Epstein-Barr virus (EBV), are similar to each other in their disease etiology but quite distinct from the alphaherpesviruses. They each cause mononucleosis in adolescents, although they may cause asymptomatic disease and go unnoticed in infants. Unlike HSV and varicella zoster virus, their mode of transmission is not sexual. It is principally by shedding of the virus into the oral cavity, a feature that has lead to infectious mononucleosis being called "kissing disease." Both

CMV and EBV infect epithelial cells at the outset of the infection, but then target different populations of leukocytes in which they establish latency. In the case of CMV, cells of the lymphocytic and monocyte lineages are preferred, while EBV takes up residence in B-lymphocytes. It is notable that these viruses from different subfamilies have evolved commonalities in their lifestyles, a primary infection followed by lifelong latency, yet they access quite different cellular niches in which to establish latency. Such differences in host cell tropism must have mandated the evolution of distinct regulatory machinery in the various subfamilies of herpesviruses in order to manipulate very different host cell types.

All in the Family *Herpesviridae*

The *Herpesviridae* are the most closely related groups of herpesviruses in terms of their genetic contents. They share a select complement of about forty conserved genes that can be considered a set of "core genes," providing the basic necessities for viral replication shared by all family members (McGeoch, Rixon, and Davison 2006). They are the genetic lowest common denominator of the family of *Herpesviridae* and they make up the majority of the seventy or so genes of alphaherpesviruses. On the other hand, the human betaherpesviruses encode a substantially larger complement of genes. For example, human cytomegalovirus (CMV) encodes 165 genes in a genome of 230 kilobases. It is indeed a sophisticated virus; a limousine "fully loaded" with optional extras. More than 100 additional genes supplement its core gene functions. The other families of *Herpesvirales* share similar basic needs for their propagation in host cells. They, too, must engineer their entry into the cell, replicate their DNA, and assemble nucleocapsids to mature into morphologically very similar virus particles. Nevertheless, any detectable amino acid sequence similarity between their proteins is lost in evolutionary time as they and their hosts diverged. Only a single instance of similarity between their proteins is discernible – for a protein associated with virus capsid morphogenesis. It is apt that the machinery for the packaging of the viral DNA genome and assembly of the virus particle is the single unifying characteristic of all *Herpesvirales*.

Evidence is now accumulating that the capsid assembly and genome packaging machinery of all herpesviruses share a common and ancient evolutionary origin in the *Caudovirales*, tailed double-stranded DNA phages that infect prokaryotic cells (McGeoch, Rixon, and Davison 2006; Schmid et al. 2012). These studies are informed by comparison of the pathways of virus particle morphogenesis together with protein sequence and structure information. Remarkably, detailed visualization of the three-dimensional structures of phage and herpesvirus nucleocapsid proteins reveals that their amino acid chains share similar spatial dispositions (otherwise known as protein "folds") despite being separated by a billion years of evolutionary history. Sometimes the first solution to a problem cannot be improved upon. Herpesvirus evolution has adopted one particular modus of virus particle morphogenesis and has never "reinvented the wheel." It is premature to suggest that herpesviruses are direct phylogenetic descendants of tailed DNA phages. Viral lineages have an uncanny ability to donate and receive genetic information, be it genes or entire genetic modules. Nevertheless, we can safely conclude that the last common ancestor of all herpesviruses, which is estimated to have existed some 400 million years ago, already possessed the machinery. That virus may have originated via speciation from a descendent of a tailed DNA phage or (more likely?) a distinct viral lineage acquired the machinery by horizontal gene transfer from a phage or other descendant of the *Caudovirales*.

Despite their genetic diversity, the genomes of all herpesviruses must encode a certain subset of proteins necessary for replication in their eukaryotic host cells. The large difference in the gene complements, even of viruses within the same subfamily, is therefore remarkable. This diversity has arisen after divergence from common ancestral herpesviruses, the result of different selective pressures sculpting the genetic composition of each virus to match its particular ecological niche. Leading experts in herpesvirus genomics and evolution, notably McGeoch, Davison, and colleagues at the University of Glasgow, Scotland, have made a detailed and convincing accounting of the family tree of the *Herpesviridae* and its component alpha-, beta- and gammaherpesvirus subfamilies (McGeoch, Dolan, and Ralph 2000; McGeoch et al. 1995). It is grounded in the phylogenetic relationships of the viruses and their hosts and is informed by knowledge of the temporal speciation of the hosts with which they have codiverged and cospeciated. For these viruses, it can be tentatively said that if one follows the branches of their evolutionary tree back in time, their common roots lay in an ancestral virus that existed some 200 million years ago. This common ancestor split into the alphaherpesviruses and a second lineage that became the common ancestor of today's beta- and gammaherpesviruses. Subsequently, the three lineages of the *Herpesviridae* emerged prior to the major radiation of mammalian evolution that occurred approximately 80 million years ago. The branching events in the evolution of these three sublineages could not be associated with any node of speciation in the common evolutionary history of mammals, birds, and reptiles. These evolutionary experts speculate that the first "speciation events" in the early history of *Herpesviridae* resulted from genetic divergence creating viral lineages capable of exploiting different ecological niches within the same host organism.

As McGeoch climbed further up the evolutionary tree of *Herpesviridae*, he discerned the evolutionary landscape of the mammalian herpesviruses with much greater certainty (McGeoch et al. 1995). In the last 80 million years mammalian herpesviruses evolved almost in lockstep with their hosts. Speciation following the transmission of herpesviruses into new but related host species has certainly occurred, but cospeciation, or codivergence of the virus with its host species, appears to be more the rule. As a result each virus becomes exquisitely coadapted to its respective host species. The primate simplex viruses represent an illustrative case study in cospeciation and codivergence. Their evolution was traced to the ancestors of Old and New World primates (*Simiiformes*), and today many primate species (e.g., human, baboon, African green monkey, chimpanzee, macaques, and squirrel and spider monkeys) can claim their own species of simplex virus. Humans are unique in being infected by two distinct simplex viruses, HSV-1 and -2. It was a topic of some debate as to whether these two viruses evolved by lineage duplication within humans or one of our ancestors, or if one or another arrived in humans via cross-species transmission from a related primate. Until recently the jury was out, but the discovery of the chimpanzee herpesvirus left little room for further speculation. It is most closely related to HSV-2 and more so than HSV-1. HSV-2 appears to have emerged in humans after cross-species transmission, while HSV-1 codiverged and evolved with our ancestors. It seems that about 1.6 million years ago one of our now-extinct ancestors became infected by a herpesvirus that originated in an ancestor of chimpanzees (Wertheim et al. 2014). Of course, this is not the only example of cross-species transmission during the evolution of the other extant mammalian herpesvirus subfamilies, but nevertheless coevolution and codivergence remains the dominant mode of evolution operative in these ancient DNA viruses.

Are recently emerged viral diseases the exclusive province of RNA viruses? These "modern viral diseases" are relatively new to our species, resulting from interspecies transmission of animal viruses from their long-standing reservoir hosts. It is true that RNA viruses are particularly well equipped to move between species and rapidly adapt to their new hosts and have done so with considerable success. The emergence of the human measles virus (originally a morbillivirus of cattle) and the human influenza A viruses (from avian influenza viruses) are examples. Nevertheless, we will meet some DNA viruses that prove the exception and suggest we should not underestimate their evolutionary prowess.

· 7 ·

EVOLUTIONARY MECHANISMS *of* DNA VIRUSES

THE FASTIDIOUS NATURE of herpesviruses and papillomaviruses is typical for viruses whose evolution has been dominated by cospeciation and codivergence with their animal hosts. It is tempting to intuit that they all share limited evolutionarily agility, or that they are evolutionarily constrained because they become so highly adapted to their particular hosts. At face value, RNA viruses are better equipped for rapid evolutionary change. The viral RNA polymerase is a consistent generator of genetic variation, creating mutations in almost every new daughter genome. Existing in each host as a quasispecies, they can utilize their rich and dynamic genetic diversity as a platform for rapid adaptive change. DNA viruses have no such luxury. The polymerases they use to replicate their genomes are DNA-dependent DNA polymerases. Herpesviruses encode their own viral DNA polymerase, while papillomaviruses rely entirely on the cellular enzymes. Each of these DNA polymerases has a proofreading capacity; it can detect and correct misincorporated nucleotides before they can be perpetuated in progeny genomes. As a result, the rate of nucleotide misincorporation more closely reflects the fidelity of the cellular machinery than the infidelity of RNA viruses. How then do DNA viruses achieve the rapid evolutionary rates necessary to compete in the arms race that exists between virus and host? The answer is complex: let me first convince you that in order to compete successfully in a Red Queen dynamic relationship with a host, a parasite (in our case a virus) need only have the capacity to *out-evolve* the host. That is, to run as fast or just a little faster than the Red Queen herself.

Evolutionary rates, whether of organisms or viruses, microbe or metazoan, are governed by several variables. The rate at which errors or other events create genetic variants, the prevailing selective pressure (purifying or positive), and the generation time. DNA viruses may be shackled by replication machinery that has greater copying fidelity than that of RNA viruses but they still soundly out-replicate their hosts. They can undergo countless replicative cycles and create large populations and many generations of viruses within the generational time frame of their hosts. Here I refer not to the time interval between host cell divisions but to the time interval between the reproduction of the host organisms. This is the time that the host needs to pass on its genes to offspring in which they are tested by natural selection. Only the mutations that occur in the gametes of the male or female hosts are inherited in sexually reproducing organisms; it is this genetic variation that drives host evolution. All viruses have generation times that far outpace their hosts. For DNA viruses, this is sufficient advantage to ensure their evolutionary success, particularly in the context of cospeciation with their hosts. Cross-species transmissions and the resulting host shift are commonplace in the evolution and speciation of RNA viruses, but have a relatively minor role in the evolutionary history of DNA viruses. Where there is evidence that it takes place, it is most likely to be between closely phylogenetically related species (this topic will be discussed in Chapter 10). The necessary adaptive agility to facilitate successful cross-species jumps is largely (albeit not exclusively) the reserve of RNA viruses.

For completeness, I do need to point out that RNA viruses do not (nor indeed do DNA viruses) have unfettered flexibility and capacity for genetic change. While RNA viruses with their error-prone replication mechanisms have a great capacity to explore genetic variation, their flexibility is limited. RNA-based replication places a significant constraint on the maximum size of the genome. The larger the RNA genome, the greater the risk of *error catastrophe*. It can be demonstrated experimentally that RNA virus genomes exist on the brink of error catastrophe. Evolution has tuned the relationship between their mutation rates and genome lengths to an optimum. In the laboratory, RNA viruses can be forced to extinction by artificially enhancing their mutation rates. This has been achieved by exposing them to nucleotide analogues that increase their mutation rate or by genetically manipulating their RNA polymerase to make more copying errors (Crotty, Cameron, and Andino 2001; Crotty et al. 2000). It seems that an error rate of about 1 per 10^4 nucleotides incorporated is close to the maximum that can be tolerated in a genome of about 10 kilobases (these experiments were performed in the laboratory with poliovirus). It follows that any change in genome functionality of an RNA virus must be accommodated without increasing the size of the genome. The mutational space that can be explored by RNA viruses must always respect their limited genome size. This constraint on coding capacity results in minimalist genomes densely packed with information. Many RNA virus gene products must execute multiple functions in the virus life cycle and can be encoded in overlapping genes. Even more important for RNA viruses, the primary and secondary structure of substantial portions of the genome play functional roles. The primary nucleotide sequence can itself constitute informational content and thus functionality. Such functions are "coded" in the particular nucleotide sequence or the folding of the RNA strand to form complex secondary structures stabilized by base pairing between self-complementary stretches of nucleotides. This "code" is not redundant in the fashion of the genetic code and there can be no synonymous mutations in such regions. Any nucleotide substitution will change the structure of the RNA and may therefore influence phenotype. Some of the flexibility of a redundant genetic code is therefore lost to the exploration of RNA viruses. They can only successfully evolve if they retain the functionality of the highly condensed genetic information that accomplishes the task of viral replication, and the command and control of the host.

On the other hand, the fidelity of DNA virus replication removes much of the size constraint on the genome. Many laboratories have estimated the error rate of replication in DNA viruses to be in the order of 1 error per 10^8 nucleotides incorporated into new genomes (Drake and Hwang 2005); they have a factor of 10-thousandfold greater replication fidelity than RNA viruses. Unless other constraints come into play, such as a small capsid size which may limit the amount of DNA that can be packaged in the infectious virus particle (Chirico, Vianelli, and Belshaw

2010), DNA viruses can evolve relatively massive genomes without any risk of error catastrophe. The genome size of some small DNA viruses, such as papillomaviruses, polyomaviruses, and parvoviruses, do appear to be constrained by their capsid size. They also have minimalist genomes that use their available capacity with remarkable economy and often encode "multiuse" proteins and overlapping genes. We will return to small DNA viruses in a later section of this chapter. Herpesviruses and the nucleocytoplasmic large DNA viruses (NCLDVs), which include poxviruses, have evolved much larger genomes. While they may not have an engine of genetic variation that equals that of RNA viruses, they need not use their coding capacity with the same economy. They exploit this luxury of genome expansion very much to their advantage. The herpesviruses utilize a variety of mechanisms to explore genetic space and opportunity for rapid evolutionary adaptation.

Gene Duplication and Gene Capture

Nucleotide substitution mutations should not be ignored as a source of genetic variation in double-stranded DNA viruses, and indeed the measured rates of mutation are still at least an order of magnitude higher than those of their host cell genes (Li 1997). Nevertheless, herpesviruses rely on their flexible DNA genome in two additional ways to create the genetic variation essential for adaptive evolution. They appear to frequently exploit gene duplication to create genomic flexibility, and they are also adept at gene capture. These capabilities become readily apparent when the genomes of herpesviruses are examined carefully; particularly so when their very different complements of non-core genes (perhaps best termed *adaptive* genes) are considered. Conceptually it might be useful to consider these as the add-ons to a base model that comes only with the essential core genes; each different herpesvirus evolves to be accessorized appropriately for its host cell. The adaptive genes are most commonly evolved as countermeasures of the host's defenses against viral infection. These include the two branches of the host immune response to virus infection: innate immunity and cell-autonomous immunity, the first lines of cellular defenses against viruses and the adaptive immune response of the host organism. It is also necessary to manipulate the metabolism of the host cell to permit replication and virulence. Adequate pools of nucleotide precursors need to be made available to support viral DNA replication, and this may require the cell to activate certain regulatory circuitry reserved for cellular DNA replication and natural cell division. By definition, these functions are those that must be exquisitely tailored to the particular host cell. They are the key battlegrounds of conflict between the virus and its host, where the selective pressures for evolutionary invention are most potent.

Genomic evolution of all domains of life employs gene duplication. It is a powerful mechanism to evolve new gene functionality. There is abundant evidence in all herpesviruses that recombination events lead to duplication of genetic material (Davison 2002; McGeoch, Rixon, and Davison 2006). The opportunity provided by gene duplication is clear; a second, and now redundant copy, of a gene is created. This copy is free to undergo adaptive evolution with no need to conserve the functionality of the original protein. The remaining unchanged copy of the gene ensures the viability of the lineage. Gene capture events are the second modality for acquiring new adaptive functionality for herpesvirus genomes. This is horizontal gene transfer which we also witness in the evolution of phages and their microbial hosts. In the case of herpesviruses, they capture genes from the cellular chromosome and incorporate them into their own genome. The precise mechanisms of gene capture remain obscure, but fragments of host cell DNA must become joined with viral genomic DNA to create a genome containing new genetic information. This new genetic material often provides beneficial phenotypic traits to the virus. The gene is no longer subservient to the selective pressures on the host organism but can evolve independently and can be tailored to perform new functions based on the random mutation and selection operating in the viral genome.

It is quite evident from the genetic composition of herpesvirus genomes that the core genes inherited from the most recent common ancestor of the herpes virus family tend to be clustered toward the middle of the viral chromosomes. Gene duplications and captured genes tend to be located near the ends. This is presumably so that recombination events between DNA molecules, be they virus with virus or virus with host, can occur here without undue risk of disrupting the central regions of the genome. Recombination events will, of course, be distributed across the entire genome, but those that disrupt the core gene complement will not result in viable recombinant viruses. If core genes were distributed equally across the genome, then it follows that a greater proportion of the random recombination events would disrupt the core genes. Natural selection has thus favored viruses with core genes clustered in the central portion of the genome. The genes that perform the essential housekeeping functions of the virus are kept out of harm's way. In an earlier chapter we discussed DNA bacteriophages that promiscuously shuffle and exchange modules of genetic information during their evolution. It is evident that this continues to play a fundamental role in the evolution of large double-stranded DNA viruses, allowing them to adapt to and dominate their host cells.

If we are to stare further back in evolutionary time, we can also perceive gene capture events as the origins of the core genes that were represented in the most recent common ancestor of these viruses. Similarities between herpesvirus genes and those of adenovirus, another DNA virus, have been observed. Large DNA virus polymerases retain structural and functional similarities to those of one family of DNA polymerases which is encoded by host cell genomes. It therefore seems highly likely that large DNA viruses have assembled much of their replicative machinery by scavenging genes from the host cell and adapting them to their needs. Nevertheless, the jury is out on this assumption: too little data is available and it remains a formal possibility that the eukaryotic cellular polymerases were scavenged from viral genomes, rather than the other way round (Shackelton and Holmes 2004). Gene capture events are prevalent throughout herpesvirus evolution. These large DNA viruses are amalgamations of genetic material. New viral functions are created after gene duplication or gene acquisition from a virus or a host, after which the new gene can be adapted to the virus's needs. These processes are evident in genes that are now common in all herpesviruses, having evolved in ancient common ancestor viruses. They can also be observed for genes that are unique even to a single species of herpesvirus today and hence were acquired after the last speciation of the lineage recently in evolutionary history.

The 2'-deoxyuridine 5'-triphosphate pyrophosphatase (dUTPase) is an enzyme that is found in all three domains of life and is necessary for the hydrolysis of 2'-deoxyuridine triphosphate to generate precursor molecules for DNA synthesis. Large DNA viruses, all of the herpesviruses (and also poxviruses), possess one or more analogues of this enzyme in their genomes (Baldo and McClure 1999; McGeoch 1990). They captured the gene from their host cells; more than likely this capture event happened multiple times in evolution. Notably, some herpesviruses have a single copy of the gene, recognizably similar to the cellular gene; in laboratory experiments it is important for the virulence of the virus in animal models of virus infection. Others have duplicates of the gene and the additional copies are substantially altered, or perhaps to be more precise "customized." Examination of their amino acid sequences reveals that they no longer have the necessary protein structures to perform the hydrolysis of dUTP. It seems then that these are examples of the *de novo* evolution of new functions for a captured host cell gene. In a surprising turn of events, some herpesviruses appear to have multiple distantly related analogues of the cellular dUTPase gene, but none of them retains the gene's original enzymatic function. These DURPs (dUTPase-related proteins) form a large family of genes in multiple *Herpesviridae*. Cytomegalovirus alone has four of these DURP genes, evolved to have independent and distinct, albeit not fully understood, functions. An elegant analysis revealed the similarity of these genes resides in the conservation of regions of the dUTPase gene that dictate the particular three-dimensional architecture of the protein (Davison and Stow 2005). They suggest that herpesviruses have hijacked a protein-coding sequence that is particularly malleable to evolution of new functions, and exploited it for their own purposes, through duplication and reinvention, to create entirely novel viral gene products.

Cytomegalovirus is the human herpesvirus with the largest complement of genes: it has taken full advantage of each of these modalities of evolution to accrue adaptive functions that allow the virus to manipulate the host cell and host organism to advantage. Among these viral gene products, a majority are proteins that modulate the host immune response and permit the virus to evade immune control and clearance from the host during both acute and persistent viral infection. Cytomegalovirus is widely prevalent in most human populations, and it infects us persistently after the initial acute infection. It is never dislodged despite a very active and lifelong host immune response. Other cytomegalovirus genes specifically promote pathogenicity; for example, during acute infection, infected cells secrete a viral protein that has genetic homology to a human chemokine. It has the potential to bind to the human chemokine receptors that mediate chemoattraction of neutrophils to sights of inflammation

(Penfold et al. 1999). It is well documented that infection of neutrophils is central to acute cytomegalovirus disease; this appears to be an elegant mechanism to attract susceptible host cells to the vicinity of virus-producing cells, accelerating the dissemination of the virus within the host organism. This strategy is seen time and time again in studies of the pathogenesis of infectious diseases, most notably those that prey on cells of the immune system itself. These are readily recruited to the locus of infection and can be used as vehicles for amplification and as transportation for the virus or infectious agent within the host from tissue to tissue.

The genome accessories that garner the most attention are the genes acquired and evolved by the virus to dampen the host immune response and to obscure the recognition and elimination of virus-infected cells by our immune system. Cytomegalovirus has a panoply of such genes and many can also be found in the genomes of other herpesviruses and other large DNA viruses. Speaking to the shared evolutionary sleight of herpesviruses, these host-derived genes can often be shown to be acquisitions of independent gene capture events in different viral genetic lineages. As part of their adopted viral genomes, they then evolved to the benefit of viral fitness and replicative success, being tailored to the individual needs of the particular virus. Among the arsenal of immune evasion proteins encoded by cytomegalovirus are those that antagonize the ability of the infected host cell to display viral antigens on its surface. Variants of host cell cytokines that naturally have immune suppressive activities and dampen the vigor of the immune response to virus infection are also recruited from the host. A case in point is IL-10, particularly notable because it has been adopted by several herpesviruses (and at least one poxvirus) in what are clearly separate evolutionary events. Evolution of the viral analogues proceeded independently within their respective genome such that they retain different subsets of functions displayed by the normal human gene product (Spencer et al. 2002).

The class I major histocompatibility complex (MHC I) on the surface of host cells is responsible for alerting cytotoxic T cells that the cell is producing foreign antigens and should thus be attacked and eliminated. Down-modulation of MHC I expression and signaling is mediated by at least two cytomegalovirus genes, their function being to cloak the infected cell, rather like a stealth bomber eluding enemy radar. Such downregulation of MHC I antigen presentation is repeatedly found as an evolved function in viruses and it is a hallmark of infected cells. Indeed, the host immune system has even evolved to recognize it as such, and cells that lack MHC I proteins on their surface are detected and become targets for destruction by natural killer cells, a distinct subset of immune cells. Cytomegalovirus, however, exhibits its remarkable prowess as a real master of immune deception. It takes its immune evasion strategy one step further; not only does it downregulate MHC I antigen presentation (a strategy that if executed alone would be detected by the host), it also deploys a decoy protein, a nonfunctional MHC I homologue on the surface of the infected cell. This fools the host's natural killer cells; they perceive MHC I function to be business as usual and the infected cells go undetected.

The proclivity to adopt host cell genes and tailor them to their own needs is evident in all herpesviruses and large DNA viruses, but cytomegalovirus is a signal example. Its genome contains fifteen gene families that each originated by gene duplication, and each family possess between two and fourteen copies. This has created the potential for multiple and diverse gene functions to evolve in parallel from each copy of the original gene. These duplications most often involve genes that have been captured from the host, and the vast majority encode proteins now employed in host immune evasion. These functions promote successful acute infection and then support the persistence of the virus in the face of host immunologic defenses. It appears then that despite having much lower intrinsic mutation rates than RNA viruses, these large DNA viruses display remarkable evolutionary guile and innovation.

Adaptive change in RNA viruses is indeed principally fueled by high mutation rates (together with rapid cycles of replication and large population sizes), but double-stranded DNA viruses exercise other options. As we have seen, they leverage many other mechanisms to support the evolutionary change necessary to wage a genetic arms race with their host. They exploit the advantages that they have: in contrast to RNA viruses, they can accommodate large numbers of genes in their genomes. While RNA viruses mutate at the nucleotide level at a high rate, they "run" to stand still in the parlance of the Red Queen; herpesviruses appear to be error-prone in their husbandry of genomic integrity. This allows promiscuous duplications and exchange of genetic material with the host cell chromosome. In effect, again to lean on the Red Queen analogy, they walk most of the time but occasionally take giant leaps, allowing them to stay one step ahead in the race of adaptive change with their hosts. These may be relatively rare or common events; it is impossible to say since we are witness only to the events that have occurred in the ancestors of today's herpesvirus lineages. Moreover, only successful genetic accidents are recorded in the genomic lineage. In any case, these recombination events provide herpesviruses with the foundation for much genetic variation, which can then be tailored by further mutational changes. Together they provide the highly nuanced adaptive evolution that is evident in the relationships of herpesviruses and their natural hosts today.

Poxvirus Evolution

The *Poxviridae*, or poxviruses, are members of the sizeable group of nucleocytoplasmic large DNA viruses (NCLDVs) that altogether make up the order *Megavirales* (Colson et al. 2013). This order includes viruses of the families *Poxviridae*, *Asfarviridae*, *Iridoviridae*, *Ascoviridae*, *Phycodnaviridae*, *Mimiviridae*, and *Marseilleviridae*. Like herpesviruses, poxviruses have large genomes, some as large as 360 kilobases in length. Their gene complements are correspondingly complex. Also like herpesviruses, the family members all share similar sets of core genes, the genetic lowest common denominator of poxviruses, conserved in amino acid sequence and passed down from their most recent common ancestor. Also familiar is that this essential set of genes takes up a central location on the viral chromosome, the genes toward each end of the genome appearing to provide additional functionality for the virus. They are the optional extras of the different models of poxvirus and facilitate their distinct and particular lifestyles. They also evolved by gene duplication and evolutionary divergence, or were captured from the genome of their host cell or another virus. Like herpesviruses, the peripheral portions of their large genomes appear to be more tolerant of genomic experimentation. Most of these accessory functions are involved in evasion of the host antiviral and immune responses. Some accessory genes, termed *host range genes*, adapt the virus for successful replication in a particular host species (Hughes and Friedman 2005; McFadden 2005).

As we discussed earlier, with a few exceptions (one of course being HSV-2, which is our own second simplex virus acquired by cross-species transmission) each lineage of herpesvirus is faithful to a single host species. Each has coevolved over millions of years with its particular host and its forebears. This is strongly supported by the absolute congruence of the phylogenetic trees that we draw to illustrate vertebrate evolution and those that can be deduced from the genetic relatedness of today's herpesviruses. Poxviruses, by contrast to herpesviruses, do not coevolve over millions of years with their hosts; their evolutionary trees do not mirror the phylogeny of their hosts. They are rather more catholic in their taste in host organism; in fact many poxviruses infect more than a single species of animal, a capacity attributable to their many accessory host range genes. Instead of coevolution and codivergence, poxviruses move between and emerge as diseases in new host species. Such evolutionary agility is reminiscent of RNA viruses, such as the human influenza virus and measles virus that invaded new species during their evolution. Poxviruses are themselves, like influenza and measles, causing acute infections that are either cleared by the host immune system or result in host death. As such and in common with other epidemic viruses, they rely on the continuous availability of new hosts to persist in a population. In Black's work on epidemic viruses, poxviruses were among those that require a host population of adequate size to persist endemically (Black 1975). They could only become endemic in human after our population densities became sufficient to sustain them. In this regard, they differ fundamentally from herpesviruses in terms of lifestyle and evolution.

Herpesviruses and poxviruses share the luxury of relatively unconstrained genome sizes and are promiscuous in their capture of new but extant genetic material by horizontal gene transfer, yet only poxviruses use this capacity to expand their host range. Expanded host range is a competitive advantage to a genetic lineage because it broadens the pool of susceptible host organisms within which the virus can be propagated. Novel host range genes permit the virus unusual evolutionary agility compared to other double-stranded DNA viruses, allowing them the opportunity to spread into new species of hosts. Instead of rapid replicative cycles, large population sizes, and error-prone replicative machinery—all tools exploited by RNA viruses to permit host switches—poxviruses leverage different evolutionary tools to support rapid evolution and an ability for host switching

It is notable that the evolutionary mechanism observed for all large double-stranded DNA viruses are similar, but the evolutionary success of herpesviruses is measured differently from that of poxviruses. Herpesviruses persist in their host after acute infection in order to enlarge the time window in which they can continue to replicate and be transmitted to new hosts. For these viruses it is advantageous to evolve a stable entente with their host to promote the propagation of their genetic information. They have minimized the necessity for rapid evolutionary change by coevolving over millennia in lockstep with their hosts. Poxviruses require that acute symptomatic infections of the host sustain their epidemic spread in the population and successful lineages expand their host range and move into new species. Poxviruses are in the evolutionary fast lane. They must remain evolutionarily agile and possess the ability for rapid genetic innovation. These qualities are essential for them to compete with their evolving hosts and to move successfully between host species, a modus operandi most often associated with RNA viruses.

Poxvirus Party Tricks

The most important field of conflict between a virus and its host is where the host defense mechanisms meet the virus's tools for immune evasion. This interface is often based in physical interactions that take place between viral proteins and cellular proteins under intense selective pressure. It is evident in positive selection on viral genes, such as the immune selection of influenza HA epitopes, and also on the genes of the host cell. In a tit-for-tat competition, the cellular genes responsible for controlling virus infection and the corresponding viral genes directing immune evasion each strive to gain the upper hand. Examination of Red Queen conflicts that have driven genetic innovation and adaptive evolution of virus and host genes alike gives evolutionary biologists a ringside seat to observe the evolutionary arms race between hosts and pathogens. No viruses are better suited for the scientific exploration of this area of evolution than the poxviruses. They have captured and evolved a disparate array of accessory genes that play roles in this virus-host interface. Many poxviruses are less than fastidious about their choice of host and exhibit a remarkable ability to infect multiple host species. Each of these species presents a different interface to the virus, and thus a different challenge, to the infecting poxvirus. Moreover, examination of poxvirus phylogenetics leads us to conclude with relative certainty that they have often crossed species barriers and must have adapted rapidly to emerge as successful disease pathogens in new hosts. For evolutionary biologists, these conclusions present a conundrum. They can be easily reconciled with RNA viruses that exist as quasispecies, replicate iteratively at a very high rate, and mutate at such high frequency that they risk error catastrophe. But for poxviruses it is less clear how they achieve the necessary rates of evolutionary adaptation to prevail in their respective arms races with their hosts and, when necessary, take up arms successfully against a different host.

One of the leaders of this area of scientific endeavor is Harmit Malik, whose laboratory is in the Fred Hutchinson Cancer Research Center in Seattle. Much of his research has involved the study of evolutionary conflicts, including those between virus genomes and host genomes, the hallmarks of which are rapidly evolving genes that display evidence of positive selection. These types of studies can throw light on the existence of ancient and now-extinct viruses that have left their footprints in the evolution of host gene sequences. We have also previously discussed how positive selection on viral genes can be inferred from studying their coding sequences. In 2012 Dr. Malik's team described a previously unsuspected mechanism that poxviruses employ to wage war with the host cell genome (Elde et al. 2012). Much of what has been learned about viral evolution under positive selection has been from laboratory experiments conducted with viruses most suited for such studies. These are viruses that replicate rapidly in culture, create vast population sizes, and have high mutation rates. Those are conditions fertile for the generation of the genetic diversity, which is the fodder of all natural selection. Viruses that meet these criteria prove suitable for the study of viral natural selection and evolution in real time in the laboratory under defined selective pressures. This might include the ability of a virus to evolve resistance to an antiviral drug, and even to adapt to replicate in a new host cell type. Such evolutionary experiments probe the ability of viruses to adapt to environmental change. In one case, the new environment contains a drug molecule that interferes with virus replication. In the other, the change is in the host cell, which has different antiviral response mechanisms that must be circumvented by the virus lest it will fail to replicate successfully. This might be representative of the arms race between virus and host that escalates with evolutionary change in the host organisms; it might also be a reflection of novel selective pressures placed on the virus when it invades a new species. These were questions that had not been addressed in poxviruses, and Malik and his colleagues sought to establish a system where poxviruses' evolutionary response to changed selective pressures imposed by the host could be recapitulated in laboratory experiments.

In order to relate these experiments in a way that the reader can interpret, I will make some basic introductions. The virus in which Malik's laboratory chose to study this phenomenon is called vaccinia virus. It is a species of orthopoxvirus commonly used in the laboratory and closely related to monkeypox and cowpox (that contracted by Jenner's milkmaids and the basis of the earliest smallpox vaccines). Each of these viruses (and indeed smallpox virus as well) possesses two particular gene products, E3L and K3L, whose roles are to antagonize the antiviral defenses of the host cell. Vaccinia virus can infect the cells of many different species, and it has been known for some years that these two gene products influence the host range of the virus (Langland and Jacobs 2002). Notably, the virus needs E3L to replicate efficiently in *human* cells; conversely, K3L is required for replication in *hamster* cells. Each of these gene products plays a role in neutralizing a key component of the cellular antiviral response. The agent in question is protein kinase R (PKR), an interferon-induced sensor of double-stranded RNA, which is a frequent by-product of many viral infections including poxviruses. Double-stranded RNA directly activates PKR to set in motion a regulatory cascade that short-circuits protein synthesis in the infected cell, causing viral replication to founder and the host cell to die. E3L and K3L usually block this antiviral response. However, the PKR of different species (e.g., hamsters and humans) is different and E3L and K3L are differentially effective at neutralizing their activities. Malik's team reasoned that should one create a vaccinia virus that lacks the E3L gene and infect human cells, they would effectively recreate a genetic conflict mimicking the cross-species introduction of a hamster poxvirus into a human host cell. The remaining K3L gene of the poxvirus is not very effective at neutralizing human PKR activity, and the new E3L-negative virus replicates only poorly in humans. Their experiment would be to ask the virus to evolve the means to grow efficiently in human cells: evolve or die!

The viruses were passaged in cell culture, a common virologic technique entailing growth of a virus in a culture of host cells followed by harvesting the progeny virus and using it to inoculate a new culture of cells. It is a particularly useful approach to discern evolutionary changes occurring in the virus population, as they adapt to the culture conditions. The results were unexpected. As with all good experiments, replicates were conducted and in each case they got the same result. After just a few passages, the virus became adapted to the human cells and grew more vigorously. A careful analysis of the virus population revealed that, in each of three independent experiments, the K3L gene locus had become amplified: viruses had between two and fifteen copies of the gene. Moreover, in two of the three replicate experiments some of the viruses also possessed copies of the K3L gene containing the same and unique amino acid substitution mutation. Viruses with amplified K3L gene sequences and viruses that had a single K3L gene sequence containing the mutation exhibited increased replicative fitness in human cells. The poxvirus employed two different mechanisms in these experiments to overcome the host cell antiviral response. Strikingly the results supported the notion that K3L gene duplication typically prefaced the acquisition of the adaptive mutational change. Gene amplification was effectively increasing the probability of acquiring mutational changes in the K3L gene. The rate of acquisition of mutations in a piece of DNA is generally related directly to its length. It follows then that a virus with ten copies of the K3L gene will have a tenfold greater probability of acquiring a mutation in K3L. Furthermore, the redundancy of the sequence provides insurance against the negative consequences of maladaptive mutations occurring in a single gene copy. Malik's team concluded that genome amplification accelerates the acquisition of adaptive changes in vaccinia virus. It is likely to be a common mechanism exploited by poxviruses, allowing them to overcome the intrinsic limitation of their mutation rates. They can thus facilitate faster than anticipated evolutionary adaptation to selective pressures. Is this an evolved mechanism? Has the poxvirus replicative machinery evolved to generate gene expansions, or are they simply a consequence of sloppy recombinogenic events occurring during poxvirus genome replication? For my money, I would place it in the same category as the error-prone RNA polymerase of poliovirus whose feasible evolution toward greater fidelity is stymied by the forces of natural selection. These observations are inspirational to evolutionary thinkers, being one small piece of the puzzle underlying how these large DNA viruses evolve rapidly to adapt to new hosts and compete in their perpetual arms race.

Malik termed the gene expansions he and his colleagues observed "genomic accordions" by analogy with the expansion and contraction of the bellows associated with the creation of music, in this case very much the music of evolution.

It is remarkable that one of the dominant forces in herpesvirus evolution, the acquisition and modification, often after duplication, of host cell genes is shared by poxviruses. The genomes of these large DNA virus families each resemble modular genomic mosaics, composed of core genes passed down from their respective last common ancestor virus and accessorized with different *after-market* bolt-on functionalities. Why is this so strange? It is so because poxviruses, although large double-stranded DNA viruses like herpesviruses, accomplish their replication entirely in the cytoplasm of the cell. Herpesviruses, on the other hand, replicate their genome in the nucleus, in proximity to the DNA of the host cell. It is intuitively easier to contemplate circumstances in which errant recombinogenic events (in which two DNA molecules come together to create a new contiguous piece of DNA) can take place between DNA strands of the host cell genome and those of replicating herpesvirus genomes. It is less evident how a virus whose lifecycle is restricted to the cytoplasm can capture nuclear genetic material at all. Even more, it appears to occur with a certain regularity, at least on an evolutionary timescale relevant to poxvirus adaptive evolution. There are clues, however: none of the poxvirus genes that have been hijacked from the host cell genome contain introns, the intervening sequences that are found in cellular genes and which must be removed by splicing of the primary RNA transcript. Poxvirus genes are *mRNA-like*, not *gene-like*. This implies that their origins lie in cytoplasmic mRNA that has been reverse transcribed in the cytoplasm to generate double-stranded DNA. These DNA fragments are then available for recombination with the poxvirus genome. This event may not be as rare as was once thought; a variety of enzymes that are up to this task exist in human cells. I will revisit this concept in Chapter 14 when we consider how viruses have been fundamental to the evolution of our own genomes.

Small DNA Virus Evolution

The herpesvirus and poxvirus families of DNA viruses owe much of their evolutionary success to the flexibility of their genome size and composition. Together with a remarkable talent for usurping host genetic material to furnish their evolutionary needs, these extraordinary adaptive capabilities make them immensely successful genetic parasites. They have adopted a wide diversity of species as their hosts, in which they occupy multiple ecological niches and enjoy a dazzling array of different lifestyles. In their own way, however, small DNA viruses are equally successful but represent somewhat of a departure from this narrative. Papillomaviruses, polyomaviruses, anelloviruses, circoviruses, and parvoviruses are examples of DNA viruses with comparatively tiny genomes. Coding for very few of their own genes, they use their genetic capacity with the frugality of RNA viruses. The coding sequences for their proteins are always packed close together, so precious is their data storage medium. Genes often overlap, the same DNA sequence being used to code for more than one gene product. The size of their genome must have been constrained in evolution by the capacity of their small nucleocapsids (Chirico, Vianelli, and Belshaw 2010). The physical structure of the virus particle is a fundamental design element of a virus, rather like the chassis of an automobile. It must be very difficult to reengineer without starting from scratch, a luxury that is available to automakers but is not an option in virus evolution. In spite of this limitation, those lineages that prevailed through evolution are efficient parasites that cause highly prevalent infections in diverse species, including ourselves.

Most of us are persistently infected by one or more of these viruses from a young age. They generally cause clinically inapparent infections or relatively benign diseases. All establish extended and usually persistent infections. They all rely on the host cell for the machinery to accomplish genome replication and the expression of their proteins. As we proceed, you will note that there is no single mode of evolution shared by small DNA viruses. We will first focus on the papillomaviruses and polyomaviruses before considering parvovirus, a single-stranded DNA virus that succeeds despite exploiting a very different lifestyle. Papillomavirus and polyomaviruses have small circular double-stranded DNA genomes, coding for less than ten genes, together with a transcriptional control signals and an origin of DNA replication. They are packaged in simple icosahedral virus particles composed of two structural proteins. The very limited complement of genes they can encode mandates that they rely heavily on the host cell for their replicative functions. They do not encode their own DNA replication apparatus, but simply encode two proteins to recruit cellular DNA polymerases and associated factors to the viral chromosome. They are minimalist in design and highly species specific, with no or few adaptive, non-core genes available to manipulate the cellular environment (DeCaprio and Garcea 2013; Krumbholz et al. 2009; Van Doorslaer 2013).

The polyomaviruses achieve tenacious asymptomatic lifelong infections in their hosts; human papillomavirus infections can be successfully resolved by our immune system, but depending on the age and immune status of the individual, they can often persist as productive subclinical infections for prolonged periods. Like rhinoviruses, HPVs exist as multiple types and reinfection by related but distinct HPV genotypes is common. These properties ensure the high prevalence and sustained transmission of these viruses in host populations.

Take for example the JC virus, first described in 1971. Between 70 percent and 90 percent of us are persistently infected. The virus takes up residence in the renal tubules of our kidneys and up to 80 percent of us excrete infectious virus particles in our urine for the rest of our lives. The success of these viruses is rooted in their ability to persist in the host undergoing continuous rounds of viral replication. They can persist and replicate at high levels in the face of our immune response. This ensures that they can successfully maintain the chain of transmission. The persistent virus is frequently passed from parent to offspring during long-term cohabitation. Furthermore, the number of genetic variants cast off from the replicating virus will be in direct proportion to the number of genome replicates synthesized. This is surely an asset to evolutionary adaptation since such variants are continually tested for superior fitness by natural selection. This rapid generation time may contribute to the relatively high evolutionary rates that have been reported for JC polyomavirus compared to papillomaviruses (Shackelton et al. 2006; Loy et al. 2012).

It may be anticipated that their mutation rates, and thus the genetic space that can be explored by these viruses during evolution, will be restricted by the high fidelity of the cellular replication machinery. In principle, their evolutionary rates can easily exceed those of the host since they have relatively short generation times. As a consequence, they continuously create copies of their genome and shed virus over extended periods of time during their persistence in a single host. This is an opportunity to create genetic variation by generating large populations of progeny virus particles. Nevertheless, the evolutionary challenge of adaptation to their respective hosts should not be underestimated. The herpesviruses can supplement their genomes with a large and varied repertoire of accessory genes that often number in the hundreds. But these small DNA viruses must accomplish their replication in the host cell using, at most, a very limited repertoire of "adaptive" gene functions.

In this regard we will take note of the papillomaviruses, encoding three genes, E5, E6, and E7, that can be considered *adaptive* genes. The members of this virus family have evolved an exquisitely balanced relationship with their hosts, infecting differentiating squamous epithelium, where they reside in a relatively quiescent state in basal cells, eluding immune surveillance (Stanley 2010). The virus takes advantage of the natural program of squamous epithelial cell differentiation to undergo replication and production of infectious virus particles that are shed from the surface of the epithelium (Doorbar et al. 2012). These genes are variably present in different papillomavirus genera and their proteins possess somewhat different properties, but they function to prepare the differentiating host cell to support viral DNA replication, antagonize mechanisms of programmed cell death, and modulate innate antiviral immune responses to the virus, permitting it to evade immune surveillance. Each of these remarkably compact proteins have multiple functions, evolved while respecting the premium on genetic coding capacity of these small genomes. The remaining papillomavirus genes are strictly reserved for the business of genome replication and virus particle formation. The relatively benign infection caused by papillomaviruses, its exquisite alignment with epithelial cell differentiation, and strict species specificity are all indicative of a long history of coevolution with their hosts. We are infected by three different genera of papillomaviruses, each of which can be found in many different primates, indicating that the papillomaviruses had diverged before our primate ancestors, probably to occupy different niches within the host organism. In addition to primates, we find papillomaviruses in birds, turtles, and other mammals. Shah and colleagues (2010) reasoned that if cospeciation were the exclusive mode of papillomavirus evolution, their phylogenies should demonstrate congruence with those hosts over the last 300 million years. This turned out not to be the case. Numerous inconsistencies in virus versus host evolution were evident and indeed some viral genes, when analyzed independently, appeared to have different phylogenetic histories to other genes in the same virus. The data substantiated the proposal that recombination between papillomavirus genomes and divergence of papillomaviruses had created new species within a single host. This source of genetic innovation may have created new species equipped to invade and occupy different ecological niches in the host. These processes may have played a significant role in the evolution of the papillomavirus lineages that we study today. Nevertheless, cospeciation is an important mechanism in papillomavirus evolution, and zoonotic transmissions of papillomaviruses between species are very rare (if they happen at all). Given the low evolutionary rate documented, it is likely that cross-species transmission may be an evolutionary barrier that these viruses are unable to readily overcome.

It is likely that the evolutionary mechanisms at play in polyomaviruses are quite similar to the papillomaviruses. It is tempting to believe that all small DNA viruses which replicate using the human DNA replication apparatus will exhibit similar evolutionary rates. However, this is not the case. The *Parvoviridae* (*parvo* is Latin for "small") represent a significant departure from these evolutionarily conservative viruses, both in their lifestyles and in their capacity for evolutionary change. They are tiny, having a mere 5-kilobase genome and encode no adaptive functions such as those of the papillomaviruses; they are absolute minimalists. They encode one protein that abducts the cellular DNA replication machinery, one for genome packaging, and two structural proteins that assemble the virus capsid. Following entry of the cell, the parvovirus must passively wait for the cell to enter into its dividing phase so that it might begin to replicate its genome. The parvovirus that we are most familiar with is the human B19 erythrovirus. It is highly prevalent among us and celebrated as the fifth diagnosed childhood infection: fifth disease. My mother referred to it as "the slaps," after the lacelike rosy rash that often appears on the cheeks of infected infants. It is a transient self-resolving infection readily transmitted in respiratory droplets, and almost half of children are infected before their mid-teens. I contracted it as an adult and experienced the arthropathy that it can cause in that setting; immune complexes accumulate in the joints leading to swelling and crippling arthritic symptoms that can persist for some time. This virus is indeed cast in a different mold to that of papillomaviruses and polyomaviruses. It does not cause persistent infections; it is a hit-and-run virus, reminiscent of the epidemic viruses such as influenza, measles, or the common cold, and lifelong immunity follows infection. Another distinction from other small DNA viruses is its rate of evolution. Surprisingly, the sequences of B19 genomes sampled over a thirty-year period revealed a rate of nucleotide substitution approximating to 10^{-4} nucleotide substitutions per site per year (Shackelton and Holmes 2006). This is an evolutionary rate comparable to RNA viruses and far from typical for DNA viruses. It signals the capacity for rapid evolutionary adaptation and an increased potential for cross-species infections.

Evidence for this is not hard to find. In 1978 a pandemic of a new viral disease spread through domestic dogs, quickly becoming endemic worldwide. Canine parvovirus emerged after cross-species transmission of a cat-specific parvovirus to dogs. Accounts of the emergence of the disease indicated that a low level epidemic of canine parvovirus in Europe was the basis for seeding of the pandemic. It is likely that genetic variants of feline parvovirus, with mutations in their capsid gene allowing them to attach to canine cells, first made the jump into domestic dogs. Over the course of just a few years prior to 1978, further and rapid adaptive evolution must have taken place to create the pandemic strain, which spread explosively across the globe. Like erythrovirus B19, this emerging parvovirus virus exhibited the same high rate of evolution more typically observed in RNA viruses. Other studies of parvoviruses uphold this as representative of all parvoviruses, and there are hints that high levels of genetic variation may occur in other lineages of single-stranded DNA viruses such as circoviruses. It remains enigmatic that viruses which use the cellular DNA replication apparatus to copy their genomes can exhibit such high rates of viral evolution, allowing them to cross species barriers and presumably exist as quasispecies within each host. Is it possible that the proofreading mechanisms of the host cell's DNA polymerase are not fully reconstituted during replication of the parvovirus single-stranded DNA genome? Are other error protection mechanisms that typically operate during cellular DNA replication abrogated in parvovirus infected cells? These remain speculations. The precise mechanisms underlying the genetic innovation of parvoviruses are still cloaked in mystery, an evolutionary conjuring trick, perfected by these fascinating, yet misleadingly simple, small DNA viruses.

· 8 ·

VIROIDS *and* MEGAVIRUSES: EXTREMES

I HAVE CONTENDED that viral identity and evolution is best captured by considering viruses as independently evolving, selfish, transmissible genetic information. Our first example was the tobacco mosaic virus, a small plant virus. I later remarked on the simplicity of the RNA bacteriophage Qβ, with its diminutive single-stranded genome and a coding capacity for only four genes. On the other end of the spectrum, we surveyed herpesviruses and poxviruses, large double-stranded DNA viruses. They infect and cause disease in vertebrates and invertebrates, and some of them wield hundreds of gene products to tailor their lifestyles to their host organisms. All meet the same essential criteria first laid down by the earliest virologists: they are filterable and transmissible infectious agents made up of nucleic acid within a protein coat, properties that derive from their physical characteristics and their ability to move between hosts, be they single cells or multicellular organisms constituting vast collectives of cells. Transmission is intrinsic to viruses. While mobile genetic elements have been described as selfish DNA, they are certainly not infectious agents. I have therefore put them aside from consideration (see Chapter 14 for a discussion of endogenous retroviruses). There are other classes of agents, however, that should not be put aside. In this chapter we will discuss the smallest and largest of viruses. They meet the criteria for inclusion in our discussion, but in different ways: they fall outside the boundaries of the simple definitions of viruses and prompt us to exercise further our imaginations.

Viroids: The Smallest

In 1967 Diener and Raymer, both scientists working for the U.S. Agriculture Research Service in Beltsville, Maryland, published *Potato Spindle Tuber Virus: A Plant Virus with Properties of a Free Nucleic Acid.* Seventy years after the first virus was described, new territory was being discovered and explored. During the intervening years, a steady trickle of research results characterized viruses as important causes of diseases in humans and in economically important animal and plant species. All of them consisted of nucleic acid within a capsid, sometimes surrounded by a lipid envelope. Potato spindle tuber virus (PSTV) infection affects the growth and foliage of the potato plant and had been demonstrated in 1923 to be a transmissible and filterable infectious agent: a virus. The pathogenic effects of PSTV on infected potato plants were often varied and difficult to detect but, after a suitable surrogate host for the virus was identified, research on PSTV took off. Five years earlier Raymer had discovered that the virus would grow on tomato plants; more important was that it caused very distinct and reproducible signs of infection. The researchers now had an *indicator* plant that allowed them to *score*, or quantitate, the infectivity of a virus preparation. They could now approach the biochemical separation and characterization of the virus. The 1967 paper was the first in a series that progressively exposed the true nature of the PSTV infectious agent. The researchers used a variety of physical and biochemical techniques to characterize the nature of infectious particles extracted from potato leaves. Their results had no precedents: infectivity in tomato plants was associated with particles that, when centrifuged in a sucrose density gradient, appeared much less dense than any other virus examined to that point. Moreover, the infectivity of the preparations resisted the effects of phenol (an agent used to denature and extract proteins from virus particles) and, when treated with nucleases, it was resistant to deoxyribonuclease and affected by ribonuclease only at low salt concentrations. The researchers remained tentative in their conclusions: "Whatever the chemical nature of PSTV, our experiments demonstrate it is a most unusual pathogen. When one considers the nucleic acid-like properties of PSTV, the ease with which it is transmitted, its remarkable stability, and high specific infectivity are astounding" (Diener and Raymer 1967). Even more curious was that spectroscopic analysis of the highly infectious extracts that they prepared had no detectable ultraviolet absorption. It was undetectable by the very technique that was routinely used by scientists to quantify nucleic acids. Moreover, the extracts could be diluted more than 1-millionfold and still retain infectivity in their test tomato plants. If the infectious agent was indeed a nucleic acid, it was present at a miniscule concentration in the extract and must have extraordinary potency. The infectious entity was later to be defined as a tiny circular single strand of RNA that in physiological conditions existed as a compact and highly folded structure. Soon it was visualized under the electron microscope; mixed with bacteriophage T7 DNA, it was but a speck on the image of the vast intertwining bacteriophage genome. This class of infectious agents, considered subviral in nature, were christened *viroids*. In 1978 PSTV would become the first eukaryotic pathogen to be sequenced in full. We might then consider that the discovery of viroids represented a watershed and ushered in the era of genomics.

Today more than thirty species of viroids have been described, belonging to two distinct families (Flores et al. 2014; Tsagris et al. 2008; Tabler and Tsagris 2004; Flores et al. 2005; Daros, Elena, and Flores 2006). They infect a variety of other plant species, among them economically important crops such as citrus trees, eggplant, coconut, and avocado, as well as ornamentals such as chrysanthemums and coleus. They have in common a single-stranded RNA genome between 246 and 401 nucleotides in length, some ten times smaller than the smallest bacteriophage genome. The genome is circular, and its compact folded nature is attributed to extensive self-complementarity of the nucleotide sequence leading to base pairing and complex secondary structures. The most remarkable aspect of viroids is that their genetic material encodes no proteins, yet they can orchestrate efficient autonomous replication and transmission between cells and even between plants. Belying their apparent simplicity, they have evolved to rely exclusively on host cell proteins to support their lifestyles. They are a compelling illustration that the ability to confer phenotype is not the exclusive domain of proteins. Traditionally we think of protein function as the ultimate arbiter of phenotype, but viroids remind us that nucleic acid sequences themselves can definitively be functional and confer phenotype. The primary nucleotide sequence of viroids "encodes" all the signals necessary to recruit cellular functions for replication and transmission; it determines the phenotype expressed by the viroid. It is this phenotype that must be subject to selective evolutionary pressures. Despite the distinction that they encode no proteins or capsid, viroids do meet all the criteria of independently evolving selfish genetic information: in my book they are essentially viruses.

The lifestyle of viroids is far from simple and differs between the two families: *Pospiviroidae* (whose exemplar is PSTVd: note the "d": it is after all not considered a virus but a viroid), and *Avsunviroidae*, exemplified by avocado sunblotch viroid (ASBVd). After entry into the plant cell, probably through plasmodesmata structures, the respective viroids make for their sites of replication. For pospiviroids this is in the nucleus, for avsunviroids it is in the chloroplast. Once in their respective target organelles, they each achieve the same remarkable feat: they recruit host cell DNA-dependent RNA polymerases to mediate their replication. In the nucleus, the pospiviroids commandeer the RNA polymerase II, which is usually responsible for the transcription of all cellular protein-coding genes. In the chloroplast, avsunviroids utilize a nuclear-encoded chloroplastic RNA polymerase (NEP). The genome is replicated by what is known as a rolling-circle mechanism that results in the creation of oligomeric replication intermediates, which are cleaved to the right length and ligated into their circular permutation. In this latter step the two different types of viroids differ again; while the nuclear-replicating pospiviroids recruit cellular enzymes to mediate genomic processing, the avsunviroids "encode" a *hammerhead* ribozyme (Cech 1993; Hutchins et al. 1986) that is responsible for autocatalytic maturation of the genome RNA. Viroids move within the plant through plasmodesmata (between cells) and are translocated over greater distances within the plant vascular system, the phloem, probably protected in complex with plant proteins. Transmission between plants is achieved by a variety of means, often through seed or pollen, but vegetative propagation is the most efficient. There are also documented examples of aphid-mediated transmission between plant individuals, and honeybees have been shown to transfer PSTV infection between tomato plants (Flores et al. 2005). Many aspects of viroid replication and pathogenesis remain mysterious, not least their ability to alter the substrate specificity of host RNA polymerases to utilize an RNA template rather than a DNA template. The mechanisms of intracellular transport are also obscure; it is not known how the avsunviroid RNA penetrates the chloroplast for example. Perhaps the most perplexing aspect of viroid infection is the basis for the very diverse disease pathologies that they induce in their plant hosts. Is this a direct result of the hijacking of important cellular functions? Probably not, since some viroids cause no adverse effect at all on the host plant. Recent work suggests that the particular pathology of a viroid is based in the corruption and redirection of the plant's own antiviral defenses by the viroid (Flores et al. 2005). What we can conclude with absolute certainty is that the genetic information to orchestrate the varied repertoire of replicative strategies and pathologies of viroids (their respective phenotypes) is captured within their 246–401 nucleotides of RNA. It does not entail encoding proteins. It appears then that nucleotide sequences themselves are at play: But how? The important properties must reside in primary nucleotide sequences of the viroid, together with RNA secondary structures induced by folding of the RNA genome into complex three-dimensional structures, in a manner analogous to the folding of amino acid chains into functional proteins.

The direct physical interaction of the genome with host proteins is evidently possible and has been proposed as a mechanism mediating the transport of PSTV RNA into the plant cell nucleus. On example of support for this comes from the direct and specific binding of PTSV RNA to a cellular bromodomain-containing protein believed to play a regulatory role in cellular chromatin remodeling (Martinez de Alba et al. 2003). The resulting ribonucleoprotein complex moves into the nucleus where PSTV replication takes place and could perhaps also influence cellular gene regulation. It is rather the viroid primary nucleotide sequence itself that is gaining favor as the central player in viroid pathogenesis. The plant host response to contain viroid replication is now known to involve posttranscriptional gene silencing (Eamens et al. 2008), an anciently evolved defense targeted to invading nucleic acids. The same processes involving the generation of nucleotide sequence–specific micro-RNAs and small interfering RNAs (siRNAs) are implicated in developmental regulation in both plants and animals (Carrington and Ambros 2003). It is known that the plant generates siRNAs to target and contain viroid replication and researchers postulate that such viroid-targeted siRNAs will play a role in viroid pathogenesis by also acting on the transcripts of host cell genes. Several such cellular mRNA targets have already been described but scientists are far from a unified understanding of the multiple mechanisms invoked by these tiny RNA genomes to manipulate the host cell (Gago-Zachert 2016; Flores et al. 2015).

Evolutionary Reliquary

Viroids are selfish RNA-based replicons, whether one considers them to be viruses or subviral (viroids) because they have not evolved to encode capsids or any other proteins. They are replicated by error-prone RNA polymerases and are therefore expected to exhibit mutation rates typical of RNA viruses at large. As a consequence, they should exist as quasispecies in the infected host. The mutation rate of viroids, as for viruses, depends on rates of nucleotide misincorporation into new genomes. It is not reflected accurately in the nucleotide substitution rate observed during the evolution of the genome because nonviable genomes are eliminated from the population. A minimalist genome such as that of viroids, capable of directing a cascade of complex biological functions, must be under extraordinary selective constraints. Larger viruses that encode proteins can tolerate more mutational changes; synonymous mutations (those mutations not affecting the encoded protein sequence) will not result in inviability. This is not the case for viroids since their genetic information is not redundant in any way; its intrinsic nucleotide sequence is directly linked to viroid phenotype. Nevertheless, nucleotide sequence polymorphisms and sequence variation have been observed *in vivo* and constitute a quasispecies of sorts, albeit one of relatively limited complexity. The underlying mutation rate of viroids was measured by Spanish researchers who employed an exquisite experimental design to capture and count the mutations arising in newly synthesized viroid genomes before selection could eliminate them from the population (Gago et al. 2009). The results were remarkable and revealed a mutation rate of 0.0025 per nucleotide incorporated, a rate equivalent to one mutation per new genome copy manufactured. This rate of mutation is orders of magnitude higher than that of the RNA viruses we discussed in depth in earlier chapters. The scientists did not resolve why this is the case, but it is speculated that the fidelity of nucleotide incorporation by the RNA polymerase might be negatively affected by several factors. Not least that viroids subvert the normal function of host DNA-dependent RNA polymerases, forcing them use an RNA template for which they did not evolve. We discussed earlier the limitation in genome size that mutation rates impose upon RNA viruses. Error catastrophe is expected to be a consequence of excess mutations per genome copy and Eigen noted that there is a direct relationship between genome size and mutation rate in different replicons be they viral, bacterial, or of larger organisms (Biebricher and Eigen 2005). It can then be speculated that given the minimal size and relatively nonredundant nature of viroid genomes, their mutation rate places a constraint on their potential for genome expansion. The evolution of increased replication fidelity must go hand in hand with any increase in genome size that would accompany the acquisition of increased evolutionary sophistication.

Comparative genomic analysis of viroids provides a compelling argument that they evolved from a common and very ancient ancestral RNA replicon (Flores et al. 2014). This holds true for both the pospiviroids and avsunviroids, despite their very different lifestyles. The nuclear pospiviroids have adopted cellular enzymes to process their oligomeric replication products into discrete circularized genome-length RNAs while avsunviroids possess a functional hammerhead ribozyme sequence that accomplishes the same task in an RNA-catalyzed reaction. For this reason, it is thought that the avsunviroids represent the most ancient form of viroids, notionally closer to the most primitive RNA-centric world. It has been proposed that they existed as replicons in *Cyanobacteriae*, which subsequently invaded the eukaryotic cell and evolved to become its symbiotic chloroplasts. Subsequently the ancestral viroid must have escaped from the organelle to invade the nucleus, diverging to form a second lineage of autonomous replicating circular RNAs that evolved to adopt the protein functionalities available within the nucleus.

The nature of viroids as strictly RNA-based replicons raises the question of their evolutionary origins: Are they ancient holdovers, representing intermediates in the evolution of pre-cellular life, captured like prehistoric insects in amber? RNA has now long been accepted as the precursor to all life; it can record information and it has the capacity to act as a biocatalyst (Cech 1986b). James Watson, the Nobelist and codiscoverer of the double helical nature of DNA intimated that as early as

1968; Francis Crick, his collaborator, had the notion that RNA might act not only as a template but also an enzyme mediating its own replication (Watson 1993). This view is now well accepted: pre-cellular life's first genetic material was RNA, and the first replicons were RNA-based polymers. Only later would the emergence of DNA and proteins relegate RNA to its predominantly subservient role in the evolution of cellular life. We can nevertheless easily imagine that it was once an RNA world. Consider the discovery of ribozymes, RNA sequences that mediate the autocatalytic excision of introns from ribosomal RNA transcripts in eukaryotic algae (Kruger et al. 1982). It has also been pointed out by Cech that modern ribosomes, our cellular factories for protein synthesis, are at their core elaborate ribozymes structured around protein scaffolds (Cech 2000). Autocatalytic introns have been widely proposed as the prototype of the earliest RNA replicons from which life originated (Sharp 1985; Joyce 1989; Cech 1986a). If this were the case, then it might follow that viroids evolved later as escaped introns. Subsequently, however, Ted Diener, the "father" of viroids and the first to describe them, made a persuasive argument to the contrary. He suggested that perhaps rather than introns being ancestral to viroids the opposite might be true; introns may have evolved from viroids (Diener 1989).

Several scientists have pondered the evolution of viroids from a theoretical ancestor, posited to be a small autonomous RNA replicon, perhaps limited in length to only a few nucleotides. Longer strings of nucleotides would not be viable if nucleotide misincorporation rates were very high. Such RNA polymers might be envisaged to become ligated to form larger "genomes"; ribozymes have been shown capable *in vitro* of ligating RNA molecules together if they are aligned on a template (Doudna and Szostak 1989). It has been further speculated that such genomes may subsequently be linked in reiterating sequence copies. There is some support for this notion; computational analysis of most viroid genomes reveals underlying periodicities in their nucleotide sequence of twelve, sixty, or eighty nucleotides (Juhasz, Hegyi, and Solymosy 1988). This may be at the root of RNA genome expansion by tandem duplication of fragments (Diener 1989); together with the formation of mosaic assemblies this could be the mechanism underlying viroid genome evolution. A further argument is made to support the notion that genome size is practically limited by replicon error rates. The adoption of a rolling circle model of genome replication by viroids allows for progeny genomes to be multiples in size of their parental genome. It is speculated that the resulting redundancy in information contained in these products of replication is fundamental to escaping the limitation in these simple genomes that have such high intrinsic mutation rates. Reiteration of the information ensures that each genome is linked to a functional form of all of the necessary genetic elements: there will always be at least one viable progeny genome (Flores et al. 2014).

Together, these conjectures place viroids at the very origins of precellular replicons. If this is truly the case, it would be remarkable if they retain recognizable features of their ancestry, such is the evolutionary distance between them. We have placed them in early cyanobacteria (hypothetically) and in the chloroplast and nucleus of eukaryotic plant cells. If they are indeed evolutionary relics of primordial RNA-based parasites, their exclusivity to angiosperms and their absence as such in other domains of life is difficult to reconcile (Koonin and Dolja 2014). It could perhaps be a consequence of extinction and replacement by successive genetic lineages with evolved lifestyles, but it is also possible that another unifying explanation will emerge. Some evolutionists favor the possibility that they may be ancestral to certain transposable genetic elements, others that they really are the ancestors of today's introns found in almost all eukaryotic RNA transcripts, but others conjecture that an explanation for their origins should be sought in their evolution after the emergence of the plant world.

A few words are called for satellite RNAs (or virusoids). These are small autonomously replicating RNA genomes (also circular) similar in nature and lifestyle to viroids. One key difference, however, is that they require a helper virus for their transmission between host cells. This lifestyle choice (perhaps artificially) places them outside my (already rather broad) definition of a virus, so I have not explored them here. I must acknowledge, however, that their evolutionary relatedness to viroids is inescapable, and they vividly illustrate how selfish genetic information can successfully explore alternative approaches to self-preservation and propagation. The reliance on other viruses (or viroids) to support their proliferation is a departure from viroids, but it opens up otherwise apparently closed doors. Hepatitis delta virus (HDV), a satellite virus of human cells, is one such example; historically it was taken under the wing of an unrelated but conventional envelope-encoding helper virus. It remains an obligate parasite of that virus, hepatitis B virus, and depends on it for its transmission. It thus seems plausible that HDV is a viroid descendant that has penetrated and persisted in a distinct lineage of eukaryotic cells as an exogenous parasitic agent.

Megaviruses: The Biggest

In 1992 Tim Rowbotham climbed to the roof of the Royal Infirmary in the City of Bradford, Yorkshire, England, to take samples from the rooftop water tower. He was a scientist on assignment from the Public Health Laboratory in Leeds, some twenty miles away, sent to track down the source of an outbreak of pneumonia. A *Legionella*-like bacterium was the suspected cause of the outbreak. The bacteria he was in search of, *Legionella*-like amoebal pathogens, are intracellular parasites of amoeba that can under some circumstances contaminate institutional water supplies. Back in his Leeds laboratory, Rowbotham attempted to co-culture the samples with amoeba in order to tease out any such bacteria in his samples. The water was indeed a source of several Gram-negative staining disease-causing *Legionella* bacterial species; it was also the source of one bacterium that stained Gram-positive and could not be identified. This new bacterium was christened "Bradford coccus" and consigned to the archives; it was, after all, not the bacterium causing the pneumonia that was rampant in the hospital.

Eleven years later the sample again came under scrutiny. Dr. Richard Birtles arrived in the laboratory of Professor Didier Raoult at the CNRS in Marseille, France. Raoult's laboratory, in one of the many national centers of research scattered across France, studied *Rickettsia* and other intracellular bacteria, and he had on hand a variety of molecular tools to characterize novel bacterial species. Birtles brought with him several samples, among which was the Bradford coccus, and they set about determining the species of bacteria in the samples using PCR amplification techniques. Since the bacterial 16S ribosomal RNA gene is highly conserved across all bacteria, it has become the go-to gene for speciation of bacterial isolates. Probes specific to this gene can be used to amplify DNA from cultured and unculturable samples containing prokaryotic DNA, yielding adequate material to perform DNA sequencing and then phylogenetic analysis. All of Birtles samples yielded to this technique, with one exception: the novel Bradford coccus. Raoult's PCR analysis could find no ribosomal genes. In an attempt to at least visualize the recalcitrant bug and determine if his procedures were successfully breaking the cells open, Raoult resorted to the electron microscope. His discovery tilted the microbiological world on its axis; what he saw under the microscope was not a bacterium but a monster of a virus. Its capsid was icosahedral, and it resembled that of the aquatic invertebrate-infecting Iridoviruses, members of the group of viruses termed nucleocytoplasmic large DNA viruses (NCLDVs), but it was much larger. It was an astonishing half a micrometer across and surrounded by an outer layer, making the virus 0.75 microns in total diameter. This was no filterable infectious agent, it redefined the perception of viruses and their possibilities and it challenged the conventional wisdom that viruses were apart from the other domains of life, inanimate and chemical in nature, not life-forms. Here was a virus that, like bacterial cells, was visible under the light microscope. Did it have the potential to be as sophisticated as some true life-forms? It was named Mimivirus, "microbe-mimicking" virus, and it opened a new arena of virus research.

Raoult was quick to work with Jean-Michel Claverie, also in Marseille, to sequence the genome of Mimivirus and to publish their findings (La Scola et al. 2003; Raoult et al. 2004). Its DNA genome was a staggering 1.181 megabases in length and encoded almost 1000 genes. The Mimivirus genome is larger than that of many species of bacteria and contains a plethora of novel genes, nearly half of which biologists did not recognize, and had no known function. Intriguingly, it also had many genes related to cellular genes not customarily seen in viruses. It contained a variety of genes associated with cellular metabolic functions and some that are part of the cellular protein translation machinery: specifically, several transfer RNAs and aminoacyl-tRNA synthases. These are functions that all other known viruses parasitize from the host cell (some large DNA viruses do have tRNA genes, but none encodes enzymes involved in protein translation). Mimivirus was clearly a strange beast. Nevertheless, it behaved like a typical virus when it infected *Acanthamoeba polyphaga* in culture. Immediately after infection, there was a characteristic eclipse phase (the period before infectious progeny virus can be detected), and regulated cascades of gene expression, protein synthesis, and viral DNA replication could be observed before the release of a burst of infectious virus after twenty-four hours.

Big and Bigger

The discovery of Acanthamoeba polyphaga mimivirus invigorated the virology community to reexplore ecosystems in search of the like. Viral metagenomics had failed to turn them up, but there was a simple explanation. Most previous environmental surveys in search of viruses used similar sampling methods; they had passed their samples through a 0.2-micron filter before analysis. This was, after all, the definition of a virus. Protists such as *Acanthamoeba* were host to the first Mimivirus, so aquatic ecosystems were reexamined. The million to trillion viral particles that are routinely counted in ocean water filtrates naturally excluded giant viruses, so a more exhaustive look was called for. The first hints that ocean water might be a rich source of new giant viruses came from reanalysis of samples collected on the *Sorcerer II* Global Ocean Sampling Expedition, whose surveys were not restricted to 0.2 micron virus filtrates (Rusch et al. 2007). Most of the nucleotide sequence they had collected was found in a fraction of seawater that escaped a 0.8-micron filter and was captured by a 0.1-micron filter. Most of the DNA sequence was of bacterial origin, but 3 percent was viral (Williamson, Rusch, et al. 2008). A reanalysis of this large dataset (representing an astonishing 4.9 billion contiguous nucleotides) revealed a multitude of gene sequences, previously dark matter, that were recognizably related to Mimivirus gene sequences (Monier, Claverie, and Ogata 2008). Soon scientists identified many relatives of Mimivirus (Colson et al. 2012). These were Mamavirus, Terra2, Moumou, Courdo 11, and the largest at that time, Megavirus chilensis. Isolated off the coast of Chile and cultured in freshwater amoeba, it had a genome larger than 1.25 megabases encoding 1,120 proteins. A more distant relative named CroV (Cafeteria roenbergensis virus) was found to grow in the marine zooplankton *Cafeteria roenbergensis*. Co-culturing water samples from the environment with various *Acanthamoeba* species has continued to turn up new giant viruses: Marseillevirus from a Paris cooling tower and Lausanevirus from the Seine River. It was becoming evident that giant viruses of single-celled eukaryotes are quite abundant in all our ecosystems.

The team that had identified Megavirus chilensis did so in the ocean water column. They elected to extend their search to aquatic sediments, speculating that the density of potential eukaryotic host organisms might be higher. They reported their findings in 2013 (Philippe et al. 2013); they had discovered two new giant "Pandoraviruses" that were isolated from marine sediment off the coast of central Chile and from freshwater pond sediment in Australia. They had genomes of 1.9 and 2.5 megabases, larger than that of some parasitic eukaryotes; the larger of the two, Pandoravirus salinas, encodes 2,500 proteins. A staggering 93 percent of them are dark matter. Neither virus of the proposed new genus has phylogenetic affinity with any other virus families including the *Mimiviridae*. These observations are remarkable but also perhaps salutary, in that they highlight just how little of the viral and microbial world has been exposed.

Virophages: Fleas upon Fleas

The *Mimiviridae* and *Marseilleviridae* are new families in the ranks of the nucleocytoplasmic large DNA viruses that appear to be descended from a single common ancestor. This notion is supported by the phylogenetic reconstruction of a hypothetical common ancestral virus that identified a group of forty or so genes as the core genes (Yutin et al. 2009). These genes are mostly conserved in some NCLDVs and provide the fundamental functions that underpin the life cycles of an incredibly diverse group of viruses. Their genomes range in size from less than 0.2 to 1.25 megabases, being either linear or circular and replicated in the nucleus or in the cytoplasm. Their hosts range from humans and birds to insects and worms, algae, zooplankton, and other phagocytic protists. Nevertheless, they must all replicate their DNA and encapsidate their genomes, and they do so with a common tool set identified by subsets of core genes; across the whole group of NCLDVs only five genes can be found in all of them. In this regard, they have much in common with the herpesviruses, which themselves share a subset of core genes. Herpesvirus core genes are also supplemented by auxiliary genes that contribute the bells and whistles to virus functionality, providing for specific virus-host coadaptation. This particularity is shared by poxviruses, themselves NCLDVs, which possess a variety of genes that influence host range and manage virus-host interactions. The size of the largest Mimivirus genome approaches tenfold that of the smallest poxvirus that encodes 130 genes. The giant viruses thus possess the luxury of a highly flexible genome with the potential to provide an extensive repertoire of adaptive functionality.

The first ultrastructural examination of Mimivirus-infected cells by Raoult and his colleagues led them to conclude that the virus was replicating and assembling in the nucleus (La Scola et al. 2003), not unlike herpesviruses. A closer examination, however, revealed that the nucleus had been relegated to the periphery of the cell, replaced by new structures. Large "virion factories" were being assembled by the virus in the cytoplasm of the cell (Suzan-Monti et al. 2007; Novoa et al. 2005). These transient structures, built by the virus in the cell, provide an organized workspace in which the infected cell marshalls the necessary resources to support viral gene expression, genome replication, and virion morphogenesis. Structures like these had not been seen before; they had little similarity to viral factories previously described for other NCLDVs. Under EM, the cytoplasmic Mimivirus viral factories resembled a membrane-bound organelle or even a bacterium in the cytoplasm of the cell. Some researchers likened its assembly to the reconstitution of a transient living microorganism within the infected cell. The seeds of doubt were sown: the research community was split: Where did these viruses reside in the evolutionary scheme of things? Could they be relics of a fourth domain of life? Not only is much of the Mimivirus genome constituted from genetic material never before witnessed in *Eukarya*, *Bacteria*, or *Archaea*, it encodes proteins that possess functions that all other known viruses depend on the host cell to supply. It has a genomic complexity comparable to many living cells, and it also appears to exist as a pseudo-microorganism during its replicative cycle within the host cell.

The viral factories of Mamavirus-infected cells were discovered to support the replication of a viral parasite named Sputnik (La Scola et al. 2008). Researchers in the field, however, were quick to point out that the Sputnik virus infected and replicated exclusively in Mamavirus-infected cells, and only within the intracytoplasmic virus factories. They asserted that it was a parasite not of the host cell but of the virus itself. This property sets it apart from satellite viruses that parasitize the host cell systems but rely on a helper virus to provide the capsid vehicle for transmission to the next host cell. Sputnik has its own capsid and morphogenesis apparatus and furthermore appeared to be pathogenic to the Mimivirus, reducing its replicative efficiency and cytopathic effects on the host cell. The class of viruses has become known as *virophages*, "virus eaters." Some investigators argue that the use of the term *virophage*, an allusion to bacteriophages infecting bacteria as an analogous biological process, is a misleading nomenclature (Krupovic and Cvirkaite-Krupovic 2011). Supporters of the term believe it recognizes the host-like quality of the immensely sophisticated, microorganism-like Mimiviruses. The antagonists suggest that designating Sputnik a satellite virus is more appropriate; giant viruses are not autonomously replicating host organisms, and despite their complexity they must rely on the living host cell to provide much of their essential machinery, just like other viruses. They are simply helper viruses for their virophages.

Within five years, more virophages were identified: Mavirus infected *Acanthamoeba* cells infected by Cafeteria roebergensis virus. Organic Lake virophage was identified by sequence homology in environmental metagenomic DNA sampled from a hypersaline lake in Antarctica; it is believed to be a parasite of NCLDVs that infect green algae. All are double-stranded DNA viruses with genomes encoding about twenty genes and, like their hosts, have certain genes in common, notably proteins involved in capsid morphogenesis, also suggesting a monophyletic origin (Krupovic, Bamford, and Koonin 2014). Some sequence similarity was noted between the virophage genes and those of a class of mobile genetic elements termed Mavericks or Polintons (Yutin, Raoult, and Koonin 2013). They are found in eukaryotes as diverse as invertebrates and protists, suggesting that they are of very ancient origin. A fourth virophage, Sputnik 2, was isolated from the contact lens fluid of a human patient with amoebic keratitis. The amoeba was infected with a novel giant virus that they christened Lentille virus. To the surprise of the investigators, the purified Lentille virus particles themselves were found to transmit Sputnik 2 to infected *Acanthamoeba* cells. The virophage DNA was integrated into the genome of the giant Lentille virus: essentially a "pro-virophage." By analogy to the phages of prokaryotes and retroviruses of eukaryotes, virophages appear capable of association with their host genome, allowing them to be vertically transmitted to daughter viruses and also to the host cells infected by those viruses (Desnues et al. 2012).

A third player also entered this act. Sequence analysis of Lentille virus identified a new genetic element, which was present as circular copies in virus particles but also as integrated elements either in the virus genome itself or indeed in its virophages. This was termed a *transpoviron*, a 7- to 8-kilobase transposable genetic element that is associated with the giant virus/virophage ecosystem. Like virophages, it has some resemblance to Mavericks/Polintons but overall it appears to be patched together as a mosaic of genetic elements acquired from a variety of unrelated sources (Desnues et al. 2012). Very much like the genomes of higher eukaryotic cells with their veritable zoo of retrotransposable elements, or bacterial cells with their populations of transposons and prophages, giant viruses have their own mobilome. Could these mobile genetic elements be part of the secret of the hugely diverse gene complements of giant viruses supporting their predisposition for genetic acquisition by horizontal gene transfer?

Chimerism

Weight is added to these speculations by several seminal contributions from the Marseilles research teams of Raoult and La Scala together with the evolutionary geneticist Eugene Koonin (Koonin, Dolja, and Krupovic 2015b; Koonin, Krupovic, and Yutin 2015; Yutin et al. 2013; Aherfi et al. 2014). They have published extensively on the evolutionary origins of the genes of NCLDVs that are proposed to constitute the new viral order *Megavirales*. Now, together with an understanding of the biology of these viruses, a picture emerges that reconciles the disparate evolutionary trajectories of this diverse group of pathogens. The discovery of Marseillevirus as a new member of the *Megavirales* served to inform us on aspects of their evolution. It retains a complement of twenty-eight of the ancestral core genes that are distributed throughout the order. The remainder of its genes have evolutionary origins in every domain of life, eukaryotes, bacteria, and even archaea. There is also evidence for gene duplication and amplification to create gene families with diverse functions, much the same as is seen in the evolution of the eukaryotic genome and herpesviruses. Eighty of the Marseillevirus genes are phylogenetically most closely related to genes in amoeba; some genes are found also in Mimivirus genomes but none of the other *Megavirales*. Taken together this suggests that these large DNA viruses acquire much of their genome complexity by horizontal gene transfer from other organisms as well as other viruses such as Mimivirus. Boyer and colleagues (2009) suggested that amoebae may represent a genetic "melting pot in the emergence of chimeric microorganisms." They note that phagocytic amoebae are infected by multiple viruses and bacteria, and routinely graze on organic matter in the 0.2-micron range. They can act as mixing vessels of microbial genetic information, allowing the evolution of novel genotypes. The potential importance of mobile genetic elements in this process should not be neglected. Indeed, virophages must play a role. The same group of scientists convincingly showed that virophages exhibit a wide host range, efficiently infecting diverse species of *Mimiviridae*. Virophages may thus be useful vehicles of horizontal gene transmission between different members of the *Megavirales*.

In a manner mirroring prokaryotic life and bacteriophage genetic diversity; the evolution of the *Megavirales* appears to have taken place through combinations of vertical gene inheritance, evident in their core gene components, and horizontal gene acquisition from gene donors in all domains of life. They are massive genomic chimeras assembled as patchwork quilts of enormous complexity and variety with their total genetic diversity comprising a *Megavirales* metagenome.

Megavirus Origins: Mavericks at Heart

Why, the virology world has asked, does a virus need to be so large and why need it carry such a large amount of genetic information? Does it call into question our views of the evolutionary origins of the *Megavirales*, viruses as a whole, and even of the other three domains of life? Absent a clear culprit for the ancestor of the *Megavirales* some researchers thought it possible that they may have arisen out of a fourth cellular life-form, now extinct, that devolved to parasitism by gene loss and subsequently diverged by the evolutionary processes illustrated above. It is certainly a tantalizing notion, as the complexity of the giant viruses flies in the face of viral minimalism and efficiency. Comparative genomics has, however, (at least for me) put this conjecture to rest. Several observations undermine this proposal, and a more concrete alternative evolutionary scenario is available. One of the most persuasive theses underlying the proposal that the *Megavirales* resulted from reductionist evolution of a common ancestral cellular life-form is that they have a variety of genes with metabolic functions typically associated with cellular life-forms. Most notable are the components of the cellular protein translation apparatus. Once a large collection of genome sequences became available, viral evolutionists could examine the phylogenetic relationships of these genes to each other and to other forms of life (Koonin, Dolja, and Krupovic 2015b; Koonin, Krupovic, and Yutin 2015). The striking observation was that these genes had their closest relatives in organisms with very distinct evolutionary origins. In other words, the genes had to have been acquired by the giant viruses on different occasions from different species. If these viruses had devolved from a common ancient cellular ancestor with its own protein synthesis machinery, then one would expect phylogenetic congruence between these proteins. As it is, they have different origins. Alternative scenarios in which multiple cellular ancestors spawned various groups of the *Megavirales* can be invoked to explain this, but there is no data to support what seems such an improbable scenario.

The most plausible explanation for the origin of *Megavirales* (all NCLDVs) resides in the recent observation that the mobile genetic elements called Mavericks (or Polintons) have homologues of viral capsid genes in their genomes. What is more, the structural particularity of their capsid protein genes places them akin to the capsids of all of the NCLDVs. Koonin has suggested that, in fact, Polintons that exist only as mobile genetic elements today may originally have been conventional viruses with an extracellular transmission infection cycle: in other words, *polintoviruses* (Koonin, Krupovic, and Yutin 2015). The evolutionary thread from polintoviruses to *Megavirales* follows the conservation of the capsid morphogenesis apparatus of the virus that is still retained in all of them today. The virophages appear to have similar origins. The evolution from Koonin's hypothetical polintovirus into today's contemporary order *Megavirales* has been far from simple and has likely involved multiple and often independent horizontal gene transmissions from other viruses, transposable genetic elements, and cellular genomes. Nevertheless, little should remain of the notion that these giant viruses were once long-lost lineages of a fourth domain of life.

· 9 ·

HIV-1: A VERY MODERN PANDEMIC

Many of our ancient diseases, those that stalked us long before the birth of our species, have evolved in lockstep with humans and share parallel evolutionary histories. Our "modern" viral diseases are the result of more recent invasions of our species. The viruses that cause them have at one time or another crossed the evolutionary divide that separates their natural host species from us. When we refer to a virus's natural host, we are usually referring to the host species to which it has become most evolutionarily adapted. Virus-host species coevolution is driven by reciprocal selective pressures. Let's examine these in turn. Natural selection on the virus judges fitness in terms of successful infection of the host and the ability of the virus to establish a chain of transmission within the host population. A virus that successfully establishes itself in a new host species (after cross-species transmission) must achieve a basic reproductive number, termed R_0, of greater than one; each infected host is the source of more than one new infection. Viruses that fail to meet this mark will not survive in the new host species: these are often termed "dead-end" infections, meaning that a chain of transmission is not established. Genetic variants that impact this success for the better are adaptive mutations and will prevail within the invading virus lineage. They may have arisen in the reservoir host and serendipitously passed through the population bottleneck to become founders of a new viral lineage. Alternatively, they may arise, again by chance, during the first stumbling infectious cycles of the virus in its new host. Now, if we look at the infection from the perspective of the new host, natural selection must favor those that survive to reproduce themselves. Those individuals that successfully reproduce despite viral infection will contribute their genes to the future gene pool of the species, enriching it with genetic traits that benefit survival. In the long term, this selective advantage may drive such variants to fixation in the population. These are the reciprocal forces typically at play in the evolutionary arms race that we have talked about and referred to as Red Queen dynamics. Each gene (and variant gene) in the virus and the host competes to prevail in their respective winning gene teams: the successful genomes of virus and host. For many virus-host relationships, such arms races appear to have run their course. These are relationships that have coevolved over extended periods of evolutionary time, reaching a stable entente, in which evolution has slowed. These are comfortable relationships in which there is little room for improvement. Examples of these highly coadapted viruses are our herpesviruses, papillomaviruses, rhinoviruses, and polioviruses, or the influenza viruses of wild birds. Even measles virus, which is definitively a modern disease, has achieved an apparent evolutionary equipoise with the human species. For each of these viruses, purifying selective pressure prevails on most of their genes because the vast majority of mutations that change the structure of their proteins are deleterious to the fitness of the genome. A quasi-stable evolutionary stasis has been approached.

Those viruses that jump from their natural host to a new host face an evolutionary challenge; they must adapt to the new host species or perish, persisting only in the security of their natural host population. There, coevolutionary pressures have created an elegant and nuanced relationship between the virus and its host cell and organism. The virus docking to its surface readily unlocks the host cell, host immune and antiviral mechanisms are parried by viral immune evasion mechanisms, and cellular signaling pathways are commandeered to benefit the virus. This carefully choreographed dance cannot hope to be faithfully recreated when a virus so naturalized to one species accidentally spills over and infects another.

The host can be thought of as the operating system of a computer, designed to provide an infrastructure that supports the proper execution of computer programs (the software) written in the corresponding language. Programs are designed to run under one operating system. Programs written for a computer running the Linux operating system will not be executable on my Mac computer. Likewise, viruses are programs designed to run on one operating system, that of their natural host; their code must be rewritten if they are to be successful in a new host running a different operating system. To extend the analogy further, to my word processing software, the program would likely still function if my computer were running a closely related, perhaps later version, of OS X. The incrementally modernized version of the operating system would likely remain compatible with my old word processing software. It follows that viruses will be more successful jumping between phylogenetically closely related species that run similar operating systems. Consider cross-species transmission between species of primates or rodents for example. We can easily intuit that crossing the large evolutionary divide between invertebrates and humans is inherently more challenging. The language of the human operating system is far more closely related to those of other primates, mammals, and vertebrates than to that of invertebrates. Their proximity on the phylogenetic tree makes them evolutionarily less distant; less rewriting of the computer code will be necessary. Evolutionary adaptation to the new host will be more feasible.

As mentioned in Chapter 7, pandemic canine parvovirus evolved after a simple mutation in the capsid gene of a feline virus unlocked canine cells, allowing it to invade domestic dogs. Although further rewriting of the cat virus code occurred during a pre-pandemic period, the virus adapted to cats with facility because the two carnivore species have very closely related operating systems, different but similar enough to be considered versions of the same operating system. Both factors paved the way for this successful cross-species transmission of the virus.

Such incursions by "foreign" viruses into humans are called *zoonoses*. They are certainly not rare events, but the evolutionary gap between species ensures that few cross-species infections result in sustained epidemics and the establishment of a new enduring virus lineage endemic in human. Most viruses, ill-equipped to jump the species barrier from their natural host to human, cause dead-end infections that do not sustain epidemic spread. In other words, they only achieve an $R_0 < 1$ and cannot sustain a chain of transmission. A dichotomy exists between those viruses that cannot sustain a productive infection in a new host and those that are simply too pathogenic for their own good. Each is a reflection of the maladaptation of the virus and the host. In one case, the virus can gain no foothold in the host because it cannot use the host cell operating system or host antiviral and immune responses effectively repel it. In the other case, the host is rapidly overwhelmed by the virus and conditions for effective transmission of the infection to secondary hosts are not met. Bear in mind that these are the two extremes of a continuum along which viral genetic variants and the genetic makeup of the individual host dramatically influence the outcome of the encounter. For illustrative purposes, consider the simian herpes virus B, until recently a hazard to laboratory animal handlers. It causes such severe and rapidly fatal illness that transmission between humans is never achieved. Absent adequate and complete cycles of infection, replication, and transmission between hosts providing the basis for adaptive evolution, no virus can adapt to a new species.

For our purpose now, we shall explore the evolutionary changes that are associated with successful zoonoses when viruses successfully make the jump across the evolutionary divide from their natural host to humans; the result: new endemic diseases of humans. Both measles and smallpox are considered "modern diseases" that resulted from zoonoses. We have established that the success of these zoonotic epidemic diseases was contingent upon an ecological variable: the existence of adequately large and concentrated human populations in the first large settlements. A disease that left the telltale pockmarks of smallpox appears to have circulated in Egyptian times, and some have implicated its emergence with the decline of the great civilizations of the Indus Valley region of India in the seventeenth century BCE (Shchelkunov 2009). Measles appears to have resulted from a successful cross-species transmission of rinderpest virus from domesticated cattle to humans, presumably in a similar time period or even later. Nevertheless, the pathways of genetic adaptation of the virus to host and vice versa have been lost to time. Viruses leave no fossil record, no cadavers, or physical trace with which to construct their genealogy with certainty. We will therefore examine a much more recent viral zoonoses in an attempt to illustrate the conditions that potentiate cross-species transmissions and the evolutionary conflicts that ensue. Our first candidate is the human immunodeficiency virus type 1 (HIV-1), the virus that ignited the most ruthless pandemic of the twentieth century. In its first twenty-five years HIV-1 infected more than 65 million people and continues to this day. More than 30 million have died of the disease and the virus is now endemic to the human population.

A New Disease and a New Virus

In April 1983 a team lead by Dr. Luc Montagnier at the Pasteur Institute in Paris announced that they had isolated and characterized a virus from a patient showing symptoms that presaged AIDS, the acquired immunodeficiency syndrome. As a scientist at the National Cancer Institute (NCI) at the time, I was well aware that Dr. Robert Gallo, the chief of NCI's Laboratory of Tumor Cell Biology, was deep in the race to discover a viral culprit for AIDS. Gallo was a human retrovirus pioneer; he discovered the first human retrovirus, human T cell leukemia virus-1 (HTLV-I) in 1981 (Rho et al. 1981; Popovic et al. 1984). His new virus was to be christened HTLV-III and its discovery would be published in the May 1984 issue of *Science* (Popovic et al. 1984). He had, in fact, re-isolated the very same virus discovered earlier at the Pasteur Institute. Montagnier shared an isolate of the virus he called lymphadenopathy virus (LAV) with the Gallo lab. The culminating events in this intense scuffle between scientific rivals are thoroughly documented by others and are not our concern here. Needless to say, bragging rights to the identification of the AIDS virus and its categorical link to the disease were at stake.

The virus was not to be called HTLV-III (it was in fact only a very distant relative of Gallo's HTLV-1 and -2 viruses), nor would it be called LAV. A compromise was struck in the name human immunodeficiency virus type 1 (HIV-1). It was the first known human lentivirus, a subfamily of *Retroviridae*. Other known lentiviruses infected ungulates, sheep, goats, and equines and caused slowly progressive diseases, many of which had some similarity to AIDS, affecting the immune and nervous systems. After acrimonious negotiation between the French and Americans, it was agreed that Gallo and Montagnier would share equal credit in the discovery of HIV-1 as the cause of AIDS: Montagnier for identifying the virus and Gallo for linking it definitively to AIDS.

The first hint of an emerging epidemic surfaced on the U.S. West Coast in 1981. It made its debut in an article, the lead author of which was Michael S. Gottlieb, in the June 1981 issue of *Morbidity and Mortality Weekly Report*, the same periodical that reports the onset and progress of the annual flu season and on unusual instances of infectious diseases across the United States. It was titled "Pneumocystis Pneumonia—Los Angeles," giving no sign that it presaged the recognition of an emerging pandemic disease. The first paragraph read, "In the period October 1980–May 1981, five young men, all active homosexuals, were treated for biopsy-confirmed *Pneumocystis carinii* pneumonia at three different hospitals in Los Angeles, California. Two of the patients died. All five patients had laboratory-confirmed previous or current cytomegalovirus (CMV) infection and candidal mucosal infection. Case reports follow. . . ." Soon there would be reports of unusual incidences of Kaposi's sarcoma, a rare malignancy of the blood vessel walls. These rare diseases in healthy individuals were to become the defining opportunistic infections associated with an AIDS diagnosis. In an editorial appearing in *Science*, Dr. Anthony Fauci of the National Institute of Allergy and Infectious Diseases observed that "the common denominator in these patients [there were now 290 cases] seems to be a profound immunosuppressed state, particularly among the patients with severe opportunistic infections." The disease was linked to T cell dysfunction; the virus, yet unidentified, was destroying the patients' ability to fend off a variety of diseases that otherwise healthy immune systems take in their stride.

The epidemic took hold initially in the gay communities of the East and West Coast cities of the United States, spreading by sexual transmission between asymptomatic and unsuspecting partners. But this was not a gay disease; heterosexual men and women fell victim, as did IV drug abusers. An epidemic raged in hemophiliacs, half of whom were infected, with thousands dying. In 1984 Secretary Margaret Heckler of the U.S. Department of Health and Human Services stood beside Dr. Robert Gallo at a news conference to announce that a virus responsible for AIDS had been isolated and a diagnostic blood test was soon to be available. The safety of the blood supply would be ensured. She also made a rash and regrettable prediction: "There will be a vaccine in a few years and a cure for AIDS before 1990." More than thirty years later, medical science has yet to deliver on her promise. The mobilization of biomedical science resources toward research on HIV/AIDS has been formidable. Academic and pharmaceutical research has delivered more than twenty-five specific drugs to battle the HIV virus. HIV-1 infection was all but a death sentence in 1981, but today it is almost a manageable, chronic viral illness.

We are however without a vaccine, and none seems imminent. The efforts to stymie the growth of the HIV-1 virus and to develop vaccines have each faced the same major challenge: HIV-1 is a master of genetic deceit. It carries with it an armamentarium of strategies to defy and destroy our immune system and demonstrates a remarkable ability to evolve rapidly in patients. It is a moving target for the immune system and antiviral drugs; to date there is not a single documented case of a patient who has successfully cleared the HIV-1 virus from his or her body.

Before delving into the origins of HIV-1 and its evolution, we will examine the lifestyle of the virus and the deadly pathogenic mechanisms that make this insidious disease tick. The clinical presentation of AIDS as it was first recognized in California signals the advanced stages of the disease caused by HIV-1. The deadly secret of the virus, which is mostly sexually transmitted, is that the infection goes unnoticed for many years before symptoms become evident. All the while, the virus courses through the victim's blood and pervades their bodily fluids. The chronic persistent infection of an otherwise healthy sexually active host provides the virus with abundant opportunities to establish a chain of transmission in the host population, passing from one unsuspecting individual to the next. The same liberated sexual mores of the post-1960s era fueled both the epidemic of genital herpes at its height in 1983 and HIV-1, which was now emerging. The route of infection of HIV-1 is predominantly via contact of infected bodily fluids with the mucosal epithelium that lines the anogenital and intestinal tract. Bloodborne transmission is also important. A group that continues to be at high risk for HIV infection is intravenous drug users, and in the early stages of the emerging epidemic, it is likely that reuse of needles in medical settings played an amplifying role in the epidemic spread of the virus.

The acute phase of an HIV-1 infection manifests itself after a few days or weeks and is signaled by a transient flu-like illness. The patient returns to a symptomatically normal state of health that belies the raging chronic infection that continues unabated. A pitched battle is being waged between the host's army, its immune system, and the rampaging viral hordes. Unlike acute infections that are usually cleared by our immune system or the latent infections of herpesviruses that are successfully controlled by our immune system, most of us do not have the capacity to substantially restrict HIV-1 replication. A small elite group of individuals, with privileged immune systems, appear to be able to exert dominance over the virus and control its replication to lower levels, but most of us are at its mercy. Our immune system never wins; this disease cannot be resolved. We can normally live and die *with* herpesvirus infections, but we live with and die *of* HIV-1 infection.

A clarification is in order here: the huge death toll that the pandemic of HIV-1 has wrought on humans results from the progressive and ultimately absolute destruction of the body's immune defenses by the virus. This acquired immune deficiency renders the body susceptible to a whole range of infections and cancers that an otherwise healthy immune system could tackle. The body is laid bare to the environment and its pathogens. The Ebola virus kills its victims crudely and savagely, taking the body by storm, attacking cells and tissues in what appears to be an indiscriminate manner, quickly leading to catastrophic organ failure and death. HIV-1 surgically lays waste to the body's immune cells. It causes an insidious creeping disease and leaves the final axe of death to be wielded by one of many opportunistic diseases. But make no mistake about it: we die because of HIV-1 infection. The success of our new viral predator is certainly due to its long asymptomatic chronic phase of infection, during which high levels of virus are continually released into the bloodstream to fuel transmission to another host.

The fact that unchecked HIV-1 infection results in death after a decade or so is a less than favorable outcome for the virus. Would it not be more advantageous for the virus and its genetic information, if infection progressed more slowly? Natural selection favors genetic variants that have more time and opportunity for sexual transmission to a new host. A healthier host leads to more transmission events. In the absence of medical intervention we might expect that this would be a natural outcome of the adaptive evolution of HIV-1 in the future, but today it is a relative novice at infecting humans. Given a long enough period of coevolution we would expect that the disease would become less pathogenic.

In 1981 there were no scientific publications on HIV-1, as it was not yet recognized. By 2014 scientific publications were appearing at a rate of more than 15,000 per year. Today almost 300,000 scientific publications document our knowledge of HIV-1. Despite its diminutive size, a virus with a genome composed of a single-stranded RNA of only 7.5 kilobases and encoding less than twenty proteins, HIV-1 has become the most intensively studied biological entity on earth.

Anatomy of HIV-1

So why does this newly discovered lentivirus cause such an intractable and insidious disease in its victims? To begin to answer this question we need to explore the molecular biology of retroviruses (family: *Retroviridae*) that have lifestyles unique in the eukaryotic virus world (Coffin, Hughes, and Varmus 1997). Although their genome is composed of RNA, they replicate via a DNA intermediate, which becomes permanently integrated into the chromosome of the infected cell. This process of reverse transcription and integration of the genome is performed by an RNA-dependent DNA polymerase (termed a *reverse transcriptase*) and an integrase that are components of the virus nucleocapsid. Reverse transcription to generate the double-stranded DNA copy of the virus genetic information occurs in the cell cytoplasm. It is then delivered to the nucleus, where it is spliced into the host cell DNA by integrase. The resulting integrated *provirus* serves as base camp for the virus in the infected cell. It is transcribed by host cell RNA polymerase II to make messenger RNAs for protein synthesis and new viral genomes for packaging into infectious virions. HIV-1 is a cytolytic virus in the majority of human cells it infects. Those not killed by the virus harbor a copy of the virus in their genome that is inherited by every daughter cell. Retroviral proviruses share some properties with the prophages of bacteriophages but are quite distinct in others. They are both inherited by daughter cells and can be a source of infectious virus upon induction, but the provirus of retroviruses is an obligate intermediate in lytic replication of the virus. On the other hand, phages need not create a prophage to replicate themselves lytically. This sets retroviruses apart from all other eukaryotic cell viruses. It is a property that has had profound consequences for the evolution of vertebrate genomes, a topic we will take up in detail in Chapter 14.

The HIV-1 virus particle is composed of a toroidal nucleocapsid within a lipid envelope, studded by copies of a single viral envelope glycoprotein (the env protein), composed of two subunits gp40 and gp120. Env is the single viral protein presented on the surface of virus particles and a major antigen against which the host immune system can direct its antibodies. The protein is made up of conserved scaffold regions and highly variable regions. The variable regions are "seen" by the host's antibodies, but continuously evolve to evade the prevailing antibody response. The functionally important parts of the protein are either folded into the interior or cloaked by carbohydrate modifications to amino acids on the exposed surfaces of the protein. In effect, the variable regions can be considered to act as decoys for the immune system. The env protein is a chameleon, its coding sequences being the most rapidly evolving gene known to evolutionary biologists (Holmes et al. 1992; Rambaut et al. 2004). It constitutes the virus's entry machinery and binds to receptors on the surface of susceptible host cells. Its cell of preference is the CD4 antigen-positive T lymphocyte (the $CD4^+$ T cell). The viral env first recognizes and binds the CD4 protein on the cell surface, but to mediate entry into the cell, env must also interact with a second protein, a coreceptor. Different viral lineages exhibit a preference for one or the other of two coreceptor proteins, R5 tropic viruses, which use the CCR5 receptor to enter the cell, or X4 tropic viruses that use the CXCR4 receptor. Some $CD4^+$ T cells have both coreceptors on their surface, but most possess mainly one type, meaning that the tropism of the virus, R5 or X4, dictates the spectrum of host cells that it can infect. The engagement of CD4 by env triggers a conformational change in the env protein, which exposes previously hidden protein structures for binding to the coreceptor. A part of the env protein called the fusion peptide moves into position and penetrates the host cell membrane, resulting in the fusion of the virus envelope with the cell plasma membrane. The viral nucleocapsid is spilled into the cell cytoplasm. These complex and tightly choreographed cloak-and-dagger tactics comprise a highly evolved strategy that shields the evolutionarily conserved viral entry machinery from exposure to host immune recognition.

The beginning of the end of an HIV-1 infection appears to be quite early in the disease, which runs its course over a decade or more culminating in the collapse of the immune system and AIDS. The virus is particularly well adapted to infect activated $CD4^+$ T cells that have abundant CD4 and CCR5 coreceptors on their surface. During the acute phase of HIV-1 infection, the virus selectively decimates the populations of activated $CD4^+$ T cells in the mucosal tissues lining the intestinal and genital tracts and the gut-associated lymphoid tissues (Hel, McGhee, and Mestecky 2006). An important aspect of the pathogenesis of HIV-1 is that it preferentially targets immune cells. Those T cells that are attracted to the site of infection and that have specificity toward viral antigens become activated by viral antigens and offer the most suitable host cells for viral amplification. The virus efficiently lures its preferred host cells to their demise. As the infection progresses, infected T cells move to the draining lymph nodes, densely populated with target cells. The systematic dismantling of the body's immune system has begun; the primary subset of cells attacked by the virus are helper $CD4^+$ T cells, which work with other immune cells supporting effective cell-mediated and antibody-mediated immune responses. The decline of a patient with HIV-1 infection is signaled by the progressive depletion of $CD4^+$ T cells and their helper T cell functions. The CD4 cell count (the number of $CD4^+$ T- cells in a milliliter of plasma) is used as a surrogate diagnostic tool to monitor the progression of the AIDS-virus infection in patients. Although the number of infected T cells during the chronic phase of infection represents only a small percentage of the total, the immune system is affected at many levels by the virus infection. The body is unable to rebuild its functional $CD4^+$ T cell populations and becomes unable to mount effective helper and cytotoxic T cell responses. The effector cytotoxic T cells become blunted in their response to viral antigens; the immune system is increasingly dysfunctional and acquired immune deficiency sets in.

HIV-1 offers particularly rich fodder for researchers who have been able to study the evolution of the virus in real time at the species and population levels throughout a major portion of its history in human hosts. For most viruses, we can only study contemporary genomes at the very tips of the branches of their evolutionary tree. For HIV-1, genome nucleotide sequences are available for viruses at all positions in the tree. Scientists can analyze genome sequences from its roots, when HIV-1 first made its debut as a human virus, and from the very tips of its branches, these being the viruses circulating today. Perhaps the most fascinating aspect of HIV-1 evolution, however, resides in the within-host evolution of the virus. Blood drawn from a patient in the extended period of chronic infection has up to 10 million viral RNA genomes in each milliliter. This high-level viremia creates 10 billion to a trillion new viral particles each day. Considering that this is sustained day in and day out over the duration of what can be decades of persistent infection, it represents an immensely powerful engine of genetic variation that allows HIV to evolve at unprecedented rates. Its RNA genome is the product of an RNA polymerase that makes on average one mutation per five genomes. The population of a trillion viruses made each day will be made up of 200 billion different viral genome sequences; in fact, during a day the virus will be able to test every single nucleotide substitution possible in its genome. The resulting quasispecies can exhaustively explore the available sequence space and provides the virus with an explosive capacity for evolutionary change and adaptation, the qualities at the core of HIV-1 pathogenesis that make it so deadly to humans.

HIV-1 is on a steep evolutionary learning curve in humans, as it must respond to our immune and antiviral defenses. Just as human influenza viruses have a higher evolutionary rate than their avian counterparts, the rapid rates of evolution of HIV-1 are characteristic of the evolutionary conflicts arising in new or recently established virus-host relationships. With enough time (and absent any medical interventions), two radically different natural outcomes of the current HIV-1 epidemic can be contemplated: the HIV-1 virus lineage and the human race will both become extinct or the virus and host genetic lineages will succeed in coadapting and hence coexist on a population level. The chain of transmission of HIV-1 in the human population is now firmly established; unchecked the epidemic is self-sustaining, the virus is now endemic. The enormous burden that HIV-1 imposes on world societies is stymieing the human potential of the most hard-hit nations in the developing world. It is of paramount importance that effective treatments and vaccines be developed. It is, of course, not a tenable option for human society to wait for the human species to evolve resistance to the virus. This outcome is in any case far from certain and human society has an obligation to intervene rather than allow the Red Queen to run her race. The very evolutionary potential of HIV-1 has presented the greatest challenges to biomedical intervention in this epidemic.

HIV in the Making

What are the origins of HIV? It took almost twenty years for the complete story to emerge (Sharp and Hahn 2011). In 1999 Dr. Beatrice Hahn and colleagues at the University of Alabama were able to definitively identify a particular subspecies of the common chimpanzee as the source of the infection. Cross-species transmission of the chimpanzee virus to humans was the cause of the zoonosis, which had developed into the pandemic. The story began to unfold as early as 1985, although the implication of the observations was not immediately appreciated. Asian macaques at the New England Regional Primate Research Center were suffering from an AIDS-like illness (Daniel et al. 1985) that was traced to an infection by a lentivirus. This was now the second primate lentivirus identified. The similar disposition of genes on the retroviral genome and the nucleotide sequence homology of this simian AIDS virus with HIV-1 indicated that it was a distant relative of the human virus, but a relative all the same. Soon afterward in 1987, under the header "Second virus linked to AIDS Peril: Study confirms West African finding and indicates second epidemic possible" the *New York Times* published the news that a virus isolated in West Africa three years earlier had been linked to AIDS-like illnesses in thirty patients. Dubbed HIV-2, it too was a genetically distant relative of HIV-1, but it was most closely related to the virus isolated from the sick monkeys in captivity at the primate center (Clavel et al. 1986). The hunt was on, and soon researchers had identified many of what were to be called simian immunodeficiency viruses (SIVs). They were widely prevalent in various species of monkey in West Africa. The virus of the sooty mangabey was particularly closely related to HIV-2 and to the virus that caused AIDS in the laboratory monkey; the geographic range of the sooty mangabey was at the epicenter of the HIV-2 epidemic (Hirsch et al. 1989).

HIV-2 was a monkey virus that had jumped the species barrier into humans; in an analogous fashion, SIV-1 in Asian macaques in New England arose from a cross-species transmission from African monkeys while in captivity. Like humans, Old World Asian monkeys get AIDS when infected with the virus, but their Old World African cousins do not. It was concluded that SIV must have invaded primates for the first time after the Old World African and Asian primate lineages diverged. That wild African simians infected by the prevalent SIV-1 suffer no outward signs of disease is consistent with the notion that a long period of coevolution between the virus and its host had taken place. The cross-species infection of foreign hosts, even as closely related phylogenetically as

Asian and African monkeys, can result in unexpectedly severe disease, and must be indicative of the degree of maladaptation of the new virus with the host.

More pieces of the puzzle soon began to fall into place. Some chimpanzees in captivity were identified as infected with viruses very closely related to human HIV-1 isolates (we will refer to these viruses as SIVcpz). Only four cases were found in all of the captive chimpanzees tested, and one of the viruses was quite genetically divergent from the others and from HIV-1 (Sharp and Hahn 2010). Hahn and colleagues were now a good way along the trail to finding the origin of HIV-1. They noted that different lineages of SIV cluster within more closely related subspecies of African green monkeys. They believed that this was probably a result of evolutionary codivergence, and they decided to investigate whether similar host-dependent coevolution had taken place in SIVcpz infections of the chimpanzee *Pan troglodytes* subspecies. There are four subspecies that occupy distinct geographic ranges in sub-Saharan Africa: *Pan troglodytes verus* in West Africa, *Pan troglodytes troglodytes* in the central region, *Pan troglodytes elliotti* in Nigeria and Cameroon, and *Pan troglodytes schweinfurthii* in East Africa. Three of the four SIVcpz-infected chimpanzees in captivity were *P. t. troglodytes*; the fourth was *P. t. schweinfurthii*. The viruses from *P. t. troglodytes* were most closely related to HIV-1; the virus from the East African chimpanzee was the outlier. The investigators proposed that SIVcpz had infected the common ancestor of these two subspecies of chimpanzees and then codiverged with each of the separate subspecies lineages (Gao et al. 1999).

The form of HIV-1 that was isolated from the vast majority of U.S. patients in the early 1980s was designated the "main" group (M) of HIV-1. But HIV-1 has several distinct lineages that were characterized later, albeit in fewer patients. Group O (group outlier) is restricted mainly to West Africa and is the cause of about 1 percent of HIV-1 infections worldwide; groups N and P are far less widespread and appear to have infected only a few Cameroonians. While the groups are clearly distinct phylogenetically, SIVcpz is the closest relative of them all. This was good evidence that several distinct transmissions of SIVcpz into human must have occurred, some being more successful than others and each seeding a separate lineage; group M HIV-1 became the pandemic virus.

Establishing that the African apes were the true reservoir of HIV-1 and the origin of the pandemic virus was a challenge. The detailed epidemiological studies that were necessary to determine the geographic distribution and prevalence of the virus in chimpanzees were impossible. Field studies involving blood draws from chimpanzees, a protected and endangered species, were not feasible. Hahn and her collaborators were resourceful and developed noninvasive techniques for sampling fecal and urine samples from the forest floor. Chimpanzee cells in these samples were used to establish the species and identity of individual chimpanzees, and sensitive immunological assays were developed to test antibodies in the samples for reactivity with HIV-1 antigens. This exercise in forensic epidemiology laid bare the HIV-1 reservoir in 10 wild chimpanzee communities across Cameroon (Keele et al. 2006). SIVcpz was widespread but unevenly distributed in different communities. Phylogenetic comparisons of these SIVcpz virus sequences confirmed that the viruses were all close relatives of the SIVcpz isolated in captive chimpanzees and HIV-1. Furthermore, the researchers observed that the viruses from different chimpanzee communities and geographies formed distinct lineages which were the result of rapid divergent evolution of the virus. With some certainty, the origin of pandemic HIV-1 group M virus was narrowed down to *P. t. troglodytes* living in southeastern Cameroon in an area bounded by the Boumba, Ngoko, and Sangha Rivers (Keele et al. 2006; Van Heuverswyn et al. 2007). In a similar manner, the group N virus lineage was also found to have emerged from Cameroonian chimpanzee communities; the origins of the other two lineages, O and P, remained enigmatic. Recently, however, SIV was discovered in West African lowland gorillas and appears to be the closely related to group O and P HIV-1. These lineages of HIV-1 thus have their origins in SIVgor, which most probably diverged after cross-species transmission of SIVcpz to gorillas (D'Arc et al. 2015). It seems that within a relatively short span of time, cross-species transmissions of primate SIV to humans had occurred not only from chimpanzees but also from gorillas.

The field studies of SIVcpz confirmed that it was only endemic to two of the four subspecies of common chimpanzee: its prevalence in these chimpanzees was highly variable, some communities having more than 30 percent prevalence and others having no infected individuals. Moreover, detailed studies of wild chimpanzee communities also revealed that SIVcpz is actually pathogenic to chimpanzees; the infected individuals live shorter lives and give birth to fewer offspring, and their offspring have higher rates of infant mortality (Keele et al. 2009). Like humans infected with HIV-1, their CD4+ T cells were depleted. Some communities were undergoing catastrophic population decline; the SIVcpz virus was causing an epidemic disease in chimpanzees too (Rudicell et al. 2010). This is in stark contrast to the nonpathogenic SIV infection of African monkeys.

The origins of SIVcpz of chimpanzee have been definitively traced back to SIV of two different species of monkey: SIVrcm from red-capped mangabey and an SIV that infects *Ceropithecus* species of monkey. Chimpanzees are known to hunt and eat monkeys, and we must assume that bloodborne virus from monkeys infected a chimpanzee, perhaps through a wound or abrasion or through contact of blood with mucous membranes. SIVcpz is actually a recombinant virus; it appears that a common ancestor of *P. t. troglodytes* and *P. t. schweinfurthii* was simultaneously infected by two types of SIV. The promiscuous viruses interchanged portions of their genomes to create what evolved to become SIVcpz. The human AIDS pandemic is the result of viruses moving from monkey to chimpanzee and then into human. I will dare to speculate that the different potential of HIV-1, SIVcpz, and SIVsmm to cause disease in their respective hosts is a direct reflection of the time that each virus-host relationship has coevolved. The evolution of simian SIVs is characterized by strong phylogenetic clustering, which shows clear congruence with the speciation of simians. This is consistent with a long evolutionary relationship entailing codivergence and cospeciation with their hosts. It would be imprudent to assert that cross-species transmission of SIV between simian species has never occurred, but it has certainly been a lesser influence in the evolution of SIV lineages. The chimpanzee virus, however, is a relative newcomer to its host, probably invading the species less than a million years ago. HIV-1 is of course evolutionarily naive to its new human host and its rough-and-ready pathogenesis is likely a reflection that evolutionary adaptation to humans is in its earliest phases.

Socioepidemiology of AIDS: A Man-Made Epidemic

In 2000 Dr. Bette Korber, of the Los Alamos National Laboratory, developed highly sophisticated computational methods to determine accurately the date of the last common ancestor of HIV-1 M (Korber et al. 2000). This would be the hypothetical virus with the genetic complement that was in the right place at the right time to make the jump from *P. t. troglodytes* to man, becoming the source of the AIDS pandemic. Several earlier attempts arrived at very recent dates for the origin of the epidemic. The earliest sample of HIV-1, however, dated back to 1959 from a patient in Léopoldville in the Belgian Congo (now Kinshasa in the Democratic Republic of the Congo). Therefore, the common ancestor must predate the sample. Positioning of this sequence in the phylogenetic tree of all known HIV-1 group M isolates suggested that HIV-1 originated in the decades before this date. Korber's very careful analysis used 159 viral *env* sequences from viral samples spanning several decades to estimate the rate of nucleotide sequence change over time. The work placed the single ancestral group M virus at a point in time between 1915 and 1941, a surprising observation since the epidemic in the United States did not become apparent until 1981. HIV-1 M viruses had clearly been circulating in the West Central African population for some time before the pandemic became apparent. More recent work has pushed this date even further into the past. Using extremely sensitive techniques, researchers extracted viral genomes from lymph node biopsy samples that had been stored for decades embedded in paraffin wax in the vaults of Kinshasa hospitals. This work established that in 1960 the HIV-1 virus in Kinshasa already existed as a diverse virus population. Incorporating these sequences into the phylogenetic analysis of HIV-1 strains placed the common ancestor of HIV-1 between 1884 and 1924, with the midpoint being 1908 (Worobey et al. 2008). At some moment after 1884, a rural African hunter, probably butchering bushmeat, became infected with the virus that would seed the pandemic. This was not a unique event; at least four cross-species transmissions of SIV are documented, those that gave rise to HIV-1 groups M, N, P, and O. It may be that such infections were not at all rare and other instances may have lead to dead-end infections that were never transmitted. Perhaps by chance or perhaps because of its unique genome sequence, the ancestor of HIV-1 M succeeded in gaining a foothold in humans.

HIV-1 circulated within the central West African population during the many decades prior to the explosion of the epidemic in 1981. The simmering zoonosis went unrecorded during this period, but the events leading up to its explosive emergence have been the subject of intense study and speculation. In his book *The Origins of AIDS*, Jacques Pépin

(Pepin 2011) makes a persuasive case that the origins of the epidemic lay in the social and political changes that took place in colonial West Africa. The establishment of urban trading and colonial administrative centers in African cities created urban pull that attracted a multitude of workers. Léopoldville quickly became the most populous city in the Belgian Congo. Prostitution was rife in the city, densely populated by workers who had left their families at home in the countryside. These factors, together with organized well-meaning medical campaigns administered by the colonial governments that spread the disease through the reuse of needles, all played into the eruption of the epidemic. It was a perfect storm. The long incubation period of the disease and deviance from traditional social norms that had governed African rural life had allowed the virus to establish its base in the human population over decades. Changes in human behavior allowed the epidemic to emerge from central West Africa with its epicenter in what is now Kinshasa, seeding the global pandemic.

Many researchers have now diligently followed the trail of clues leading to the United States and the rest of the world. Epidemiological studies and phylogenetic analyses of viral isolates separated in geography and time have allowed the emergence of the global pandemic to be mapped (Gilbert et al. 2007). A pandemic "clade" of HIV-1 M termed subtype B traveled in an infected individual from Africa to Haiti, probably around 1966. Although there is evidence of several transmissions of this virus outside Haiti, it was a single introduction of the virus into the United States that became the founder event of the North American epidemic. Interestingly, this may have occurred as early as 1969. The virus must have spread in the U.S. population for more than a decade, undetected and at a low level. Perhaps it was limited to the heterosexual community before it spread to the most high-risk population, males who have sex with males. Here its mode of transmission was far more efficient, and the emergent pandemic virus spread more rapidly and was noticed in the gay communities of the major metropolitan areas of the United States. From there to the rest of the world were direct flight paths. A sexually transmitted virus causing a long asymptomatic illness in which the individual harbors infectious virus is a potent vehicle for pandemic transmission. Both in Africa and beyond, changes in human behavior provided the opportunity that zoonotic pathogens grasp.

Within-Host Evolution: A Very Personal Arms Race

Before medical science succeeded in developing an armamentarium of antiretroviral drugs, HIV-1 infections were a death sentence for the vast majority of its victims. Some progressed from the asymptomatic disease to AIDS more rapidly than others, and only a few lucky individuals could hold the disease in check (Yue et al. 2015). There is compelling evidence that the outcome of an infection is influenced by the genetic makeup not only of the virus but also that of the infected individual and the individual who transmitted it (the virus donor) (Alizon et al. 2010). In other words, not all transmitted viruses are created equal, and we are not all equals in our ability to counter an infection. The most important conflicts between a virus and its host occur at the level of the host cell antiviral responses and the host adaptive immune response. In influenza virus infections, the battle between our adaptive immune response and the virus is one that we witness being played out on the population level and in the evolution of new epidemic strains. Immune escape variants of the influenza virus antigens provide a selective advantage in their ability to infect new hosts and seed new epidemics. HIV-1, on the other hand, in the face of a vigorous immune response causes a chronic infection with sustained high rates of genome replication. The resulting selective pressures on the replicating virus population dictate that within-host evolution plays a profound role in the natural history of the infection in the individual. Within-host evolution of immune escape variants of viral antigens presents a continually changing face to the immune system. These escape mutants have a selective advantage because they result from changes in immune epitopes, the small segments of viral proteins that are recognized and targeted by the immune system in its effort to control the viral infection. Viruses with variant epitopes frequently emerge and evade the specificity of extant antibodies, helper T cells, or cytotoxic T lymphocytes, the warriors of adaptive immunity. The HIV-1 viral envelope protein, env, is a veritable shape-shifter. It is not the only viral protein whose evolution defends the virus from the host immune response, but it is certainly on the front lines of the battle. It is the principal target of neutralizing antibodies as well as cell-mediated immunity. In a longitudinal study of HIV-1-infected patients, the rate of disease progression (measured by declines in CD4 cells) was compared with the number of adaptive mutations in the env protein (Williamson 2003). It became clear that CD4 cell counts in patients with more immune escape mutations in the virus *env* gene declined more slowly. The ability of those individuals to mount a broad antibody response that exerts strong immune selection on the virus was slowing the progression of the disease.

The framework of the env protein is composed of conserved scaffold regions and four hypervariable regions, one of which, V3, is the immunodominant epitope. As we discussed, the functionally critical portions of env—the workings of the entry machinery—are conserved and largely invariant but hidden from view of the antibody response. On the other hand, the variable regions are exposed to the immune system and are under high levels of positive selective pressure. This is plainly evident in the high ratio of nonsynonymous to synonymous mutations that are observed within them. Mutations of epitopes that are the targets of immune responses become rapidly fixed in the viral population by selective sweeps that replace the targeted sequence with one no longer recognized by the immune response. It has been estimated that one new adaptive mutation is fixed within the HIV-1 env protein every 3.3 months (Williamson 2003; Bar et al. 2012). Many research groups have contributed to our understanding of env immune escape by examining the antibody response in patients at different time points after infection. Successive changes in the virus followed by changes in the immune response are observed: cycles of control followed by escape.

In a recent study, Bar and colleagues (2012) published a detailed analysis of three patients who became infected by HIV-1 and seroconverted. Sensitive DNA sequencing technology allowed the researchers to examine the individual genome sequences of the very first viruses that began to replicate in each patient; from this data they could deduce the sequence of the founder virus genome transmitted between individuals. With this information in hand, they sequenced individual viral genomes in the emerging quasispecies that were under selection pressure by the immune response. Within two weeks of seroconversion, neutralizing antibodies were detected in the patients. It was a relatively weak initial antibody response, but the antibodies rapidly selected for escape mutations in the *env* gene. Detailed characterization of the individual escape mutant genomes in each patient revealed a disparate set of escape pathways in *env* sequences. Immune escape in one patient resulted from mutations in the immunodominant V3 loop and then in V1; multiple mutations in V2, many of which affected glycosylation of the protein, were evident in patient 2, and mutations in patient 3 mapped to the outer domain of the env protein. The investigators remarked upon the very rapid evolution of immune escape mutations despite the weak selective pressure of the initial immune response. This serves to illustrate that the large population size, the rapid cycles of viral replication, and turnover of the virus population can leverage even very mild selective pressures to drive fitness optimization of the HIV genome.

It makes sense that HIV-1 exploits the same repertoire of evolutionary tools that influenza uses to evade the host immune response. The coding sequences of the immunodominant, hypervariable regions of env display a preference for volatile codon usage compared to other regions of the protein; codons that tend to mutate nonsynonymously to small hydrophobic residues are favored (Stephens and Waelbroeck 1999). The HIV-1 genome is not segmented, and therefore reassortment is not in the HIV-1 arsenal, but recombination between viral genomes of the quasispecies replicating in the host is commonly observed in HIV-1 (Zhuang et al. 2002). Contrast this with influenza in which recombination is rarely (if ever) observed, perhaps because its mode of genome replication provides less opportunity for this type of genetic exchange. Recombinant lineages of HIV-1 subtypes are frequent in the pandemic and they may also play a significant role in disease pathogenesis within the host. It appears that recombination, particularly when individuals experience very high levels of viremia, can be an important source of genetic novelty to optimize viral fitness (Levy et al. 2004; Wei et al. 2003). An additional strategy has evolved in HIV-1 that has not been documented (at least not yet) in other viruses: early studies of HIV-1 antibody neutralization and escape recognized that the virus employs a rapidly evolving "glycan shield" that generates immune escape variants (Wei et al. 2003). In and of itself this is not a unique strategy of immune evasion; pathogens often alter the glycosylation patterns of their surface glycoproteins to evade immunity. Scientists in Lausanne, Switzerland, recently published an analysis that revealed motifs of repeating nucleotide triplets in the env protein hypervariable regions (De Crignis et al. 2012). These repeats have the facility to generate genetic diversity and alter the glycan shield of the virus. The cryptic trinucleotide RNY (where R is a purine and Y is a pyrimidine and N can be either) is found at significantly higher frequencies in *env* variable regions than mere chance can account for; stretches of RNY repeats are also commonly found and they are typically "in frame" with the amino acid code. The Swiss team previously observed that variable region 4 in viruses from the same patient often accumulated variants with insertion or deletion mutations of three or multiples of three nucleotides. In their study, such variations of RNY repeats were seen in four of five variable regions of the *env* sequence. The amino acids encoded within the affected triplet repeats most frequently included an asparagine, the amino acid residue of proteins that is most commonly the site of N-glycosylation. It is likely then that these sequence peculiarities provide a mechanism for the virus to shuffle the envelope glycan shield, creating a barrier between the protein epitopes and host antibodies. That natural selection can maintain these cryptic nucleotide repeats in the *env* gene appears to constitute a prospective preference for error-causing hypermutable sequences. As for the selection of volatile codon bias, the selection of hypermutable sequences must be dependent on a history of balancing selection at these sites in the protein. Both "forward" and "reverse" mutations by duplication or deletion are each the most probable beneficial mutational events at these sequences. Just as duplication will occur frequently and confer a fitness advantage under one immune selective pressure, deletion or duplication of the same nature at the same site will commonly occur and offer benefit under changed selective pressure. These deletion or mutation events will commonly change the patterns of asparagine glycan modifications, which will frequently be beneficial for immune escape. So, no causality or evolutionary forethought is at play; this mechanism of sequence variation arises from the recognized processes of natural selection (which, of course, act on extant phenotypes and in the moment).

The particular epitopes targeted by the immune response of one infected individual may be different in another due to genetic differences in their respective immune repertoires. It is the result of selection for improved viral fitness within that unique individual, but it does not guarantee that the newly evolved genotype will possess a selective advantage in the next individual host it encounters. The genetic information of the founder virus of the infection carries the indelible imprint of the immune system of its former host. The most influential host genetic factors (at least to our current understanding) in the within-host arms race between the virus and the immune response are the HLA class I alleles. These dictate the capability of the immune response to recognize viral cytotoxic T cell (CTL) epitopes. Viral protein-derived peptide fragments are differentially displayed for T cell targeting depending on the HLA alleles carried by the host individual. The possession of select HLA haplotypes have been found to be beneficial to some individuals, allowing for more effective control of viral replication, while others are not. Peptide CTL epitopes from the main structural protein Gag of HIV-1 protein are major targets of cell-mediated immunity and hence subject to strong selective pressure for immune escape by mutation. This protein is highly conserved and has numerous important structural and functional protein domains. It is thought that CTL recognition of peptides within critical structural features of this protein is an advantage to the host. Immune escape mutations that change these epitopes are expected to reduce viral fitness and result in reduced replicative capacity. Indeed, this appears likely; individuals who mount an effective and broad CTL response to Gag CTL epitopes appear to maintain lower viral loads (Edwards et al. 2002; Ramduth et al. 2005). These responses seem to be particularly effective in patients who possess the beneficial HLA I alleles in HLA-B.

Much research, particularly by teams led by Dr. Eric Hunter of Emory University, Georgia, has sought to unravel the most influential factors governing the intensity of the early viremia that emerges in newly infected individuals. This measure may be indicative of viral virulence and has been strongly implicated in determining the rate of early disease progression (recall that early stages of infection may be pivotal in destruction of host immune competence) as well as transmission rate. This enterprise entailed a detailed study of transmission pair cohorts, in which the virus in both the donor and linked recipient are available, and the HLA haplotypes of both hosts are known. The fitness (replicative capacity) of the respective viral isolates can be determined in the laboratory, together with an analysis of the immune escape mutations that develop in the donor. These parameters can be compared with the viral loads in the donor and the newly infected partner. Several themes emerge (Prince et al. 2012): there is a significant concordance between viral load in the donor and the recipient, regardless of their HLA haplotypes. This is consistent with the important influence of the inherent virulence of the virus.

Furthermore, founder viruses with multiple escape mutations often have a lower replicative capacity and replicate at lower levels during early stages of infection. Finally, the levels of early viral replication have a persistent effect on disease progression over the long term. These results are consistent with the hypothesis that viruses that accumulate multiple escape mutations have attenuated virulence in the next host and that the outcome of the disease is most strongly influenced by the early stages of infection. The HLA I status of the donor individual determines what CTL escape variants are selected in the viral genome. Certain beneficial HLA alleles have been associated with slower disease progression as they select for variants with reduced virulence. It is therefore of particular interest to assess the outcome after transmission of such viruses to individuals with mismatched HLA I haplotypes (Chopera et al. 2008). In this focused study, individuals infected by a virus from a patient with the beneficial HLA I alleles (and thus with the associated attenuating escape mutations) controlled their disease better. Moreover, as expected, the escape mutations associated with reduced viral fitness underwent reversion during within-host evolution. Nevertheless, these patients appeared to retain the initial benefit of being infected by viruses with attenuated virulence and progressed more slowly, despite not possessing the beneficial HLA haplotype. The outcome of the game of HIV-1 disease may be largely won or lost in the first period of play.

Shortsighted Evolution

Evolution in and of itself is a process that has no objective. An entity that can sustain its own replication and that creates variants which compete with their parental genotype will be subject to the laws of natural selection and, in principle, can evolve. The variants of HIV-1 that compose the quasispecies in the chronically infected individual are tested against each other, based simply on whether they can sustain their own replication in the host at that moment. The relative fitness of each genotype under the prevailing selective pressures is determined by its replicative capacity compared to others in the population. HIV-1 replicates so rapidly and exists as such a diverse swarm of variant genotypes that, within the chronically infected host, it can evolve to explore almost limitless genetic opportunities. Despite existing as a quasispecies during chronic infection, each new infection is typically initiated by a single founder virus (Keele et al. 2008). This first virus in the host enters and replicates in CD4⁺ T cells at the site of infection. Notably, this virus is always R5-tropic and uses the CCR5 receptor protein as coreceptor to enter the cell.

The tropism of the virus for activated $CD4^+$ T cells and the coreceptor that it uses are determined by sequences within the *env* gene of the virus. As chronic infection in each patient proceeds, the envelope is under strong positive selection by the immune system. In response, successive immune escape genotypes evolve. In fact a form of natural selection is also at play in the host immune system: the B and T cells that have specificity for the antigens displayed by the virus and virus-infected cells are selectively activated. The viral antigenic identity is the natural selective pressure that provides the cognate immune cells with a proliferative advantage. The immune system is modeled to exploit natural selection at the cellular level. Other selective pressures also influence the pathogenesis of the virus infection. Scientists can cultivate the virus from infected patient plasma on a variety of different cell types and lineages in culture. Some viruses grow on cells that represent different lineages of lymphocytes or myeloid cells that display one of the two viral coreceptors. In this way the tropism of the virus (R5-tropic or X4-tropic) for different cell types at different stages of infection can be described. Early in infection R5-tropic virus that infects primarily activated $CD4^+$ T cells prevails, but later in the infection viruses of distinct tropism phenotypes emerge (Arrildt, Joseph, and Swanstrom 2012; Gorry and Ancuta 2011). These can significantly influence the pathogenesis of the disease. In almost 50 percent of patients with advanced disease, X4-tropic viruses with altered cell tropism emerge as the dominant phenotype. At this stage of infection, the disease and associated erosion of the host immune system advances more rapidly, the virus now infects naive X4-positive T cells and often has a more rapidly destructive cytopathic phenotype. It remains unclear whether the X4-tropic virus is instrumental in the acceleration of immune destruction or simply an evolutionary sideshow. Viral variants with a capacity to exploit this new niche, an abundantly available cell type (X4 positive T cells) may simply be more successful at this time in infection. It is noteworthy that the evolution of coreceptor-switched viruses is not essential for the disease to progress to full-blown AIDS; some patients with AIDS have no X4 tropic virus. Indeed, while the most prevalent group M subtype B virus of the epidemic frequently evolves changed coreceptor selectivity, subtype C viruses rarely, if ever, make the switch. The precise genetic changes to the envelope that allow coreceptor switch have been the subject of numerous studies. Although it is commonly associated with changes in the V3 variable loop of the protein, additional poorly defined amino acid changes also appear to be necessary. It has been postulated that the "genetic distance" between R5-tropic and X4-tropic *env* is greater for subtype C than subtype B viruses (Coetzer et al. 2011). Perhaps the stepwise accumulation of mutations for a successful change in tropism is not readily accessible via viable intermediate mutants in subtype C env proteins.

It appears that although these within-host evolutionary trajectories do influence disease pathogenesis, their raison d'être may be more because the virus *can* than because the virus *must*. Immune escape is central to the persistence of the infection in the host and must be a relevant factor in the ability of the virus to be transmitted successfully to a new host. However, the evolution of altered cell tropism is characteristic of the later stages of the disease and may offer little advantage for successful transmission of the genetic lineage. There are two possible reasons for this: First, transmission rates are likely highest in the early stages of the disease when the infected host is still in good health and making the necessary sexual contacts for transmission. Second, but perhaps more importantly, the viruses that prevail later in disease are rarely transmitted. In fact, the virus passed to the recipient during sexual transmission is more closely related to the ancestral viruses circulating early in infection than it is to the contemporaneously circulating viruses (Redd et al. 2012). An acute bottleneck occurs during HIV-1 transmission and a single virus (or a very limited number of viruses) becomes the founder of the new viral population that expands in the new host. In HIV infections, this virus is always R5-tropic and infects activated $CD4^+$ T cells, the cell type infected by the virus at the outset of disease. It is never X4-tropic nor is it M-tropic (viruses with tropism for macrophages that are commonly isolated from HIV-1 infected patients). These products of evolution are never the vehicle of transmission, and appear to be simply by-products of a system that is compelled to evolve.

It is worth examining this phenomenon in more detail, as it reveals a paradox in HIV evolution. The rate of HIV-1 evolution in each patient is extremely rapid; it has been estimated that the virus fixes one adaptive mutation every 3.3 months (Williamson 2003; Rambaut et al. 2004). Despite this, the tempo of epidemiologic evolution of the virus is substantially slower (Rambaut et al. 2004; Lemey, Rambaut, and Pybus 2006). There is a disconnect; the viral evolutionary clock within the host runs at a higher pace than the evolutionary clock governing evolution of the virus among hosts. It is as if each time the virus is transmitted to a new host the clock is turned back. The rate of nucleotide sequence diversification by any measure, be it nonsynonymous or synonymous nucleotide changes, is faster in every host and cannot be reconciled with the much slower rate of evolution evident in the epidemic. Within the host, evolution is characterized by the hallmarks of positive selection, mostly driven by strong selection for immune escape. Evolution among hosts, on the other hand, appears relatively neutral and is characterized by the co-circulation of multiple variants over extended periods (Rambaut et al. 2004). Unlike within-host evolution, mutations are not successively fixed in the population by selective sweeps. Their distribution appears to be largely governed by chance geographical and temporal factors. It appears then that it is not necessarily the most replication competent and evolutionarily adapted virus in the quasispecies that succeeds in becoming the founder genome for a new infection. Fitness for replication must necessarily not equal fitness for transmission. Some hypotheses have been advanced to address this paradox (Lythgoe and Fraser 2012). One among them resides in the unique transmission of R5 virus. The slower tempo of evolution among rather than within hosts could be due to low rates of within-host viral evolution early in infection (due to low levels of selective pressure), coupled with high rates of transmission during this period. In this scenario R5-tropic virus would still be dominant and few other genetic variants would have the opportunity to arise and become fixed in the population. This, however, appears not to be the case, and indeed there is evidence to the contrary. Selective pressure on the virus is already very high early in infection (Bar et al. 2012). Moreover, the viruses that are transmitted even at late stages of infection appear to be most representative of the genotypes that prevailed earlier in infection rather than those that evolve in later stages. The favored hypothesis has thus been termed "store and retrieve" (Lythgoe and Fraser 2012; Vrancken et al. 2014).

A particularity of the retrovirus life cycle is that the genome of the virus becomes integrated into the host chromosome. Although HIV-1 kills most cells that it infects, some cells, in particular resting memory T cells, can become latently infected. These cells form a permanent archive of virus genotypes laid down starting very early in the course of the disease. Furthermore, latently infected cells reactivate periodically to make virus that can be representative of the genotypes circulating much earlier during the course of the disease. Is it likely that the virus which is transmitted to the next host and founds a new viral population is an archived viral genome selected for its favorable properties? If this is the case, it would explain why the virus that is transmitted from one infected host to a new naive host is much more similar to its ancestral founder virus than we should expect. Consequently, although immune escape throughout persistent infection is critical for lineage survival within the host (and immune escape variants are indeed often transmitted), we must conclude that a great deal of the unbridled genetic variation and opportunistic evolution that the virus experiences during advanced disease is a sideshow. Much of it is neither necessary nor useful for the survival of the genetic lineage at the population level and maintenance of the chain of transmission. It is evolution without consequence.

Adaptive Evolution: An Evolving Relationship

I should not leave the impression that HIV-1 does not evolve among hosts in response to selective pressures. Indeed, the transmission of immune escape variants is commonly observed and can have a profound influence on the outcome of subsequent infections. Whether these transmissions play a large part in shaping the pandemic genetic lineage on a population level is, however, less clear. I have consistently promoted the theme that over time, natural selection promotes the coevolution of a virus and its host, and that one outcome of such evolution can be reduced pathogenicity. Our own trudging rate of genome evolution denies us the possibility of witnessing substantial evolutionary changes in ourselves in response to our new virus within our lifetimes. We can, however, expect that HIV, currently a novice in the human population, will be under extreme selective pressure to optimize its replication and transmission in the face of host-determined selective pressures. Despite the quite modest tempo of directional evolution of pandemic HIV-1 at the population level (compared to within hosts), the evolutionary clock of HIV-1 still ticks many orders of magnitude faster than that of humans. If we are lucky then, we may be able to witness the HIV-1 lineage embark on an adaptive evolutionary trajectory. Generally speaking, it seems that the pathogenicity of HIV's ancestor viruses is inversely correlated with the time that they have infected and coevolved with their respective hosts. SIV is minimally pathogenic in West African monkeys, its longtime natural hosts, but it is highly pathogenic and causes AIDS in Asian macaques, which it has only recently infected (in the laboratory). The SIVcpz is mildly pathogenic in chimpanzees that it has infected for an intermediate period, likely some 2 million years. We might venture then that given adequate time, primate lentiviruses which infect chimpanzee or humans will diminish in virulence (disease-causing potential). There is obviously a trade-off that the virus must make if it is to become less virulent. More virulent viruses cause rapidly progressive disease, but replicate at higher levels and are therefore more readily transmitted during a sexual liaison. On the other hand, a virus with reduced virulence may extend the life of the host but replicate at lower levels resulting in fewer successful transmissions per liaison. In 2007 Fraser and colleagues published their epidemiologic analysis of the Amsterdam seroconverters cohort. Homosexual males in Amsterdam had been prospectively recruited starting in 1982 to generate systematic observations on the natural history of HIV-1 infections (Fraser et al. 2007). Their study specifically explored the epidemiological effect of differences in viral loads of untreated patients with the duration of their asymptomatic infectious period and upon their infectiousness. The results confirmed that there was a trade-off between the two parameters as virulence diminishes. The analysis predicted that a viral load of 4.8 log10 viral RNA copies per milliliter (approximately 30,000) would maximize the reproductive number (R_0) of the virus and optimize epidemic transmission. In fact, the average viral load of the cohorts under study matched this number: Is it the result of viral adaptation? The authors were quick to point out that although this makes intuitive biological sense, there is no absolute proof that this is anything other than coincidence.

Evidence for adaptation of HIV-1 to humans has, however, begun to emerge. In a seminal study of almost 3,000 patients in cohorts across five continents, an international team addressed the adaptation of HIV-1 to HLA class I restriction (Kawashima et al. 2009). I have already introduced you to the importance of HLA class I alleles as central players in the immune control of HIV-1 and key drivers in the selection of immune escape variants. It makes sense that one of the major battles in the arms race between virus and host will be fought over immune recognition by the adaptive immune system. Of the most beneficial alleles of HLA class I, those in HLA-B, the allele HLA-B*51 was selected for intensive study; a particular immune escape variant of a viral protein termed I135X (a substitution of isoleucine by a different amino acid) is commonly selected in individuals carrying this HLA allele. This particular mutation is exclusively found in the context of the HLA-B*51 allele and is not prone to reversion in hosts that have a different HLA haplotype. This is a strong indicator that this immune escape mutation does not carry with it any loss of viral fitness. One cohort was particularly informative; HLA-B*51 has a high prevalence in the ethnic Japanese cohort, being present in more than one fifth of the patients. More than half of these patients carried virus with the escape mutation; an analysis of all cohorts confirmed a statistical correlation between HLA-B*51 prevalence and the escape mutation. On a population level, the adaptive mutation was arising in patients with the HLA-B*51 allele. In an elegant comparison, the researchers compared the frequency of occurrence of the adaptive mutation in HLA-B*51-negative Japanese hemophiliacs, who had been infected in 1983 (before the blood supply was made safe); only 21 percent carried virus with the mutation. In more recent years, between 1997 and 2000, more than 70 percent of HLA-B*51 negative patients were infected with the virus carrying the mutation. Early in the epidemic it was beneficial to carry the HLA-B*51 allele as the disease progressed more slowly. Today the allele has no protective effects because the prevalent epidemic viral genotypes now carry the adaptive mutation.

With these observations in mind and the knowledge that many HLA-driven immune escape mutations reduce viral fitness, scientists set out to examine whether this type of viral adaptation might be associated with reduced virulence. Payne and colleagues (2014) chose to study the epidemic in South Africa and Botswana, two severely affected countries in which the HIV-1 epidemic began at different times. They reasoned that if viral adaptation were occurring it would be most evident in populations with such a high prevalence of disease. The results were striking and revealed that HIV-1 is evolving remarkably rapidly in response to immune selective pressure exerted by HLA class I. Surveillance of known immune escape mutations in HIV in the two epidemics confirmed that the virus was adapting: the frequency of adaptive immune escape variants was significantly higher in Botswana, where the epidemic has raged for longer, compared to South Africa. Although several of the most beneficial HLA-B alleles remained protective in South Africa, where their possession correlated with reduced viral loads, they no longer had protective effects in the Botswana epidemic. There was, however, a bright spot in the data; a systematic evaluation of replicative fitness of viruses currently circulating in Botswana and South Africa confirmed that the HLA class I–driven evolution of the epidemic virus was associated with reduced virulence. Although Botswanan subjects were often infected by viruses that had immune escape mutations for their respective HLA class I alleles, it appeared to be somewhat offset by the reduced virulence of the circulating virus. The transmission of viruses burdened by immune escape variants has resulted in circulating viral strains with reduced replicative capacity. This in turn is associated with slower disease progression and decline in CD4 cells.

The results of these studies prompted Payne and colleagues to speculate on the potential of antiretroviral therapy to accelerate the adaptive evolution of epidemic HIV-1 populations with attendant reductions in viral virulence. Since 2010, antiretroviral treatment of HIV-1 infected subjects is typically started only when CD4 cell counts sink below the critical level of 350 CD4 cells per mm^3. Treatment with antiretrovirals is therefore skewed toward those patients with more advanced disease. The Payne team's train of thought ran thus: if the selection of HLA-driven adaptive viral mutations is associated with reduced virulence and hence slower progression of disease, then over time, antiretroviral therapy will be initiated in a growing proportion of subjects who are infected with more virulent (and, therefore, less adapted) HIV-1 isolates. Since antiretroviral therapy effectively blocks viral transmission, the propagation of the more virulent lineages enriched in the treatment group will be stopped. Viruses with lower virulence will be preferentially transmitted. Antiviral therapy should result in attenuation of the epidemic virus. The crux of this argument resides in the presumption that more viral transmission events occur in later-stage disease than in early-stage disease. If the majority of transmission events were to take place in early infection, before initiation of antiviral therapy in any infected patients, the window of opportunity for pretreatment transmissions would not be affected. The "selective advantage" for attenuated viruses resides in the opening of a time window between which virulent infections have been suppressed by antiviral therapy while subjects with immune adapted (yet attenuated) virus infections remain untreated and infectious. The most recent WHO guidelines recommend even earlier initiation of antiretroviral therapy (when patients have CD4 cell counts < 500 cells per milliliter). Will this facilitate a more potent acceleration of adaptive evolution by antiviral interventions and attendant loss of virulence? Can the average viral load set point of patients be forced below 30,000 copies per milliliter, reducing the R_0 and slowing the epidemic? It is a seductive thought! Nevertheless, evidence of such effects has been hard to come by in North American populations, possibly as a result of the far greater genetic diversity in our multiethnic society.

The Red Queen rules over all host-parasite coevolution. It is a reciprocal affair, but from our moment in the continuum of time it is the adaptive evolution of the parasites, in our case viruses, that is most evident. The selective pressures for coevolutionary change of a virus and its host are at their height during the earliest stages of species invasion by the virus. We have already witnessed evolutionary changes in HIV-1 genotype prevalence in some regions of the HIV-1 epidemic. Moreover, Payne showed that in just a few decades of the Botswanan epidemic, evolutionary adaptation to the human population had resulted in measurable attenuation of viral fitness (Payne et al. 2014). These changes are evident within a single human generation, just a moment on the timescale needed to document reciprocal human evolutionary change on a population level. It must be expected that adaptive evolutionary change in humans in response to the selective pressure of HIV infection would necessitate many millennia to become evident on a species level.

Outrunning the Red Queen

Projections regarding adaptive evolutionary change in humans in the face of HIV-1 infection are artificial constructs that assume our battle with the pandemic virus will play out only in our genes and those of the virus: business as usual for the Red Queen. It is well established that a polymorphism in the CCR5 receptor protein, known as the CCR5Δ32 mutation, can be found in 10 percent of Caucasians in the United States. Individuals homozygous for this variant allele are resistant to HIV, and show no signs of the disease (Dean et al. 1996). They fail to produce a particular cell surface protein that constitutes the essential coreceptor recognized and bound by HIV during its entry into the host cell. These rare individuals exhibit no adverse effects from the lack of the protein and appear to be healthy. This is a de facto genetic trait associated with resistance to HIV. Could this be a beneficial polymorphism that would spread through the human population rendering us resistant to HIV-1 infection? Imagine a computer-generated *in silico* model of the human population, in which males and females interbreed randomly, and the population is under the selective pressure of HIV infection. We can expect that such a resistance gene mutation would progressively spread through the population as a polymorphism, providing a fitness advantage to offspring who have two of the resistance alleles. The survival of the fittest would lead to increasing prevalence of the alleles in our imaginary human population. Such a course of events could in principle render the species resistant to the disease. Of course, the rate at which this occurred would be influenced by myriad factors; the resistance allele would, for example, spread more rapidly if individuals possessing a single copy of the mutation reaped some phenotypic benefit (in fact, heterozygosity for CCR5Δ32 has been associated with more slowly progressive disease). It is also possible that the allele would be associated with some unforeseen disadvantage, in which case it will spread more slowly (or not at all). In any case, as the computer game moves through its iterative program of breeding and natural selection in the face of the *in silico* HIV epidemic, the population would necessarily experience a severe contraction. A substantial portion of the individuals with wild-type alleles, initially the overwhelming majority of the population, would die of the disease before reproducing. Increased prevalence of the resistance allele would necessarily be associated with a severe bottleneck in the population, before the fitter genotypes can emerge and dominate. If you have a good working knowledge of population genetics, you must forgive this gross vulgarization of the science. I have ignored some factors to simplify my argument; among them the effect of frequency-dependent selection on resistance alleles and the certainty that in nature the virus too will evolve.

Nevertheless, I think all readers will get my point. It is that if a species is to adaptively evolve in the face of a fitness challenge, like a fatal disease, it will do so only after severe attrition of its population.

Today's methods of genome-wide association studies have identified many "protective" alleles that circulate in human populations, conferring differential sensitivity to HIV disease progression (Passaes et al. 2014; An and Winkler 2010; O'Brien and Nelson 2004). In principle, over the long term, these polymorphisms that render carriers at a competitive advantage increase in prevalence. Perhaps I can also take a moment to convince you that the Red Queen has left her imprint on certain primitive human populations that may have been visited by zoonotic SIV. Scientists at the University of Illinois reasoned that since several cross-species infections with SIV of chimpanzees (and gorillas) were at the root of the HIV pandemic, then historically, Western African populations living in the geographic range of *P. t. troglodytes* may well have been exposed to SIV repeatedly before the twentieth century pandemic (Zhao et al. 2012). They sought evidence that the genomes of some modern African populations might have been shaped by such prior experiences. They chose to study Biaka Western Pygmies of the Central Africa Republic whose communities have historically been in the forested geographic range of the chimpanzee. The frequencies of single nucleotide polymorphisms in the pygmy genotypes were compared with those of Mbuti Eastern Pygmies who have never lived in proximity to chimpanzees (and hence were not at risk of SIV zoonosis). Their results were striking and intriguing. The frequency of certain alleles of genes related to the innate antiviral response and immunity were found at higher frequencies in the pygmy populations that have shared territory with chimpanzees. There was evidence for selection of variants that would be predicted to be "protective" against zoonotic SIV infections. It is possible then that over a long period of historical time, previous zoonotic immunodeficiency viruses may have engaged in a genetic arms race with primitive human societies.

These coevolutionary scenarios are based in ruthless Darwinian survival of the fittest. This must have been the mechanism by which retroviruses and their many different vertebrate hosts established long-standing relationships, and indeed how the "naturalization" of SIV to simian hosts evolved. It will not, however, characterize the developing relationship between zoonotic SIV (HIV) and its new modern human host. It is simply not acceptable from the standpoint of human society: we do not wait for changes in our genotype to solve public health crises.

Medicine at the Virus-Host Interface

We have already examined the very personal arms race that takes place between our immune systems and HIV, driving selection for mutations in the viral genome that allow it, at least temporarily, to evade our immune responses. This aspect of our ongoing arms race with HIV is very much akin to that fought between other vertebrates and their viruses. We can expect to detect evolutionary change in the virus, which undergoes iterative rapid replicative cycles and frequently makes replicative errors that create immense genetic diversity. Nevertheless, the HIV virus today experiences a distinct set of selective pressures that influence the success of viral lineages. First and foremost, our awareness of the disease on a societal level and our understanding of its mode of transmission have permitted adaptive changes in our behavior. The communication of risk factors across populations and the practice of safe sex are clear behavioral adaptive changes that have direct impact on the basic reproductive number of the virus, which must continue to exceed unity if the epidemic is to persist. Breaking the chain of transmission in this way is rather definitive and it is difficult to foresee evolutionary change in the virus genome that might circumvent this idea-based barrier to HIV transmission.

Perhaps the most instructive example of adaptive change in the human host can be found in one of the greater achievements of medical science to date. Within a single generation of the human species, society has responded with the development of more than a score of antiviral medicines. Taken chronically in combination, these drugs can halt the replication of HIV-1 in its tracks. For those segments of our global community that have access to these drugs, the disease is no longer an irrevocable death sentence. The success of this endeavor illustrates a pinnacle of achievement for a species fighting a new disease. It was not an easy pathway to tread and the virus countered the treatments with some success. In an escalating arms race, the virus deployed its evolutionary skills while the human hosts benefited from ideas, the invention and distribution of new antiviral drugs, and their use in effective combinations. What I find most fascinating about these drugs (and forgive me, because these are an almost lifelong source of passion for me) is their mechanism of action and what they represent to the virus. When these antiretroviral drugs are administered to a patient, they permeate all of the living cells of the body. They are "seen" by a virus that infects an individual cell as an integral part of the virus-host cell interface; antiviral drugs can be viewed as an extension of the virus-host interface. We have literally modified the cellular environment and provided another antiviral response for host cells. If the virus is to prevail in a host whose cells contain the drug, it must exploit its capacity for creating and exploring genetic diversity.

Resistance Is Futile

Antiviral drugs are typically small organic compounds designed to specifically recognize and bind to particular viral proteins, disrupting their function and interrupting the viral replicative cycle. The rate of viral replication in an infected individual is such that the extant genetic diversity of the virus quasispecies is more than likely to explore genetic variation at every nucleotide position of the genome. The discrete molecular interactions that antiviral drugs make with their protein targets places them at risk for development of viral drug resistance. Their presence in the cell provides an exquisitely targeted selective pressure, favoring genetic variants that are viable, yet subtly changed, at the site of drug interaction. These so-called resistance mutations were evidence of the genetic arms race between HIV and the extended host interface presented by the cell, collectively the products of human culture. Indeed, at the outset, resistance to the first antiretroviral drugs developed quickly, and the virus prevailed by evolving out of its constraints. It was human ingenuity that provided the counterpunch to the evolution of resistance, and it is widely attributed to Dr. David Ho and colleagues at the Aaron Diamond AIDS Research Center. They realized that should antiretroviral drugs be given in combination, then two independent mutations would be required to confer drug resistance on the virus. Furthermore, the existing genetic diversity of the viral quasispecies in an individual patient might not be adequate to contain within it a virus that has those two necessary mutations in the same viral genome. A typical patient's quasispecies would not contain any virus that possessed both resistance mutations and that could circumvent the host cell interface, modified as it was by the presence of two different inhibitory drug molecules. Medical science rapidly built upon these observations. New classes of drug were developed in the space of a few years and multiple combinations of three antiretroviral drugs, termed highly active antiretroviral therapy or HAART, were mobilized.

More and more HIV-infected patients throughout the world have access to these molecular tools that allow human host cells to present an inhospitable interface to the HIV virus and stifle its replication. Human society is successfully mobilizing its collective culture in the arms race with zoonotic HIV. The adaption of the human species to HIV is not based upon heritable change in our genome and has occurred within a single generation. Human society need not fall back on evolutionary change in its battle with zoonotic viruses.

· 10 ·

CROSS-SPECIES INFECTIONS: MEANS *and* OPPORTUNITY

THE NARRATIVE of the HIV-1 pandemic is one compelling and salutary example of a viral zoonosis with irrevocable global consequences for humanity. A more comprehensive consideration of the variables that influence the outcomes of cross-species virus infections will allow us to more fully appreciate the future risks of viral zoonoses. These variables include genetics, ecology, and behavior and they dictate the probability that a virus will jump the species barrier. We will think of these in terms of the *means* and *opportunity* for a virus to commit the crime of zoonosis. Can the virus successfully bridge the species divide and establish a successful and durable genetic lineage in a new host?

Opportunity within the rubric of the crime of zoonosis comprises ecological and behavioral parameters. The cross-species transmission of a virus from its natural species to a new host species requires contact with infectious material. The two host species must occupy the same or overlapping geographical areas. They must also exhibit behaviors that create the opportunity for the infection to occur. We can draw on our knowledge of the cross-species transmissions of SIV in western Africa and the emergence of the AIDS epidemic. It is evident that the ultimately zoonotic SIVcpz lineage had moved between different primates: from African monkeys to chimpanzees (and to gorillas) and then to humans. Each of the cross-species transmissions took place in regions where the territories of the respective primates overlapped. Phylogenetically related species often occupy similar ecological niches and exhibit similar behaviors, a factor that also promotes the circumstances in which their paths cross. The conflicts and predation that occur between primate species (e.g., chimpanzees hunt and eat monkeys) and the hunting and butchering of bushmeat by humans provided the remaining pieces in the opportunity puzzle for zoonotic SIV. All the criteria for creating a host zero fell into place. Behavioral factors may also influence the likelihood that an epidemic chain of transmission can be established in the new host: to wit, the emergence of the HIV-1 pandemic and its promotion by social and economic upheaval in French and Belgian colonial African nations.

The *means* to commit the crime of zoonosis is equally complex; it is impacted by genetic factors in both of the hosts and the virus. The probability of successfully establishing an infection in a new host is certainly favored if the natural host and the new host species are closely related phylogenetically. The cellular surface proteins that serve as viral receptors and allow the virus to dock with and enter the cell are more likely to be evolutionarily conserved and hence structurally and functionally similar. The same holds for other host cell infrastructure with which the virus must interact. These constitute the operating system of the cell. The more closely related the two operating systems, the more likely the invading virus will be able to establish replication in the new host. Though these are broad generalizations that make many assumptions, it is reasonable to predict that species jumps over greater evolutionary distances will be more challenging. Contrast the outcome of SIV infection in Old World African monkeys and HIV-1 infection in humans. SIV is essentially benign in its natural monkey host but HIV-1 infection in humans irrevocably leads to death in almost all infected individuals. Similarly, the African monkey virus causes AIDS in Asian monkeys. These are dramatically different disease outcomes, despite the close evolutionary and phylogenetic relatedness of African and Asian monkeys. These differences illustrate the profound species barriers that viruses must overcome to move successfully between even closely related primate lineages. Although the African and Asian monkeys run their cells on broadly similar operating system platforms, coevolutionary adaptation between the virus and its natural host introduced incompatibilities that influence the pathogenesis of the virus profoundly.

The *opportunity* for zoonotic transfers is abundantly available to viruses, but the *means* to establish themselves as endemic parasites of the new host population presents a greater barrier. What is the evidence for this? It is impossible to accurately quantify failures of cross-species transmission. Taken to its logical extreme, this is a pointless exercise: bacteriophages do not and cannot infect eukaryotic cells; there is no hint that inter-kingdom viral zoonoses are a real threat. In a recent paper, researchers from Aarhus University in Denmark describe a plant virus that appears to have infected honey bees and mites (Francis, Nielsen, and Kryger 2013). This represents a remarkable example of cross-species, even cross-phyla, transmission, but some in the field suggest that the researchers fell short of definitively proving that the virus is actively replicating in the insect hosts. This indeed is rarefied territory; successful species jumps and zoonoses are exceptionally rare and the underlying basis for success or failure is more often than not obscure. It is more fruitful for us to fall back on clinically evident zoonotic infections with different outcomes in man. SIVcpz and SIVgor each appear to have achieved at least two zoonotic transfers (HIV-1 M and N, and HIV-1 P and O, respectively); SIVsmm of the sooty mangabey has spawned at least eight lineages of HIV-2, each of which resulted from a separate cross-species transmission. How many cross-species infections by primate lentiviruses go unnoticed because the infection in host zero ($host_0$) is not successfully established, or is undiagnosed and not transmitted? HIV-1 group M is the source of the global AIDS pandemic, and HIV-2 groups A and B are the only lineages that successfully spread among humans but remain predominantly a disease of West Africa. These different zoonotic viral lineages experienced differential success adapting to the human host. The nature of the RNA virus quasispecies dictates that every cross-species transmission has a genetically distinct founder virus (or a limited population of founder viruses). As a consequence, not all variants that enter a new host will have the same adaptive and, hence, epidemic potential.

One to several sporadic cases of swine influenza virus infections (usually in farmworkers) are reported each year in the United States. Workers in the swine and poultry industries have a higher probability of zoonotic infection by swine and avian influenza viruses. In a recent study of farm residents in eastern China, antibodies directed against swine influenza were detectable in more than 10 percent, suggesting that zoonotic transmission events are far from rare events (Yin, Yin, Rao, Xie, Zhang, Qi, et al. 2014). On the other hand, onward transmission between human individuals, and its attendant risk of seeding a flu pandemic, appears to be an extremely rare event. A minor but dead-end epidemic of swine flu (probably related to the 1918 pandemic virus) emerged in Fort Dix, New Jersey, in 1976 among army recruits. It killed 1 person but infected 230 others before disappearing and never reemerging (Kilbourne 2006). These examples, whether of zoonotic transfers of primate lentiviruses or swine influenza viruses, serve to illustrate the abundant opportunities for zoonoses to occur. However, even if successful cross-species infections take place causing disease in that individual host, there is a spectrum of outcomes. Chance is a major factor (it must always play some role, particularly in what are likely infrequent events compared to the infections of natural hosts). Nevertheless, much of the variability is attributable to genetics: the genetics of the emerging founder virus, the genetics of the new species, and indeed the genetics of the individual, $host_0$. Some viruses that infect new hosts possess the particular variant genotype that allows them to establish an infection and to adapt to the new host: they have the means to be successful while others do not.

A Rogue's Gallery of Emerging Viruses

From where will the next pandemic come? The likely favorite is a virus. Of the six classes of pathogens that cause disease in humans (bacteria, fungi, protozoans, helminths, viruses, and prions), viruses constitute the majority of recently identified new pathogens (Jones et al. 2008). In the last few decades and first years of the twenty-first century, the rogue's gallery of newly emerged viruses grew at an alarming rate. HIV-1 and hepatitis C surfaced in the twentieth century as worldwide pandemics; hantavirus pulmonary syndromes appeared in the United States and continue to cause periodic outbreaks; Ebola, Marburg, and Lassa fever haunt the African jungles. More recently atypical viral pneumonias caused by coronaviruses, emerged in man. The SARS (severe atypical respiratory syndrome) epidemic of 2003 almost reached pandemic status, but was stalled due to prompt implementation of effective global containment measures. Today Middle East respiratory syndrome (MERS) is simmering at low levels on the Arabian Peninsula and has caused an outbreak in South Korea. Vector-borne arboviruses such as West Nile virus and chikungunya virus have invaded North America, and Rift Valley fever epidemics have exploded. Newly discovered paramyxoviruses, Nipah and Hendra have emerged as threats to man, killing more than half of those who become infected. The list goes on.

Some of these viruses are newly discovered genetic entities; others are familiar viruses that have undergone evolutionary change and become more widely transmissible and sometimes more virulent. Some are highly contagious and present real threats to global health, while others (at least in their current incarnations) cannot be construed as such. Nevertheless, the disturbing regularity with which new diseases are threatening humans serves as a reminder that our knowledge of the diversity of viruses with the potential to invade the human species is quite limited. Almost daily headlines paint a grim picture; for example, in March 2014, *Science* magazine proclaimed, "*New killer virus in China?*" referring to a publication from China (Wu et al. 2014). Following the death of two miners who worked in a derelict copper mine in Yunnan Province, scientists went in search of an infectious culprit. They found a new paramyxovirus related to Hendra virus, a known human zoonotic virus. The new virus infected bats and rodents in the cave system. Was this the source of their mystery illness? It seems that if you look, you will find a candidate human viral pathogen. How many such viruses lurk undetected and uncharacterized; how many human zoonotic infections go undiagnosed and unnoticed?

We live in an era of unprecedented global change. Human populations on all continents are growing, pushing up against territories that were once the exclusive domains of wildlife. Ever larger international metropolitan centers are globally interconnected by airlines—we are all, just a little over a day away from each other. Industrial farming of animals, poultry, and fish intensifies the potential for sustaining epidemic viruses, creating opportunities for spillover virus transmission between species and to ourselves. This is not unlike the circumstances that provided the opportunity for the emergence of measles virus from our domesticated cattle in early civilizations, but it is occurring on a far grander scale. To date only one human viral disease, smallpox, has been successfully eradicated by vaccination—a remarkable accomplishment. Despite effective vaccines, many other viruses still flourish in pockets of undervaccinated populations (the measles virus and poliovirus are signal examples). For some viruses, neither vaccine nor drugs are available, and many of these are dramatically expanding their ranges across the globe.

Mosquito-borne viruses are making headway into North America: in 2012 the CDC reported 2,000 cases of chikungunya in the United States, a trifling number compared to the 2 million infections that plagued South America, but a disturbing trend, nevertheless. Dengue virus and recently Zika virus have also gained a foothold, and West Nile virus has become endemic. Each has taken advantage of climatic trends and the movement of their insect vectors to exploit previously naive human populations, ripe for epidemic spread.

I would be remiss if I did not allow that other changes are at hand; our ability to rapidly recognize and diagnose emerging infections for one. We cannot discount that many of these viral infections just went unnoticed in former times. Nevertheless, the preponderance of evidence points to a real phenomenon: the increased emergence of viral threats, provoked by global social and climatic changes. Never before has the opportunity for zoonoses been higher. But what of the *means*?

Adaptive Evolution in Zoonosis

A quick survey of zoonotic viral infections readily reveals that RNA viruses cause the overwhelming majority; they are uniquely equipped with extraordinary evolutionary agility. Their error-prone genomic replication dictates that they exist as quasispecies, clouds of genotypic variants that most densely occupy genetic space associated with optimal fitness. The means for zoonosis are twofold and fundamental; the virus must be able to access the cell of the new host. For this to be possible, there must be a functional receptor on host cells. The virus must also be able to sustain adequate replication within the cell to achieve onward transmission. In some instances, simple genetic changes may be all that are required to seed infection of a new species, but in others the necessary genetic adjustments may be complex and mysterious. The probability of surmounting these obstacles is quite evidently higher when the evolutionary divide between the natural host species and the new species is narrower. Nevertheless, every virus has undergone substantial coevolution with its natural host, creating a remarkably high species barrier over which a virus must cross. Strong positive selection for those individuals that survive viral infection drives highly specialized adaptive genetic changes in the host genome, which quickly render virus-host relationships exclusive and highly specialized. These coadaptations in the host are not shared with closely related species that have not been exposed to the virus. The virus itself also undergoes reciprocal selection for genetic variants that are specifically adaptive in its natural host. These changes may be maladaptive for other species, even those that are closely related phylogenetically. Coevolution of a certain entente between a virus and its natural host reinforces fidelity and the jump from one to another new species is no trivial feat—a fact for which humankind should be grateful.

I have previously stressed (in Chapter 7) that viruses, particularly RNA viruses, possess the means to evolve rapidly but do not have the flexibility to exploit this in an unfettered fashion. This is particularly germane to a discussion of cross-species infection. The durability of viral genetic lineages and their capacity to adapt via genetic variation is certainly remarkable and for RNA viruses has its roots in high mutation rates, allowing them to exhaustively explore available genetic space. Only in this manner can they stumble upon the right genetic variant to counter the natural restrictive measures of the host and optimize viral fitness. RNA viruses evolve under severe genetic constraints in large part because their genomes are limited in size by the poor fidelity of their replication machinery and they must encode a great deal of necessary functionality in just a few viral proteins and RNA sequence.

Several hundreds of human cellular genes have been implicated as essential to the replication of the HIV-1 virus in a cell (Bushman et al. 2009). When one considers that only fifteen proteins encoded by the virus must mastermind their exploitation, the complexity of functional networks and the inevitable multivalence of viral protein functionality become evident. To make matters even more challenging for the virus, the innate and adaptive arms of the host immune response each represent challenges to the genotype. Our own genome dedicates about 1,000 genes to immune defenses (Lander et al. 2001), and since our genomes are diploid, many of them are present in different allelic forms on our two chromosomal copies. A simple genetic change in the virus that may be adaptive for one function is more often than not maladaptive for another. A mutation in the viral polymerase that changes an immunogenic epitope to one no longer recognized by host cytotoxic T cells might offer the virus an advantage in terms of immune escape. It may, however, come with profound negative consequences for replicative efficiency; a mutation that evades one arm of the immune response may render the virus susceptible to the antiviral action of an alternative cellular system. Adaptive mutations then often "rob Peter to pay Paul." Only viruses that find a way to negotiate these obstacles and adapt while minimizing such trade-offs can survive as successful lineages. Particularly for the minimalist genomes of RNA viruses, the mutational space that can be explored to achieve this goal can be remarkably restricted and virulence is often necessarily mitigated.

Fitness Landscape

Before we explore how these challenges play out in the situation of cross-species transmission and zoonosis, I will lay down a conceptual framework that helps me think about the relative fitness of viral (or indeed any) genotypes that exist in a particular environment, a particular habitat to which a virus must adapt. For our purposes, environment can denote a particular host species or even a newly evolved genetic variant of that species. It is a simplified way of conceptualizing Eigen's genetic space that he described in terms of multi-dimensional clouds of gas coalescing around favored fitness optima. Instead, we will think of viral fitness in this host in terms of a landscape featuring mountains and valleys. This three-dimensional relief map is reminiscent of the old papier-mâché museum exhibits that displayed geographic and geological strata in miniature. Each location or map reference and its altitude on the landscape represents a single genotype and its aggregated fitness phenotype expressed as the basic reproductive number (R_0). The fitter the genotype, the higher the R_0 and the more successful the genotype. The fittest genotypes will occupy higher elevations; the valleys represent genotypes with diminished fitness. We might now think of particular contours at each elevation connecting distinct genotypes of equivalent fitness. For conceptual ease I will flood the landscape; the water level will rise to a point that represents $R_0 = 1$. Viruses that have an $R_0 < 1$ cannot sustain a chain of transmission, while viruses with $R_0 > 1$ can successfully spread in the host population. Now we have created a landscape with mountains and lakes; some of the mountains have steep sides and are surrounded by water, others may be separated by a high pass at sufficient altitude to allow passage across to a neighboring mountain. If one thinks of each point in this landscape as the fitness of a single genotype removed from its closest neighbors by a single mutational change, then one can picture how the landscape determines the mutational flexibility of a virus and the adaptive pathways that it can follow in that particular environment. In principle, the landscape may be rolling with broad valleys and rounded hilltops or severe with deep, narrow valleys and craggy mountainous summits. In the former, the difference in elevation (fitness) between adjacent points on the landscape can be quite modest, while in the latter case the steep mountainsides signify sharp gradients in fitness, where discrete mutational changes will dramatically influence fitness. Genotypes located in a flooded area on the landscape will not be viable. Adaptive mutations must steer a path to higher ground, but sometimes to make this passage the genotypes must cross lower ground between mountaintops; it cannot, however, pass through water.

The model serves to illustrate that different genotypes may be viable in particular environments, but the virus can be constrained as to whether it can find a path to higher ground. In reality, the fitness landscape of viruses is likely very rugged, tolerating little mutational change, without suffering substantial trade-off in fitness. As a result, for most viruses and most parts of their genomes, purifying selection dominates. This is particularly so for viruses that are well adapted to their hosts (recall the avian influenza virus). Most mutational changes that affect the protein sequence are detrimental, moving the genotype downward toward the valley. Given the complex coding needs of the compact RNA virus genome, even synonymous mutations can be deleterious to the fitness of the genotype. The mutational space that can be explored by the virus in a constant environment can, therefore, be relatively small and the fitness peaks have steep sides. It is important to clarify that while real-life observation and experimental measurements of RNA virus mutation rates all come in at between 1×10^{-3} and 1×10^{-4} mutations per site per generation, the nucleotide substitution rate is much, much lower. Rather, it reflects the rate at which nucleotide changes are fixed in the viral lineage. The virus quasispecies that takes form during replication of the virus within a host is by its very nature a representation of the diversity of mutational changes. As a consequence it possesses a great deal more genetic variation (and spectrum of viral fitness) than can survive the filters of purifying selection.

A Shifting Fitness Landscape

Let's use these concepts to look at a shifting landscape that represents environmental change, specifically when a virus moves from its natural host to infect a new species. Here we have to overlay our first landscape that represented the fitness of virus genotypes in their natural host, with a second landscape, that of the new host. This new fitness landscape can be very different; points with the same grid reference may be on top of a mountain (and represent a very fit genotype) in one landscape but located in a valley (and hence be less fit) or underwater (unviable) in the other. The inverse can also be the case: genotypes unviable in the first host (and therefore inaccessible) may be fit in another. The virus that spills over to infect a new host finds itself in a new and hostile landscape. Its code was optimized for a different operating system; it may be more susceptible to antagonism by the immune defenses of the new host, and it cannot effectively exploit the essential network of cellular cofactors that it needs. It must rapidly evolve and find higher ground in the new landscape and with it, increased genotypic fitness. This is essential if it is to establish a chain of transmission. Otherwise, it will be a nonproductive or dead-end infection. Most cross-species infections are indeed just that; the adaptive pathway to an $R_0 > 1$ proves too circuitous or is prohibited by the features of the landscape. The genotype cannot traverse the terrain via mutational intermediates of adequate fitness and it cannot achieve the necessary fitness to replicate and be transmitted efficiently. It is condemned to the valley of death and extinction in that host species.

While the topology of the fitness landscape for two closely related host species may be quite similar, that of a more distantly related species may be radically altered. The location of fitness peaks in the genetic landscape may be quite different and require genetic changes that cannot be made successfully without passing through an inviable fitness minimum. It is, therefore, self-evident that if a virus invades a new (but closely related phylogenetically) species it has a higher probability of gaining the high ground via a relatively straightforward mutational pathway that avoids valleys of diminished fitness.

Sympatric species occupy the same geographic range and are often closely related. They may be the result of speciation without geographic separation of the descendent species. There is accumulating evidence that species jumping by RNA viruses, particularly between closely related sympatric species, has occurred frequently and with some facility in evolutionary history. The theory has been advanced (Charleston and Robertson 2002; Bohlman et al. 2002) that such cross-species transmissions may have played a significant role in the establishment of what we recognize as distinct "species" of viruses in distinct host species today. The phylogenetic congruence of many RNA viruses with their host species also strongly suggests that there is a substantial role for virus-host codivergence and cospeciation similar to that invoked for the herpesviruses (Holmes and Zhang 2015; Woo et al. 2012). Charles Ruprecht and colleagues published a seminal study on the evolution of rabies viruses in bats. An often neglected but notoriously multi-host zoonotic virus, rabies is almost certainly one of the deadliest of the viruses that infect humans. Rabies infections in humans are pure spillover infections, resulting mainly from human encounters with infected domestic pets and sometimes bats. While rabies is never transmitted beyond the first infected individual, it is lethal. Ruprecht and colleagues surveyed 372 rabies viruses in twenty-three different bat species in the wild in North America and found eighteen distinct bat lineages (Streicker et al. 2010). They were able to reconstruct the phylogeny of the lineages and relate them to the phylogeny of their respective host bat species. This revealed that cross-species transmission and establishment of separate lineages had taken place often. It was most strongly influenced by the existence of opportunity, when the geographic distributions of the species overlapped, and by phylogenetic distance between the two host bat species. Cross-species transmissions had been most commonly successful when the two bat species were more closely related phylogenetically. Evidently such close evolutionary relationships offer a lower barrier to infection by the novel rabies virus compared with more distantly related bats.

Other viruses have jumped a species divide on more than one occasion. They offer a unique opportunity to identify the key footprints along the pathway to adaptive change. In several instances, a striking observation has been made. Comparison of the genetic changes associated with independent cross-species transmission events reveal an identical adaptive change in each lineage. We may hypothesize then that evolutionary adaptation to the new species has an absolute requirement for this mutation and natural selection reveals and favors this particular mutant genome.

This underlines the limitations that viruses face in successfully adapting to a new host and highlights their genetic resourcefulness in finding the needle in the haystack. The most familiar example of this phenomenon draws on the cross-species transmissions of SIVcpz and SIVgor to humans. Scientists have documented four independent transmission events from chimpanzees or gorillas to humans, and in each case have reconstructed the genome sequences of the viruses before they made the jump to a human and afterward. In each case, the emerging HIV-1 lineage contained a single mutational change in the viral matrix protein that caused the amino acid residue at position 30 to change from methionine to arginine (Sharp and Hahn 2010). Tellingly, when researchers took the emergent HIV-1 strain containing this mutation and reintroduced it into chimpanzees, the virus gene underwent reversion of the same amino acid residue. Moreover, if a simian virus was engineered to contain the human virus–associated mutation it grew more efficiently in human cells. These are two very convincing demonstrations that this was a real adaptive change selected during the zoonotic event. Furthermore, it is a strong endorsement of the opinion that the mutational pathways for adaptation are often very restricted and may be limited to a single option. Successful adaptation requires that the virus traverse a narrow territory of critical genetic changes to successfully achieve cross-species transmission.

We can contemplate the following question: Was the founder SIVcpz virus that infected the hunter as he butchered the disease-carrying ape already in possession of the necessary adaptive mutation for success in man, or did it emerge as the virus expanded to form a quasispecies from which to select new adaptive variants? It is impossible to tell.

The Paradox in RNA Virus Evolution

The RNA viruses circulating today are most certainly descendants of ancient parasitic replicators that are believed to have evolved in the pre-DNA era. A bewildering variety of RNA viruses infect bacteria, archaea, and eukaryotes. RNA viruses have a multitude of lifestyles and have found their greatest success in eukaryotes, where they are the most diverse members of the eukaryotic virome. They are viruses that use genomes of negative polarity, positive polarity, or double-stranded RNA; some have exterior icosahedral capsids, others have enveloped virions, but all share a genome of relatively limited size (the largest being about 30 kilobases in length). They are limited in genome complexity because of the error-prone nature of their RNA-based replication strategies; larger genomes would accumulate too many mutations in each round of replication, resulting in their extinction. The contrast with double-stranded DNA viruses is stark; these viruses could exploit their DNA-based genetics to evolve highly complex genomes (as we saw earlier, some exceeding 2 million bases). A single common thread connects all RNA viruses: the possession of an RNA-dependent RNA polymerase (RdRp). The amino acid sequences of these RNA virus RdRps and more recently the comparison of their three-dimensional atomic structures root them back to a single monophyletic origin. The enzyme, once invented, created the RNA viruses and became the single prototype gene from which all subsequent RNA virus replicases evolved (Koonin, Dolja, and Krupovic 2015b). The reverse transcriptases are also evolved from this prototype. That is not to say that RNA viruses can be plotted simply on the radiating branches of an evolutionary tree; they cannot. As for DNA viruses such as the NCLDVs and herpesviruses, which emerged much later, their origins and evolution are best thought of as a network of intermeshed branches. While a simple phylogenetic tree illustrates vertical inheritance during evolution, a branching network reveals a substantial role for multiple horizontal gene exchanges taking place between viral lineages and between viruses and their hosts. Notwithstanding the limitation in genome size faced by RNA viruses, it seems a reasonable proposition to assert that they owe much of their evolutionary success to their error-prone genome copying and the evolutionary adaptability that comes with the continuous creation of genetic variants.

The herpesviruses are an example of viruses that emerged from the DNA bacteriophage world when eukaryotes arose and ultimately formed long-standing relationships with their invertebrate and vertebrate hosts. Their phylogenies are congruent with those of their host organisms as they codiverged, cospeciating with their hosts, permitting the establishment of harmonious and mutually acceptable lineage coexistence. Our discussion of cross-species transmission and zoonotic infections established that from the perspective of biological and genetic feasibility, cross-species transmissions of viruses are most successful between phylogenetically closely related host species. Now I will argue that it therefore makes "evolutionary sense," once a long-standing relationship between virus and host has been established, that the codivergence of the virus and its now "natural" host species will always be the most probable outcome during evolution and speciation. That is, of course, unless the relationship is perturbed by changes in the external environment. RNA viruses have proved themselves particularly adept at jumping between host species. RNA viruses are the most common cause of zoonotic infections; indeed, the emergence of measles virus, SARS, and HIV-1 have figured prominently in our dialogue. In each case the movement of the ancestor virus from the natural species was facilitated by environmental changes, whether they were in the domestication of farm animals, the introduction of the virus into dense population centers, or in societal changes. These changes provided the opportunity for the viral lineage to be introduced into a new environmental niche: a new host. Regardless of the evolutionary agility of RNA viruses in making them uniquely capable of cross-species infections and creation of distinct viral species in a new host, the originating lineage remains unchanged in the first host. For both of these virus lineages it is reasonable to contend that should their existing host undergo a speciation event, in all likelihood they will undergo cospeciation with their respective hosts. This might be thought of as the evolutionary path of least resistance; the one which requires the least genetic distance to be traversed on the fitness landscape.

So despite empirical evidence that RNA viruses have frequently speciated by cross-species transmissions, if my contention is plausible, there should be ample evidence that RNA-viruses codiverge with their natural host species as a predominant evolutionary pathway. Indeed, this appears to be the case, and evolutionary virologists have accumulated multiple examples of marked congruence between RNA viruses and host species lineages. Striking examples are the New World hantaviruses or the coronavirus family; for hantavirus, a detailed view of the phylogenetic tree of the virus species and their multiple bat and rodent hosts is consistent with a substantial role for virus-host codivergence deep in their phylogenetic histories. In the case of coronaviruses, which today have been found in many bird and mammal species and have caused troublesome zoonotic infections in human (SARS and MERS), there are also strong indicators that their natural hosts are multiple species of birds and bats. The coronaviruses and their hosts share deep evolutionary roots over many millions of years, and species distributions bear all the hallmarks of codivergence and coevolution of virus and host lineages (Wertheim et al. 2013). Consistent with the evolutionary aptitude of these viruses, there is also abundant evidence for multiple cross-species transmissions giving rise to the current virus lineages in each respective species (Guo et al. 2013; Holmes and Zhang 2015). Cross-species jumps are more readily achievable between phylogenetically closely related sympatric species and should be expected to have a high probability of occurrence between the numerous coexisting species of rodents, bats, or birds that often exhibit similar behaviors and occupy closely associated ecological niches. These are the conditions that provide the means and the opportunity for cross-species transmissions that were the subject of earlier passages in this chapter.

It appears then that, as with DNA viruses, codivergence and coevolution played a significant role in RNA virus evolution. DNA viruses create further opportunities for evolution and speciation through horizontal transfer of genetic information and by collecting new gene functions in their highly flexible genomes. RNA viruses do not have the luxury of expanding their genomes, but exploit their inherent genetic instability. It is this aptitude that has allowed host jumps, most frequently into phylogenetically closely related hosts, to be an influential factor in RNA virus evolution and the creation of new species.

The currently accepted model of RNA virus evolution acknowledges both the roles of host cophylogenetic divergence and cross-species jumps, in the relatively recent (in evolutionary terms) emergence of some RNA virus species. Nevertheless, it is natural to believe that, as for DNA viruses, they have extremely ancient roots. After all, the prevailing opinion is that life started out as an RNA world. There was, however, a glitch: given what virologists knew of the rates of evolutionary change of RNA viruses, measured by rates of nucleotide substitutions, they attempted to estimate the age of the RNA virus families circulating today. A paradox emerged: none of the RNA virus families could reasonably be estimated to be older than 50,000 years. If this were true, they are younger than our own species (Holmes 2003).

RNA Viruses and Molecular Clocks

Molecular clock estimates of the rates of viral evolution are based on the rates of nucleotide substitutions becoming fixed in virus genomes over time. Many factors can confound these estimates. Measurements of nucleotide substitution rates in double-stranded DNA viruses such as herpesviruses are as low as 3×10^{-9} substitutions per site per year, while the RNA genome of human influenza virus incorporates substitution mutations at a rate of 4×10^{-3} per site per year, six orders of magnitude faster. This latter rate is not atypical of RNA viruses, but they too can evolve more or less rapidly depending on their circumstances. Avian influenza virus in its native bird hosts is said to be in relative evolutionary stasis and has very low rates of nonsynonymous substitutional changes. In contrast, epidemic human influenza A virus undergoes much more rapid genetic change with nonsynonymous mutations accumulating a much higher rate. This difference is accounted for by the intense, diversifying selective pressure exerted on the human virus by our immune response. Synonymous nucleotide substitutions, however, that do not affect the structure of viral proteins accumulate at similar rates in both avian influenza (replicating in its long-standing and natural host) and human influenza viruses. These rates are also comparable with those of most other RNA viruses.

Simian foamy virus (SFV) is a retrovirus that infects primates. It has done so for millennia, and the phylogeny of foamy viruses and their host primates demonstrates the congruence typical of codivergence and coevolution. Since accurate fossil records document the speciation of their primate hosts, the rates of SFV evolution can be definitively measured as 1.7×10^{-8} substitutions per site per year (Malik 2005). Such a sluggish rate of viral evolution is unprecedented among retroviruses and other RNA viruses which all have evolutionary rates several orders of magnitude faster. The fundamental mechanisms of retroviral replication in SFV are unchanged compared to other retroviruses, so we must assume that mutation rates per genome replication will be the same. The explanation for this particular discrepancy can only be in lifestyle: SFV infections are extraordinarily lethargic, and the virus remains latent in the host genome for extended periods. With fewer iterative cycles of replication per unit time compared to other viruses, they simply evolve more slowly. Lifestyle has other significant effects on viral evolution. As we shall see later in this chapter vector-borne viruses are constrained in their evolutionary flexibility due to the necessity for meeting the requirements of infecting alternating host organisms. Furthermore, recent studies have illustrated that there is a relationship of host cell generation time on rates of viral evolution; it seems that viruses that infect cell types that divide more rapidly, themselves can evolve more quickly (Hicks and Duffy 2014). Viruses that infect epithelial cells generally have the capacity to evolve most rapidly.

There is now a general understanding of the biases inherent in the casual use of molecular clock predictions of viral evolutionary dynamics (Duchêne, Holmes, and Ho 2014). The origin of most uncertainty arises from the fact that scientists have necessarily been "tip dating" viruses. Because we are limited to samples from recent evolutionary time, we necessarily measure the length of the terminal twigs of the evolutionary tree and extrapolate that to the oldest interior branches. This process is fraught with uncertainty. Rate estimates are always biased toward faster rates when measured in contemporary time (the very tips of the branches). Estimates that take into account longer sampling periods, measuring some of the more mature branch lengths as well as the terminal twig lengths, yield evolutionary rates progressively slower. Two factors are most influential in these biases: purifying selection and substitution saturation. When contemporary virus genomes are compared, the nucleotide substitution rate—the rate at which mutations are fixed in the population—is an overestimate. It is a measurement of the combination of mutation rate and substitution rate since some deleterious mutations may not have had adequate time to be purged from the genome by purifying selection. Moreover, the constraints upon minimalist RNA genomes place a strong emphasis on purifying selection; there may in fact be few sites in the genome that have the flexibility to undergo nonsynonymous changes, while the vast majority of the genome evolves extremely slowly. Mutational saturation at some positions will also be a significant and systematic driver introducing bias into short- versus long-term mutation rates. It is reasonable to expect that those nucleotide positions in the genome that have the flexibility to be changed may undergo repeated substitution. Although these events are captured in short-term measurements of nucleotide substitution, over the long term multiple changes at the same site are only counted once.

There is, therefore, a reasonable basis for believing our eyes when biological observations tell us that codivergence and coevolution of viruses within natural host species are more the rule than the exception. Two examples of recently published research serve to substantiate this view and highlight the limitation of molecular clock approaches.

Scientists continue to refine computational models to determine more precisely the actual age of RNA virus lineages. Early assessments of the age of the coronaviruses using molecular clock calculations placed its age at no more than 10,000 years, a result that flew in the face of the geographical distribution of distinct but evolutionarily related coronaviruses in different bat species ranged across multiple and distant continents. New methodology to account for the bias that is evidently introduced when extrapolating genomic data from present-day viruses into deep history suggests that ancestral coronaviruses have infected birds and mammals for almost 300 million years (Wertheim et al. 2013). Strong purifying selection appears to have masked millions of years of coronavirus evolutionary history. The same can be expected to be true for other RNA viruses whose histories have been artificially truncated by uncorrected tip dating. It seems that the RNA viruses that circulate today are most likely the direct descendants of their ancient counterparts. Their potential for evolutionary change is effectively bridled in their natural hosts to which they become evolutionarily adapted. It can, however, be unleashed with a vengeance when they are subject to changed selective pressures such as those presented by the hostile environment of a foreign host. Coronaviruses are truly an ancient viral lineage that may have infected bats and birds for tens or hundreds of millions of years. The world's bat species appear to be natural reservoirs for coronaviruses (and many other viruses). The have coevolved during this extended relationship into chronic asymptomatic infections, which if transmitted to a new species can be the source of rapidly evolving viruses that are truly modern diseases. A recent example is that of SARS which emerged to pose a global health challenge.

The recognition of the emergent HIV-1 and HIV-2 viruses as human epidemic viruses led to an explosion of research into lentiviruses. Soon, almost forty different primate lentiviruses, all simian immunodeficiency viruses, were catalogued along with their natural hosts in the wild. Scientists wanted to know more about SIV because it spawned the zoonotic lineages that caused the AIDS outbreak. They tracked down the approximate dates when SIV had jumped from monkeys to chimpanzees and then into humans to ignite the HIV-1 epidemic and when SIV from sooty mangabeys emerged in humans as HIV-2, but fundamental questions remained. Answering those questions might have important consequences for the development of future treatments or vaccines. How long had lentiviruses been infecting primates? Why was HIV-1 fatal in Asian monkeys yet replicated to similar levels in African monkeys, leaving them unscathed? The difference was clear: African monkeys had adapted to the virus of African origin; they were its natural hosts. The mechanism, however, even today remains only partially resolved. Evolutionary virologists estimated that a long period of coevolution, perhaps over millions of years, was necessary for this mutual accommodation between virus and host to develop. The samples used to date the origins of HIV in human were collected over several decades and spanned a very significant portion of the time over which the virus had been evolving in man. These samples allowed very accurate dating of the evolutionary divergence of SIV-HIV since the sequences available to scientists were representative of the whole period of evolutionary divergence. Very little extrapolation was needed; the scientists were not tip dating. To probe the origins of SIV lineages themselves, a greater extrapolation of a contemporary set of data into the past would be necessary to assign a date to these older lineages. Unfortunately, the evolutionary scientists could not fall back on endogenous retrovirus (ERV) fossil evidence; no lentivirus ERVs could be found in primate genomes. Lentiviruses had only recently been discovered as ERVs in the distantly related rabbit (subsequently they have been found in Malagasy lemurs and ferrets). Evidently lentiviruses were indeed an ancient virus lineage infecting (and endogenizing) a variety of mammals, but what of their association with primates?

Worobey and colleagues (Wertheim and Worobey 2009) were some of the first to explore the origins of SIV lineages that subsequently emerged from chimpanzees and sooty mangabeys as HIVs. They sought to identify the most recent common ancestor of these two viruses by backward extrapolation of their sequence divergence over time using an advanced computational technique known as "relaxed molecular clock dating." Their results flew in the face of expectations. It appeared that SIVcpz, the virus that became HIV-1 in humans, had infected chimpanzees for just 500 years; SIVsmm, that emerged as HIV-2, had infected sooty mangabeys for only 200 two hundred years. When the complete phylogenetic tree of primate lentiviruses was examined as a whole, they found it to be rooted just a little over 1,000 years ago. These data were certainly not in line with the expectation of evolutionary biologists. They believed that apathogenic SIV infections in simian species must be the result of an extended period of virus-host coevolutionary adaptation. If the earliest primate lentivirus infections were in the order of 1,000 to 2,000 years ago, it is necessary to posit that either their ancestor viruses were not pathogenic to primates or that primates adapted to pathogenic lentivirus infections in a remarkably short period of time. These scenarios are certainly not consistent with the radically different pathogenicity exhibited by SIV in natural and nonnatural primate hosts (see Chapter 9).

The resolution of this conundrum came in the next year or so when Worobey and his colleagues undertook a phylogeographic analysis of some SIV species and their host simians that allowed them to firmly plant markers in time and definitively calibrate the SIV evolutionary clock (Worobey et al. 2010). The team was inspired to take advantage of a geographic change affecting the coast of Equatorial Guinea 10,000 to 12,000 years ago. Rising sea levels created an isolated landmass off the African coast, today called Bioko Island. The scientists reported the sampling of six different species of monkey on Bioko from which they isolated four different species of SIV, three of them novel. The fourth virus, SIVdrl-Bioko, infected an island mandrill, that is known to have diverged from its mainland counterpart during the 10,000 years of population isolation. The corresponding mainland drill was found to be host to a phylogenetically distinct but closely related SIV. The researchers could therefore conclude the common ancestor of SIVdrl-Bioko and SIVdrl (the contemporary mainland virus) infected the mandrill population at a time before sea levels rose to isolate the island population. With this knowledge in hand the researchers deduced that the nucleotide sequence divergence between the island virus and the mainland virus had taken (at least) 10 millennia. The nucleotide substitution rates were calculated as 10^{-6} substitutions per site per year, a rate three orders of magnitude slower than that originally derived by the computational methods available. The revised phylogenetic analysis of SIV placed the most recent common ancestor 76,000 years before present, a result that supports the ancient origins of primate lentiviruses and the long-term coevolution necessary to establish the existing pathogenic entente with natural simian hosts. The researchers end on a down note: this being true, the expectation that the human-HIV relationship will evolve to become apathogenic in the near future is very unlikely.

Arboviruses: Vector-Borne Viruses

We will temporarily travel into the territory of arboviruses, transmitted by arthropod vectors. Remarkably they make up almost 40 percent of known pathogenic viruses (Rosenberg et al. 2013). We will discuss examples of *Togaviridae* and members of the genus *Alphavirus*. Our chosen ones are not newly discovered pathogenic viruses but they have assumed expanded geographic ranges; in recent decades they have emerged as ever more important mosquito-borne human diseases. In the context of our discussion of the principles that govern virus-host coevolution and restrict cross-species transmission, the alphaviruses are particularly fascinating. They are mosquito-borne RNA viruses with a genome of some 12 kilobases. Their lifestyle requires that they infect and replicate in both their mosquito vectors and animal hosts; they are not simply passengers. The spread of the disease between animal hosts needs the virus to achieve adequate levels of viremia in the blood, such that it can be taken up and transmitted in the blood meal of a mosquito. It must actively replicate, first in the midgut and then the hemocoel and ultimately the salivary gland of the mosquito. It is transmitted when the mosquito bites the next susceptible host. The requirement for the alphavirus to replicate successfully in both a mammalian and insect host is believed to exert particularly stringent requirements on the viral genome and is certainly a singular achievement of viral adaptive evolution. In effect, the virus software must be engineered to operate under two very different operating systems.

Alphaviruses in the Americas cause encephalitis in humans and equids (we will focus on Venezuelan equine encephalitis virus, VEEV), while in the Old World they are mainly diseases associated with rash, arthralgia, and fever. We will turn first to an Old World virus, chikungunya. Its first reported outbreak occurred in 1952 on the Makonde Plateau in the Southern Province of Tanganyika (present-day Tanzania). It was presented to the Royal Society of Tropical Medicine and Hygiene by Marion Robinson of the Lulindi Hospital Universities Mission to central Africa (Robinson 1955). The epidemic was explosive, with 60–80 percent of the population of some villages affected. Robinson reported that joint pain of "frightening severity," rash, and fever were characteristic of the disease. It became known as chikungunya, a Makore word that roughly translates into "that which bends up." Its spread was favored in locales that provided breeding places for mosquitoes and could be attributed to the species *Aedes aegypti*. The virus exists under ordinary circumstances in African and Southeast Asian nonhuman primate hosts as an enzootic infection, one that is prevalent among animals in a particular geographic region. It is during the rainy season when mosquito populations swell that it spills over into rural human communities. It periodically emerges into urban settings where human-to-human transmission is mediated by indigenous mosquitoes. Although in the last century recorded outbreaks of chikungunya virus were typically quite limited, since 2000 extensive outbreaks have been recorded and have caused some 5–10 million infections. The virus now affects not only Africa and Southeast Asia but also Europe and the Americas. The worldwide emergence of the disease has its basis in the evolutionary adaptation of the virus to a different *Aedes* species vector whose geographical range is rapidly expanding.

In 2004 a large outbreak of chikungunya in Kinshasa was the origin of an epidemic that spread to the Indian Ocean Islands. There, genetic variants of chikungunya emerged. A single mutation in the envelope protein of the virus that changed the amino acid alanine to a valine residue (E1-A226V) permitted it to replicate well in *Aedes albopictus*, a different species of mosquito as its vector. *A. albopictus* has a much wider geographical distribution than *A. aegypti* and can be found in more temperate regions. It is currently endemic in Africa, the Middle East, Europe, and the Americas, including the United States, where it most probably arrived as eggs traveling in rainwater trapped in used automobile tires. The means for the new genetic variant of chikungunya to become a global health concern are thus at hand. It is remarkable that a single amino acid change has been so influential in the global spread of the virus and that its emergence has been recorded on at least four separate occasions (Tsetsarkin et al. 2011). Despite the apparent simplicity of this mutational pathway to improved epidemic potential, it has been observed that in some strains of the virus, notably endemic Asian strains, the mutation has never been selected in epidemics. This appears to be an example of another type of constraint that can limit the flexibility of viruses to evolve: some mutations are tolerated and even advantageous in the context of one genotype, but are deleterious in another. Such epistatic interactions are known to strongly influence the available pathways of adaptive evolution. If we relate this to our relief map of the fitness landscape, epistasis contributes to its ruggedness and for some genotypes the routes available for adaptive evolution are more difficult or impossible to traverse without passing through regions of genome inviability. For others, a short high-mountain pass may be crossed easily. Today the Indian Ocean lineages continue to evolve and find higher ground on the fitness landscape. E1-A226V variants are accumulating additional mutations that result in yet greater selective advantages for the virus in the widespread *A. albopictus* mosquito, and we must expect that such variants will emerge as the dominant epidemic virus in the future.

Our second alphavirus, a cousin of chikungunya, is Venezuelan equine encephalitis virus (VEEV). It will also teach us about genetic variants that contribute to creating human epidemics. For the past century, explosive epidemics of VEEV have occurred in the Americas; a recent epidemic in 1995 afflicted 100,000 people. Human epidemics appear to arise as a result of spillover infections of the virus into horses, mules, or donkeys. The disappearance and sporadic recurrence of the disease was an enigma until virologists discovered VEEV was enzootic, circulating in small rodents in forest and swampland using the *Culex* spp. mosquito as its vector. The virus that circulates in these reservoir hosts is typically avirulent to equids, but researchers found that the emergence of each epidemic virus consistently corresponded with the acquisition of the same mutation in their envelope gene (Brault, Powers, and Weaver 2002). These variant viruses were more virulent in equines and circulated at higher levels in their bloodstream, leading to more efficient transmission after a mosquito blood meal. Consequently, the efficiency with which the disease was transmitted to humans, particularly those in close contact with equine species was dramatically increased. Here, a novel mechanism is at work to cause zoonotic human infections; variants with increased virulence in an epizootic host have a selective advantage and are more readily vectored from the amplifying host to humans.

It is notable that the *Aedes taeniorhynchus* species of mosquito is implicated in equine to human transmissions of the virus. This is not the same mosquito species that customarily shuttles the virus between its natural rodent hosts in the swampy forest environment. VEEV appears to have the remarkable capacity to infect both a variety of mammals as well as insect hosts. Nevertheless, as for chikungunya virus, its choice of mosquito vectors seems to be an important factor influencing the emergence of epidemic and epizootic VEEV. A recent isolate of VEEV from an outbreak of encephalitis in horses in Mexico revealed a surprise. While the virus caused substantial viremia in equines, it could not infect *Aedes taeniorhynchus*, the necessary vector for efficient infection of humans. In a creative set of experiments, Brault and colleagues (2004) probed the underlying basis for these observations. They were able to demonstrate that the ability of the virus to infect *A. taeniorhynchus* could be engineered into the Mexican VEEV isolate using sequences from the genome of strains that had previously caused human epidemics. It therefore appears that the emergence of genetic variants that can be transmitted with higher efficiency in a new vector species creates the potential for VEEV to cause human epidemics.

Evolutionary Compromise

Both of our examples illustrate alphaviruses adapting to new hosts and achieving increased virulence. This is despite the fact that they must evolve under strict selective constraints to maintain their chain of transmission, alternating between vertebrate and invertebrate host species. Our understanding of virus-host coevolution suggests that alternation of host species would be detrimental to the optimal adaptation to each host, and, therefore, a fitness trade-off would likely to be seen. Can the consequences of these multiple and perhaps conflicting selective pressures be perceived in alphavirus evolution? It seems the answer is yes. Arthropod-transmitted RNA viruses have been shown to exhibit significantly lower nonsynonymous versus synonymous nucleotide substitution ratios, indicative of a reduced extent of positive selection compared to RNA viruses transmitted by other means (Woelk and Holmes 2002). In elegant experiments that I will refer to as "cutting out the middleman," scientists at the University of Texas Medical Branch Center for Tropical Diseases in

Galveston and their collaborators made a direct assessment of whether the alternation of host in the transmission of an arbovirus might constrain its evolutionary adaptation to each host organism (Coffey et al. 2008).

To test this experimentally, they passaged serially different natural isolates of VEEV, either in rodents or in mosquitoes, or passaged them alternately in rodent then mosquito. Serial passage in rodents was achieved by directly inoculating infected blood from one animal into another. Serial passage in mosquitoes, as you might imagine, was a little more difficult, and entailed feeding mosquitoes artificial blood meals containing the virus. The virus was allowed to grow for ten days in the mosquito at which time it was harvested, mixed with synthetic blood, and fed to the next batch of mosquitoes. Alternate passage of the virus was achieved by allowing cohorts of mosquitoes to feed on infected animals. After ten such passages the viruses from each of the three arms of the experiment were recovered and their virulence compared to the starting virus. The results were remarkably clear: the virus that was passaged alternately in rodent then insect was unchanged in its virulence, while virus passaged only in rodents had become more virulent in rodents and had reduced virulence in mosquitoes. For virus passaged from mosquito to mosquito, its virulence in rodents declined. Although it did not grow to higher titers in mosquitoes, it did appear to be more easily transmitted (that is, more of the mosquitoes that fed on the blood meal became infected). So, in each case of serial passage the virus appeared to become better adapted to the respective host, while the virus that experienced the natural cycle of vector-mediated transmission between two species remained adaptively unchanged.

These results broadly support the assertion that being vector-borne arboviruses comes at the significant cost of a fitness trade-off. The virus must be an evolutionary generalist and, as a result, cannot optimally adapt to either (or any) of its hosts. A mere ten encounters with an exclusive host were enough to select for an increase in virulence of VEEV that could not be achieved in nature. These constraints are likely to operate on all arboviruses: in each successive alternate host one can expect that adaptive genetic variants accumulated in the previous host may be maladaptive and subject to purifying selection. Experience tells us, however, that both discrete genetic changes, as well as ecological changes, can influence the success of emergent variants which can become dangerous human pathogens. Such variants, of course, are themselves evidence that positive selection still has a role to play in these viruses and may influence their evolutionary trajectories in unexpected ways.

Host Restriction

The fundamental chassis of a retrovirus genome is constituted by the essential viral proteins encoded in the *gag*, *pol*, and *env* gene sequences. Lentiviruses' genomes are among the most complex retrovirus genomes, and possess a series of accessory genes encoding proteins whose principal roles are played at the virus-host cell interface; they are involved in evasion of antiviral and immune responses and strongly influence host range (Malim and Emerman 2008). Like *env*, they are rapidly evolving, and perhaps not by coincidence they are clustered together in the genome. It is speculated that this separation of essential, and therefore obligatorily conserved, genes from those that are necessarily rapidly evolving due to their association with immune and antiviral evasion may be advantageous to evolution. In this way, genetic variation in the accessory and envelope genes can be explored by mutation and also recombination without jeopardizing the fundamental machinery of viral replication. This bears analogy to double-stranded DNA viruses, in which the non-core host range genes are placed toward each end of the linear DNA chromosome, creating an evolutionary sandbox for genomic experimentation. HIV-1 and SIVcpz have four accessory genes that have been named *vpu*, *vif*, *vpr*, and *nef*; their simian immunodeficiency virus relative that infects the African green monkey and must have shared a common ancestor is, however, lacking the *vpu* gene. The SIVmac virus that emerged when SIVsmm infected macaques is also lacking *vpu* but has acquired a novel gene termed *vpx* that appears to have arisen by duplication and divergence of the *vpr* gene. It is unknown whether a similar mechanism, perhaps in deeper recesses of evolutionary time, was responsible for the creation of the other lentivirus accessory protein genes. The remarkable diversity of accessory genes belies their shared purpose: they are all designed to bind physically to cellular proteins. They are neither structural proteins nor enzymes, but rather highly adaptable protein ligands, whose job is to counteract the activity of host cell restriction mechanisms and modulate host functions that interfere with viral replication. Perhaps because of the minimalist coding capacity of the genome, these proteins are very small, often fewer than 150 amino acids in length, yet most can bind multiple structurally unrelated cellular proteins. The evolution of accessory protein functions in the context of primate lentivirus evolution and cross-species transmissions reveals them to be highly flexible interfaces with the host cell. They are evolutionary rapid-response systems that adapt quickly to alterations in the host cell defenses. It is almost as if their evolution has assembled them according to a distinct set of structural principles that create malleable chameleon-like proteins, inherently plastic and capable of redesign under altered selective pressure (McCarthy and Johnson 2014). They are available at all times in the viral quasispecies as promiscuous shape-shifting protein ligands, available for repurposing. The extent to which this is true of the accessory genes and not other viral gene products is uncertain; it would seem to be an asset if it could be exploited more widely in viral genomes. Nevertheless, the rapid evolution of accessory genes and portions of the envelope gene place them apart from other viral genes that are under purifying selective pressure to retain their conserved and already well-tuned functions in the viral life cycle.

Let's look at the molecular mechanisms exploited by cell-encoded virus restriction factors and examine how virus proteins are deployed to respond to them. These cellular proteins have evolved to be participants in the first line of a multifaceted defense system that protects cells from viral infections. Cells have evolved a network of pathogen-sensing systems that alert the cell to foreign invaders. These proteins are collectively termed pathogen-recognition receptors (PRRs). Some have evolved to detect the molecular signatures of viral infections, typically the presence of foreign nucleic acids; single-stranded and double-stranded RNAs and double-stranded DNA are all under surveillance. Some PRRs are stationed on the plasma membrane or in the cytosol, others in the membranes of endosomes, allowing them to sample the contents of these membranous vesicles that are a route of entry commonly used by viruses (Rustagi and Gale 2014). Though they comprise a diverse group of proteins, they all respond to their respective stimuli by initiating signaling cascades that trigger interferon responses and other arms of the innate immune response. Viral restriction factors are all type 1 interferon-induced proteins and have evolved to restrict viral infection at a number of levels (Altfeld and Gale 2015; Malim and Bieniasz 2012). Typically, they target processes shared by many viruses or groups of viruses and thus they can be effective against a wide array of viral infections. Viruses that are adaptively evolved to a particular host species have usually evolved counter-defense mechanisms that neutralize host restriction. During virus-host coevolution, much of the genetic arms race plays out in the conflict between restriction factor genes and viral counteraction mechanisms.

We will now examine how primate lentiviruses have evolved to counteract cellular restriction mechanisms. These lentiviruses are particularly interesting since we can consider viruses that have highly evolved or lesser evolved (more recent) relationships with their hosts. SIVagm has coevolved for millennia with Old World monkeys, while HIV, SIVcpz, and SIVmac are the result of recent cross-species transmissions (Sharp and Hahn 2010). It has been empirically observed in the laboratory that HIV can replicate efficiently in human cells in tissue culture but that its replication in nonhuman cells is severely constrained by restriction factors. This is a strong indicator of the importance of cell autonomous restriction on viral host range. Cross-species transmission requires this first line of cellular defenses to be eluded so that the virus can replicate in the new host cell with sufficient success to propagate itself (and continue to evolve) in the new host species. The evasion of viral restriction mechanisms is, therefore, a prerequisite of foremost importance for successful zoonotic transmission. It is quite likely that the existence of a preadapted genetic variant in the infecting virus can be the single pivotal event that separates a successful cross-species transmission from one that is abortive. Lentivirus restriction factors have a variety of mechanisms, but those studied (and these probably represent a fraction of the cell's repertoire of antiviral weapons) often have quite general mechanisms and can therefore target many diverse viral infections.

We will consider three examples of cellular restriction of lentiviruses and the evolution of lentiviruses to neutralize them. The recent cross-species jumps of the lentiviruses under consideration focus our attention on the restriction factor-lentivirus relationship that has evolved most recently, but in fact cellular restriction factors are shaped over eons in arms races with many different retroviruses and other virus families. Lentivirus accessory proteins typically affect host restriction mechanisms via direct physical binding to restriction factor proteins. This allows evolutionary and molecular biologists to deduce the site of protein-protein interaction; these sites of interaction have been under repeated bouts of strong positive selective pressure as they compete to avoid the viral proteins that target them (Emerman and Malik 2010). This is evident in the markedly high levels of nonsynonymous mutations that accumulate in these positions. The three modes of viral restriction we are to consider have radically different mechanisms but all bear these distinctive hallmarks of Red Queen dynamics.

TRIM5α (tripartite-motif-containing 5α) is a host cell restriction factor that is one of a large family of related proteins and is unique in its mechanism of viral restriction. Upregulated after the interferon response is triggered, it binds directly to retroviral capsid proteins, interfering with capsid disassembly and preventing reverse transcription of the genome (Stremlau et al. 2006; Malim and Bieniasz 2012). A carboxyl terminal domain of the protein is highly variable in sequence between species and displays a high ratio of nonsynonymous versus synonymous nucleotide substitutions when the genes in different species are compared. These signatures betray the rapid evolution of this region of the protein under positive selective pressures. It assembles into multimeric complexes that bind capsid in a multivalent fashion. This multivalent binding provides avidity to binding interactions and may facilitate TRIM5a activity despite the modest intrinsic binding affinity of monomers. This feature perhaps also allows it to readily evolve alternative specificities when the species is infected by a new retrovirus. It is noteworthy that while the other viral restriction factors I will discuss are in genetic conflict with retrovirus accessory proteins, TRIM5α is not. Susceptibility of a retrovirus to restriction by TRIM5α is directly attributable to the sequence of the viral capsid. TRIM5α restriction is a formidable barrier to cross-species transmission of retroviruses that have capsids bound by the restriction factor. Viral host coevolution typically results in the selection of capsid proteins that elude binding of the host restriction factor allowing ordered disassembly of the capsidated viral genome in the infected cell.

The HIV accessory protein vif is essential for growth of HIV virus in human $CD4^+$ T cells in the laboratory. It was discovered that its role was to neutralize a virus restriction factor expressed in human cells (Sheehy et al. 2002). The factor involved was ABOBEC3G (apolipoprotein B mRNA editing enzyme, catalytic polypeptide-like 3G), a protein originally discovered to be responsible for cytidine (C) to uridine (U) editing of messenger RNA. It has sequence homology to prokaryotic cytidine deaminase, but appears to have evolved the ability to bind and edit cytidine residues in nucleotide polymers. It exerts viral restriction by becoming incorporated into virus particles and being associated with the reverse transcription complex in infected cells, where it causes deamination of cytidines in the negative-stranded DNA that is then converted into the provirus. The net result is that guanosines are substituted by adenosine in the viral RNA genome: the restriction factor is mutagenic and can edit one in ten cytidines in the genome. The HIV vif protein neutralizes the restriction by binding to APOBEC3G and directing its proteolytic degradation by proteasomes, that are the major cellular pathway of protein turnover and homeostasis. An amino acid subdomain in the vif protein interacts with ABOBEC3G in a species-specific fashion. HIV-1 vif binds and neutralizes human APOBEC3G but cannot neutralize the homologous protein of the Old World primate African green monkey. Conversely, vif of SIVagm can neutralize the simian APOBEC3G, a difference attributable to a single amino acid change at position 128 which defines the binding interface between vif and APOBEC3G protein (Malim and Bieniasz 2012).

ABOBECs, being cytidine deaminases with the potential for G to A and C to T mutations in DNA are potentially a source of genetic variants. Evidence is accumulating that despite antagonism of ABOBECs by the HIV-1 vif protein in infected human cells, there is a substantial residual mutagenic effects on the viral genome during infection (Sadler et al. 2010; Kim et al. 2010). It appears this viral restriction mechanism has the potential to accelerate adaptive evolution of viruses through sublethal mutagenic effects on the viral genome. Notwithstanding these observations with the human adapted HIV-1, the potency of ABOBEC's viral restriction activity was recently laid bare in research described in the laboratory of Dr. John Coffin. They were exploring the pathogenic potential of XMRV, a novel virus isolated from human cells (xenotropic murine leukemia virus-related virus). The virus was under investigation in many laboratories for its reported link to prostate cancer and myalgic encephalomyelitis/chronic fatigue syndrome. These observations were ultimately debunked and the bona fides of the virus were discredited. It was merely a recombinant virus lineage that arose in tissue culture and was not a naturally evolved virus lineage. Notwithstanding this lamentable scientific sideshow, reviewed elsewhere in detail (Delviks-Frankenberry et al. 2012), the Coffin laboratory experiments stand on their own and provided biologically informative data (Del Prete et al. 2012). They grew the virus in tissue culture and inoculated pigtailed macaques seeking to determine its pathogenic potential. The virus failed to establish a vigorous infection and no chronic virus infection was evident. Failing to find significant levels of viral RNA, the team examined cell-associated DNA, looking for XMRV proviral DNA. They found the expected proviruses, characteristic of retroviral infection, but all showed evidence of hypermutation of G to A. The macaque APOBEC-mediated restriction had systematically inactivated the foreign virus genome. Virus restriction in action!

Our final example of restriction mechanisms affecting primate lentiviruses is equally curious. It is a protein called tetherin that exerts broadly acting restriction of enveloped viruses. It is a uniquely adapted protein that literally tethers budding viruses to the cell membrane, preventing their escape from the host cell (Perez-Caballero et al. 2009). What is most instructive for us is to understand how primate lentiviruses have evolved to neutralize its activity. Remarkably, in a vivid illustration of evolutionary dexterity, different retroviruses employ different viral genes to antagonize tetherin's activity. It is neutralized by direct binding of a viral protein; the African green monkey tetherin is targeted by the protein produced from the *nef* gene, as is the tetherin of chimpanzees, but the SIV nef protein is unable to bind and neutralize human tetherin. HIV-1 possesses the novel accessory protein gene *vpu* (originally evolved in SIVcpz) and it is this protein that has evolved to target tetherin in HIV-1 infections (Malim and Emerman 2008). The human tetherin protein clearly represents a novel structural challenge that simian lentiviruses cannot address by selection of variants of the SIV *nef* gene. In lieu of this feasibility, HIV-1 has evolved an alternative ligand in vpu. It is noteworthy that the zoonotic HIV-2 virus descended from SIVsm of sooty mangabeys, evolved yet another distinct solution to the problem of human tetherin. In HIV-2 the envelope glycoprotein has assumed the role of tetherin antagonism (Le Tortorec and Neil 2009; Bour and Strebel 1996). It seems that the possession of multiple accessory genes and genes with the capacity to evolve rapidly and exhibit considerable structural plasticity is a great asset to primate lentiviruses as they successfully coevolve with new host species despite the natural resistance posed by host restriction.

Our species barrier is owed in large part to human-specific host restriction mechanisms, key elements of our species-specific innate immune response. The examples we have discussed have considered species barriers to cross-species transmission of primate lentiviruses. Despite the phylogenetic relatedness of the viruses and the hosts in question, the host restriction mechanisms diverge remarkably in a short space of evolutionary time, a hallmark of the strong positive and diverging selective pressures at play in virus-pathogen genetic conflicts. Host restriction is our best defense against the emergence of new pandemic pathogens, but some viruses do successfully bridge that divide with devastating consequences, and as we shall see, a host of them lurk outside our walls replicating benignly in their natural reservoir species. They are patiently testing genetic variants that might coincide with the necessary circumstances to create both the means and the opportunity for them to circumvent host restriction and become a human virus.

· 11 ·

FUTURE PANDEMIC INFLUENZA: ENEMY AT THE GATES

Our examination of virus evolution has by necessity focused on trying to understand and explain past events. Even looking backward, the evolutionary virologist is forced to make assumptions: save for the fossil record of dateable and morphologically informative remains of plants and animals (but lacking for viruses), there are few data points in evolutionary history to hang our hypotheses on. Nevertheless, we are driven to extrapolate our acquired knowledge of virus evolution to predict the future. Is this an exercise in futility? Threats by definition refer to potential future outcomes, and there is, of course, an overwhelming need for human society to assess and understand the risks that viruses pose to human health in the years ahead. We rely on those we have elected to govern, and the public health agencies in their service, to gauge the steps that can be taken to mitigate the consequences of newly emergent pathogens. These may be in public health surveillance (and early detection), the prospective development of vaccines and drugs, or even in changes to human behavior through dissemination of information, establishing regulations or enacting laws.

At a Pentagon news conference in 2009, U.S. Secretary of Defense Donald Rumsfeld laid out a thought paradigm he felt was applicable to deliberations on U.S. military strategy in the Middle East: "There are known knowns. These are things we know that we know. There are known unknowns. That is to say, there are things that we know we don't know. But there are also unknown unknowns. There are things we don't know we don't know." I contend that the *known unknowns* concerning the future of viruses and their relationship with their host species far outweigh the *known knowns*. Furthermore, recognizing the unpredictable genius of viral evolution, we must also confront the more haunting concept: the *unknown unknowns*. These factors cannot be quantitated and by their nature are intangible and unpredictable and cannot be assessed at this time. Newly emergent viruses that turn up at an accelerating pace fall into this category (Jones et al. 2008; Woolhouse et al. 2008; Kuiken et al. 2003; Woolhouse, Scott, and Hudson 2012). Absent evidence of their existence, they do not weigh into our risk assessment.

Let me lay the groundwork for a discussion on the future of our relationship with viruses by expanding our earlier analysis of influenza virus. The threat of pandemic influenza is ever present. Like earthquakes, we can be sure they will occur, but when and how severe they will be is anyone's guess. Historically, over the last century or so we have recorded a flu pandemic every ten to forty years. Indeed, known knowns include earthquakes and the certainty of another pandemic of influenza in the future. The shape that that pandemic will take, however, the severity of the disease it will cause, and the threat it will pose to global human health reside firmly in the territory of the known unknowns. Seasonal epidemic human influenzas and past pandemic viruses have all taken their antigenic identities from a small subset of the genetic diversity of hemagglutinin and neuraminidase genes theoretically available to influenza viruses. They are limited to three HA subtypes (H1, 2, 3) and two NA subtypes (N1, 2). The combinations of these envelope proteins used by the virus represents a great diversity of antigenic properties. Each of the HA and NA subtypes themselves comprise substantial diversity, being the products of individual and unique evolutionary histories. Circulating human influenza viruses employ HA and NA genes acquired by reassortment between viruses that have circulated and moved between humans, wild birds, poultry, and often pigs, and have consequently been subject to a variety of distinct selective pressures. Moreover, this handful of HA and NA subtypes represents only a fraction of the available genetic diversity of the HA and NA subtypes that circulate in avian species. Importantly, we should not ignore the fact that the particular constellation of eight virus gene segments making up an influenza virus "gene team" defines the properties of the virus and its pandemic potential. It is reasonable to assume then that the potential of the influenza virus metagenome continues to evolve; it has not been exhaustively explored, neither by reassortment of gene segments between viruses nor by the evolution of the resulting viruses by mutation and natural selection.

Real and Present Danger

In a review appearing in the *Science*, Richard Webby and Robert Webster reflected on events early in 2003. An outbreak of a severe respiratory disease in Hong Kong was attributable to human infections caused by H5N1 avian influenza A. It led the WHO to immediately declare pandemic alert status. Shortly thereafter, a rapidly spreading flu-like respiratory illness emerged on the mainland of China, leading all to fear that the H5N1 virus was spreading epidemically among humans. It was not; in fact the outbreak marked the emergence of the widespread epidemic of SARS, an unrelated and hitherto unknown human virus—an unknown unknown. This recollection serves to remind us that in all likelihood the greatest imminent threats to human health are highly pathogenic and transmissible viral infections. Moreover, although new zoonotic viruses will challenge us on occasion (e.g., the SARS outbreak), the emergence of a pandemic human influenza virus is justifiably our greatest preoccupation. In the past two decades, we recognized highly pathogenic avian influenza viruses as the most likely culprits of a future pandemic. Initially emerging in 1997 in Guangdong Province in China, avian H5N1 influenza viruses caused devastating viral pneumonia, leading to death in 30 percent of the individuals infected (Xu et al. 1999). The H5N1 virus emerged again in 2003 and 2004, causing widespread outbreaks in poultry across East Asia and infecting humans in several countries. At this point, it became evident that migratory wild birds were spreading the virus and it was endemic in domestic poultry throughout the region (Li et al. 2004). It thus gained an important foothold from which to potentially launch genetic variants with pandemic potential into the human population. The H5N1 virus appeared to take full advantage of the available talent to draft its gene team: it co-circulated in birds with influenza viruses of distinct genetic makeup and acquired many of its internal gene segments from an H7N9 virus circulating in quail (Guan et al. 1999). Even more worrisome, human influenza H3N2 was now endemic in pigs throughout China, raising the specter of further genetic intermixing of H5N1 with these viruses should it invade the pig population. Disquiet looms even larger today. Other new reassortant avian influenza viruses—notably H7N9, H9N2, and H10N8—are infecting poultry or pigs (and sporadically humans) across Asia and throughout the world.

The primary mode of transmission of avian influenza viruses to humans appears to be via infected poultry. The imposition of certain controls and measures such as monthly "clean days" on wet markets across China has certainly helped reduce the potential for a pandemic outbreak of an avian influenza strain. Nevertheless, the most significant impediment to the virus seems to be one of genetics.

The H5N1 avian influenza transmitted to humans in 1997 was not a reassortant between a human virus and avian influenza virus. It was of avian origin and particularly virulent in humans. A significant majority of patients suffered particularly severe classical influenza symptoms, but a few patients betrayed evidence of more systemic involvement of the infection outside the respiratory tract, to which the virus is typically restricted in humans (Horimoto and Kawaoka 2001). It has long been recognized that avian influenza viruses display different degrees of pathogenicity in birds. Indeed in turkeys, the most commonly afflicted species of poultry, virulent viruses cause fowl plague, a systemic and fulminant multiorgan disease. Avirulent avian influenza viruses, however, lead to only mild or asymptomatic infections, restricted to the respiratory and gastrointestinal tracts. A significant body of research has now documented that highly pathogenic avian influenza viruses carry variant HA genes. This variation is, however, not based in differential antigenicity of the protein that can signal the potential for a virus to become a pandemic influenza. It is a functional difference relating to viral cell tropism and the role of HA in mediating virus entry into the host cell.

The HA proteins of influenza viruses are produced as a single polypeptide, HA0, that is posttranslationally processed by proteolytic cleavage into two subunits, HA1 and HA2, which undergo conformational changes but remain associated. The cleavage releases the amino terminus of HA2 that possesses a hydrophobic string of amino acids termed the *fusion peptide*. The exposed fusion peptide penetrates the endosomal membrane within which the virus has been taken into the cell, mediating fusion with the viral envelope and entry of the virus nucleocapsid into the cytoplasm. Cleavage of HA by host cell proteases is essential for virus infectivity, and it is believed that the necessary proteases are restricted to the respiratory and gastrointestinal tract of birds and human. This explains why avirulent avian influenza virus replication is typically restricted to these tissues. But what of the highly pathogenic viruses that can infect other tissues and cause systemic disease? The explanation resides in the HA1-HA2 junction amino acid sequence. Avirulent avian influenza viruses have a single basic amino acid residue at the amino terminus of HA2, an arginine or a lysine. This site is recognized and cleaved by serine family proteases secreted by cells in the respiratory and gastrointestinal tract (Horimoto and Kawaoka 2005, 2001). Highly pathogenic avian influenza viruses have cleavage sites composed of multiple basic amino acid residues that are recognized and cleaved by ubiquitous proteases present in many if not all tissues. Often, human influenza virus and avian influenza virus strains are grown in cultured canine kidney cells. Here, it is routinely necessary to supplement the culture medium with a protease, typically trypsin, to accomplish the processing of HA and release the infectious potential of the virus particles. These "nonnatural" host cells do not produce the necessary protease to do the job. Highly pathogenic avian influenza virus variants with a stretch of multiple basic amino acids at the cleavage site grow perfectly well in the absence of any added protease. Both H5 and H7 HA proteins can have cleavage site sequences recognized by ubiquitous proteases, rendering the viruses highly pathogenic. Suarez and colleagues analyzed the nucleotide sequence of H5N1 viruses isolated from the 1997 outbreak in China. The viruses that caused the spectrum of severe respiratory and systemic infections were confirmed to possess a stretch of basic amino acids at the HA1-HA2 cleavage site, the hallmark of highly pathogenic avian influenza viruses (Suarez et al. 1998).

It should be stressed that highly pathogenic avian influenza virus variants with virulence in poultry (and in man) are not themselves considered capable of causing a human pandemic. It should also be noted that pandemic influenza viruses need not possess an HA protein with a stretch of multiple basic amino acids at the HA cleavage site. The 1918 H1N1 pandemic influenza virus is an excellent example: it is believed that it is of avian origin (Chapter 5) but its HA protein has a single basic residue at the border of HA1 and HA2. Despite its high virulence in man, its replication was restricted to the respiratory tract (Chaipan et al. 2009).

It seems that viruses such as H5N1 and H7N9 circulating in poultry are not readily transmitted among humans. They cause sporadic human infections, hence they have the opportunity, but as yet they have not acquired the means to become epidemic human pathogens. Experts are divided on the potential for these influenza viruses to become true pandemic threats that might equal the 1918 influenza outbreak in severity. Some argue that if such a virus has not yet emerged, it is unlikely to do so in the future; others argue that we should not be so sanguine. The genetic changes that highly pathogenic avian influenza viruses must acquire in order to achieve a genuine capacity for epidemic transmission among humans remain obscure and certainly warrant exploration. It is a topic of unusual scientific complexity and it has triggered a firestorm of political controversy that divides the research community and brings science and politics into collision.

Pandemic Threat Level

The avian H5N1 influenza virus has to date caused only limited outbreaks among humans. Although highly virulent in an infected individual, triggering severe lower respiratory tract illness and often death, it is not easily transmitted from person to person. Most infections meet a dead end, which is often the outcome of interspecies jumps. The deficit in H5N1 that results in poor transmission between humans can be explained. Avian influenza is an enteric virus, replicating in the gut of birds where the susceptible cells of the intestinal mucosae express on their surfaces abundant cellular viral receptor molecules. On gut cells these are constituted of abundant oligosaccharides that have a terminal sialic acid residue linked via an α-2,3 linkage to galactose (I will refer to these as α-2,3 receptors). This particular moiety is bound preferentially by the HA protein of the avian virus, allowing it to recognize and enter host cells in the gut. Respiratory droplet transmission of influenza virus in human requires that infectious virus reaching the oropharyngeal cavity of the new host finds susceptible cells with appropriate receptors for attachment and infection. In humans, however, the ciliated epithelium that lines the upper respiratory tract does not display α-2,3 receptors; those particular receptors are abundant only in the lower airways. Rather, in the human upper respiratory tract oligosaccharides with α-2,6 galactose linked sialic residues are most prevalent (α-2,6 receptors). It appears that the scarcity of susceptible cells with α-2,3 receptors in the upper airways limits respiratory transmission of avian viruses between humans. Person-to-person transmission of avian H5N1 influenza depends on infectious virus gaining entry to the lower airways where susceptible cells with α-2,3 receptors are abundant (Cauldwell et al. 2014). This necessitates intimate and extended exposure of the individual to the virus and evidently occurs only rarely. Human influenza A viruses, on the other hand, are adapted to recognize α-2,6 receptors prevalent in the upper respiratory tract and therefore are readily transmitted by the aerosol route.

This raises the obvious question: Why has the avian H5N1 virus not acquired the necessary mutational changes to become human-adapted and easily transmitted between humans? What is the secret of the limited repertoire of HA and NA subtypes that have been uniquely successful in viruses that have adapted to human transmission? There is no simple answer, and it remains elusive, a known unknown, but it is worth exploring here because it is central to assessing the relative risk of an emerging pandemic avian influenza virus. Many researchers around the world have attempted to get to the heart of this problem.

The Pandemic Phenotype

Despite the fact that avian H5N1 viruses infect bird and poultry populations around the world, they have gained no significant foothold in human or mammalian populations. According to the WHO Global Influenza Program, the virus caused less than forty human infections in 2013 (WHO 2015d). It is not easily communicated between persons. If a simple mutational change or reassortment would suffice to create a readily transmissible variant with pandemic potential, it would most likely already have emerged. It seems that H5N1 is genetically distant from a genotype with pandemic potential—getting from here to there must represent a substantial evolutionary challenge. Given that global preparedness for future emergent pandemic threats is based on surveillance of circulating virus lineages, it is reasonable to ask whether there are genetic signatures that we should be on the lookout for. What might signal that a dangerous constellation of genetic changes is accumulating in circulating H5N1 viruses, such that they pose a significantly increased risk for the acquisition of the "pandemic phenotype"? With no understanding of the genetic changes that can confer this dangerous phenotype, surveillance can offer no benefit. There can be no assessment of increased threat level, except after the fact: after the pandemic virus has been minted and is making its way ruthlessly through our global community.

Seminal work on the lethally virulent 1918 H1N1 pandemic influenza virus has confirmed that its very particular phenotype is defined by the totality of its gene complement (Tumpey et al. 2005). Using the nucleotide sequence of viral RNA recovered from a victim of the 1918 flu, whose corpse had lain buried in the Alaskan permafrost, allowed Tumpey and colleagues to reconstruct in the laboratory the same virus genotype that proved so devastating almost a century earlier. The investigators made multiple reassortant viruses by mixing and matching different combinations of the 1918 H1N1 virus gene segments with those from a recent H1N1 virus of lower human pathogenicity. They assessed viral virulence by infecting mice. They found that the 1918 H1N1 virus HA and polymerase gene segments conferred substantially higher virulence on the contemporary H1N1 virus genotype (a result that confirmed their earlier work). Nevertheless, a virus with all eight of the 1918 virus gene segments was by far the most virulent human virus that they had ever assessed in their laboratory. This was experimental evidence that the 1918 pandemic virus had indeed assembled a true gene dream team. No individual star players were sufficient to explain its pathogenicity; it was the product of recruitment and training together as a team. Reassortment had brought together the gene segments, but their coevolution under selective pressure acting on the entire genotype expressed in concert must have refined the phenotype.

Together with the drafting of gene segments with appropriate properties, the development of subtle adaptive and complex epistatic interactions between gene segments is central to creating an influenza virus with optimal fitness that can succeed as a highly pathogenic pandemic virus akin to the 1918 H1N1 strain. This notwithstanding, a pandemic avian H5N1 virus must acquire the ability to bind and infect cells in the human upper respiratory tract. Several groups of investigators explored the potential of the H5N1 genotype to adapt for transmission by the respiratory route in humans. The availability of many HA gene sequences from both human-adapted and avian influenza isolates has allowed investigators to understand precisely what amino acid changes will effect a switch in HA receptor preference from avian-type to human-type receptors in the upper respiratory tract. Avian viruses engineered to contain these mutations did indeed exhibit altered receptor preference and bound to the α-2,6 receptor expressed in the upper respiratory tract in humans. Nevertheless, these viruses replicated poorly and showed no increase in transmission compared to the original, unmodified parent virus (Maines et al. 2006; Maines et al. 2011). A breakthrough of sorts came in 2012 in the form of results from what would come to be called "gain-of-function" experiments, which caused a great deal of controversy. Two independent groups of scientists sought to adapt H5N1 virus to airborne droplet transmission between ferrets housed in neighboring cages (Herfst et al. 2012; Imai et al. 2012). Herfst and colleagues engineered an H5N1 virus to contain several mutations in HA sufficient to switch its specificity toward binding of α-2,6 receptors prevalent in the upper respiratory tract of ferrets (and humans). They also introduced a mutation into the viral polymerase gene, PB2, that confers increased virulence to avian viruses in human cells. The resulting virus caused disease in ferrets but failed to infect animals in neighboring cages through airborne droplet transmission. Employing a classic technique used by virologists, the investigators took the virus shed in nasal washings from the ferrets and infected new ferrets. This was repeated several times in what is termed a "serial passaging" experiment. Soon the virus acquired the capacity to be transmitted between ferrets in different cages. They had created an aerosol transmissible H5N1 virus. The virus acquired additional adaptive mutations during its replication in the serially infected the ferrets. These variants were those transmitted between cages. But was this new genotype representative of a virus with pandemic potential?

The respective academic institutions made celebratory press releases, but the international press saw these accomplishments in a different light. They pulled no punches and hailed the results with horror; the public were led to believe that scientists were creating pandemic killer viruses. Soon afterward an article in the *Independent* newspaper on December 6, 2013, proclaimed, "Experts warn research into H5N1 bird-flu virus could lead to deadly pandemic." A group of fifty-six eminent scientists came out against their colleagues' experiments, warning in an open letter that the benefits of the research in their opinion did not outweigh the danger of accidental release of such a pandemic pathogen. The scientific world was thus divided; it had run afoul of its own hubris, fear and politics fueled by inflammatory coverage in the press. Despite the scientific importance of the gain-of-function experiments and the critical need to understand human-pathogen interactions, many took the side of the fifty-six scientists, deeming the investigations too dangerous to undertake given what they considered to be their dubious value. In October of 2014, after being apprised of a series of unrelated biosecurity gaffes in U.S. government research facilities, the White House released a policy statement entitled "U.S. Government Gain-of-Function Deliberative Process and Research Funding Pause on Selected Gain-of-Function Research Involving Influenza, MERS, and SARS Viruses" (DHHS 2014). The research was to be starved of funding in the United States; further progress was stalled. But how far had they come in understanding what might make H5N1 a more serious threat to human health?

Absent from the hoopla following the publication of the key scientific papers was any mention that while airborne droplet transmission of mutant H5N1 viruses had been achieved, the modified virus suffered from vastly diminished virulence compared to the parental virus. What went unmentioned in the press and unnoticed in the abstracts of each of the peer-reviewed articles was the explicit statement that the virus exhibiting airborne transmission between ferrets had substantially reduced virulence. In fact, the virus caused only mild signs of disease and did not kill any of the animals infected by the airborne route. Lethality was only observed when the high titers of the variant virus were artificially inoculated into the ferret trachea, an unnatural route of infection. The virus was attenuated. It had evolved by "robbing Peter to pay Paul"; the selection for airborne transmission had resulted in a gene constellation with reduced overall fitness. It had lost its potential to cause severe disease. There was a high expectation that the phenotype of the virus in ferrets would resemble that in humans, therefore it is more than likely that the virus would pose no serious threat to humans: the researchers had in fact created a dud.

This was far from the smoking gun that reveals H5N1 is but a few mutations away from acquiring true pandemic potential. Far from it, the genetic distance between this H5N1 genotype and a virus with true pandemic potential remains uncertain. It is hubristic to believe that we can predict the particular amino acid changes, their order of accumulation, or sequence of random (by definition) mutations in the influenza virus genome that might pave the way toward the pandemic phenotype. It may require one, more likely many, mutations together with reassortment of gene segments with different influenza virus strains to realize the necessary evolutionary change. In short, a complex and convoluted process requiring multiple serendipitous genetic changes, the probability of which may be negligible, but of course cannot be ignored. The genotype of avian H5N1 is clearly a substantial genetic distance from that of a virus with a human-adapted pandemic phenotype. The virus lineage is undergoing a cross-species transmission and it must overcome the species barrier between bird and man. Thinking in terms of a fitness landscape, we might envisage the H5N1 virus genotype residing in a remote genotypic space in the landscape, distantly removed from the high ground of optimal fitness. It may have no available mutational pathway that leads to high ground and adequate fitness in the human species, without passing through a fitness nadir of inviability where it is either poorly transmitted or of inadequate virulence to sustain epidemic spread. On the other hand, given the diversity of the influenza metagenome in the avian species, it should not be denied that the circulation of H5N1 places us in real and present danger of the emergence of a pandemic strain. The probability of such an event occurring is a known unknown, but it is not zero.

The emergence of swine-origin H1N1 influenza virus was the cause of the most recent influenza pandemic in 2009. Recent work has demonstrated a high degree of genetic compatibility between that emergent virus and the avian H5N1 influenza strains that are endemic in East Asian poultry. Viable and virulent genetic reassortant influenza viruses arise when gene segments are exchanged between the two strains in the laboratory (Schrauwen et al. 2013). Perhaps we should be concerned that these naturally circulating viruses may have the opportunity to recombine in nature, establishing a new evolutionary trajectory for H5N1 avian variants. But then, of course, there are also the unknown unknowns.

Outbreak

In early 2014 Ebola virus emerged from the forest in the Democratic Republic of the Congo (DRC). Sixty-six patients contracted Ebola hemorrhagic fever; forty-nine died, a mortality rate of 76 percent (CDC). The genus *Ebolavirus* has five species of virus: Zaire ebolavirus, Bundibugyo ebolavirus, Sudan ebolavirus, Taï Forest ebolavirus, and Reston ebolavirus. With the exception of the Reston ebolavirus, all are pathogenic in man. This was an outbreak of the Zaire strain (EBOV), the most feared and virulent of the Ebolaviruses that spill over into human populations in Africa. The EBOV has emerged in DRC on several previous occasions, as recently as 2007 and 2008. Its repeated emergence there led scientists to reason that it must be endemic or was spreading epidemically in wild animal populations in the region. This particular species of the virus caused the first recognized human Ebola virus outbreak in 1976 in Yambuku, Zaire (now DRC), and it had been one of the worst to date. Hindered by ignorance of the nature of the disease (which had never been seen before) the virus spread rapidly by person-to-person contact and the use of contaminated needles and syringes in the nearby clinic: almost 9 out of 10 of 318 infected persons succumbed to the Ebola hemorrhagic fever. Remarkably, we have now become somewhat inured to periodic outbreaks of the virus. Lessons learned during the early outbreaks in Zaire, Sudan, Gabon, and the Ivory Coast seem to have prepared local communities to be alert to the disease. If the necessary measures are put in place promptly, outbreaks are typically contained. Despite the highly transmissible nature of Ebola virus (it is spread through contact with the bodily fluids of symptomatic individuals or with contaminated fomites), each of the outbreaks, horrific though they are, have constituted medical emergencies of only local dimension. During the recorded outbreaks, the zoonotic viral strain of Ebola has never been allowed to achieve a degree of sustained person-to-person transmission that might fuel adaptive evolution to the human host. They have been dead-end infections, and the emerging viral lineages were lost when the outbreaks were extinguished. The outbreak of EBOV in DRC in 2014 followed the same pattern and was quickly extinguished. Elsewhere in Africa, in Guinea, another outbreak of EBOV was taking place, but it unfolded very differently. EBOV Makona ignited an epidemic of devastating proportions, reawakening society to the potential global consequences of emerging zoonotic viral diseases on remote continents.

· 12 ·

EBOLAVIRUS

The EBOV Makona outbreak, tragic though it was, provided a unique opportunity for scientists to accumulate empirical observations of Ebolavirus biology and evolution. Hard data could now replace mere speculation. We will begin by laying the groundwork and discussing the family *Filoviridae* that is made up of three genera: *Ebolavirus*, *Marburgvirus*, and *Cuevavirus*. Among these genera and among the five recognized Ebolavirus species, there is a wide spectrum of pathogenicity for humans and nonhuman primates. The storied Reston ebolavirus achieved repute after it caused a deadly outbreak of Ebola disease in monkeys housed in a U.S. animal facility. The fear of a deadly hemorrhagic fever virus threatening to emerge in the American heartland was tangible but ill-founded: the virus was not pathogenic in humans. The outbreaks that have come to our attention in central Africa are all associated with severe, often fatal disease in humans and in primates. Human outbreaks have been linked to wildlife die-offs (Leroy, Rouquet, et al. 2004). The outbreaks are so feared because of the ruthless efficiency with which the virus replicates and spreads systemically, apparently unchecked, resulting in renegade inflammatory responses causing the filovirus-typical disease symptoms: extremely high fevers, vascular leakage, and bleeding disorders. It is ironic that the rapidity with which the infection takes down its victims and the horrific symptomology of the disease are the very factors that have restricted its capacity to establish extended chains of transmission in man. It is simply too pathogenic to cause a pandemic. Direct physical contact with body fluids of the infected victim is required for transmission, so the combination of effective quarantine measures, contact tracing, and rigorous deployment of protective gear to health care workers has routinely brought a halt to outbreaks. Such measures would not be as successful as they are if the virus were to be transmitted via an airborne route, by a cough or sneeze. Such is the scenario for the next pandemic influenza virus or other respiratory diseases such as SARS that can be transmitted via the respiratory route in small particle aerosols and spread globally in a matter of days or weeks.

Highly pathogenic filoviruses appear to owe their success (at least regarding disease-causing potential in humans) to a first-rate program of immune evasion and a proclivity for infecting a broad swathe of cells and tissue types (Zampieri, Sullivan, and Nabel 2007). The natural systems of antiviral defenses that our bodies and individual cells use to detect and respond to viral infections appear to be virtually sidelined by pathogenic filoviruses, leaving the body helpless in their sway. The virus has evolved mechanisms that prevent the cell's pathogen recognition receptors from sensing and responding to foreign viral ribonucleic acid in the cytoplasm of the cell. Absent these cellular early warning systems, they are unable to mobilize the critical interferon response to viral infection. Infected cells also include both $CD4^+$ and $CD8^+$ lymphocytes, the very cells that are fundamental to the generation of antibodies and cell-mediated immune responses. Filoviruses also appear to posttranscriptionally edit their mRNA transcripts in a manner commonly seen in other RNA viruses of the order *Mononegavirales*. In other words, the information encoded in the messenger RNA can differ from that encoded in the gene (Volchkov et al. 2001). This mechanism allows the virus to make multiple different forms of its envelope glycoprotein from a single gene; one form is not tethered to the virion envelope, but floats free. This soluble form of the glycoprotein (sGP) is thought to act as a decoy, drawing antibodies away from virions and infected cells. The free-floating protein further blunts our capacity to control the infection; it displays unusual epitopes that direct the immune response to make non-neutralizing antibodies that have no effect on the infectivity of the virus itself. The virus infects macrophages and dendritic cells that infiltrate tissues and organs of the body, disseminating the virus widely. The cytolysis of dendritic cells also interferes with the innate antiviral response and precludes the presentation of viral antigens to the adaptive arm of the immune system. Such an armamentarium permits the virus to replicate to extremely high titers in tissues, and bodily fluids of infected patients and direct contact with such fluids provide an extremely efficient vector of transmission.

Small-particle aerosol transmission of Ebola virus appears to be rare or impossible. Perhaps the airborne virus particles are unstable, or the infectious inoculum that can be vectored in small particle aerosols is inadequate to successfully establish an infection in the new host. Alternatively, the oropharynx and lung may not be optimal portals of infection. Nevertheless, it is salient to observe that the Reston ebolavirus can be transmitted experimentally by the oral-nasal route, and respiratory transmission of the virus to human laboratory workers in Reston, perhaps by small-particle aerosols, could not be absolutely ruled out (Marsh, Haining, and Robinson 2011; WHO 2009). The genetic diversity that can be explored by the genus Ebolavirus, therefore, may be sufficient to embrace both a broad spectrum of virulence in humans and perhaps differential potential for transmission by the respiratory route between nonhuman primates (Miranda and Miranda 2011). The genetic differences that underpin these distinctive phenotypes remain a mystery. It is unclear what genetic distance separates an Ebola species lineage that can be transmitted readily in small-particle aerosols from one that requires relatively intimate contact with much larger amounts of bodily fluid to be infectious (and one can only speculate as to whether it would retain high levels of pathogenicity).

The singular phenotypic difference distinguishing a highly pathogenic Ebolavirus species (such as EBOV) from one that is benign (the Reston ebolavirus) in humans is clearly evident. It almost certainly originates in differences at the virus-host interface that modulate the ability of the viruses to corrupt our innate and adaptive immune responses. Although divergent in nucleotide sequence the precise genetic differences that encode the differential pathogenic qualities of these Ebola virus lineages remain mysterious, obscured in complexity and different evolutionary histories. It is not surprising that these different virus visitors to the human species exhibit varying degrees of virulence, having evolved independently over long periods, perhaps in different species of reservoir hosts.

EBOV Makona

Notwithstanding our generally successful efforts to stymie cross-species Ebola jumps, almost 2,500 individuals were infected and more than 1,000 succumbed to the disease between 1976 when it first appeared and 2013. In human terms, these are not insignificant numbers but they betray no hint that Ebola virus might cause an epidemic of global proportions or that, like HIV today, it might become endemic in human populations. Reminiscent of zoonotic SIVcpz, which seeded the HIV pandemic in humans, the cross-species transmission of Ebola virus is most frequently a result of direct contact with infected primates (or primate carcasses). This was reported to be the case for the 2014 outbreak in the DRC and reassuringly, following the familiar pattern of past Ebola outbreaks, it was quickly contained. The other contemporaneous outbreak in Guinea was unusual from the outset. With the exception of a single isolated infection, all previous outbreaks of Ebola were in central Africa. That one case was in Ivory Coast and was caused by a novel subtype of Ebolavirus (Taï Forest ebolavirus) contracted from an infected chimpanzee and the virus has never reemerged. It was an unwelcome surprise to discover that EBOV was circulating in animal reservoir species in Guinea so far removed from the sub-Saharan equatorial region of central Africa. The most disturbing development was the unprecedented scale of the outbreak, challenging the firmly held conviction that Ebola would never become a global phenomenon and would never successfully evolve into an endemic human disease.

Starting as an unrecognized communicable disease in a remote area of Guinea, it took some time for the outbreak to catch the attention of the international community in March 2014; it undoubtedly began in December 2013. Patient zero was a two-year-old child in Meliandou Village, Guéckédou. She displayed the symptoms that were to be typical of this outbreak of the disease: fever, blood in the stool, and vomiting. Historically Ebolavirus infections have been referred to as Ebola hemorrhagic fever (EHF) but this term was recently replaced by Ebola virus disease (EVD) to acknowledge that the spectrum of symptoms can be quite distinct in different outbreaks. External bleeding (hemorrhaging) was not commonly observed in 2014 EBOV epidemic.

The first report of the outbreak appeared in the scientific literature in October of 2014 (Baize et al. 2014). Isolates of the emergent virus had been characterized by full-length genomic sequencing, allowing their phylogenetic relationship with those of previous outbreaks to be determined. It was a distinct clade of EBOV and not a lineage descended directly from previously isolated viruses. The rapidity of this analysis signaled another difference of this outbreak: scientists could, for the first time, analyze the genotypes and evolution of the epidemic EBOV in near real time. Analysis of patient isolates allowed assembly of epidemiological data on disease transmission and, in retrospect, researchers could walk in the footsteps of the spreading virus. Infection was traced to the infection and death of family members of patient zero, a midwife who traveled from village to village, and mourners at funerals who subsequently brought the infection back to their home communities. The outbreak quickly spread through communities on the main road leading to Conakry, the Guinean capital and a major local urban center, where cases emerged by March 2014. It then began its international spread: two viral lineages initiated the epidemic in Sierra Leone after mourners at a funeral in Guinea were infected and carried the viruses across the border. At the height of the epidemic, EBOV was epidemic in Guinea, Sierra Leone, and Liberia. Outbreaks in Nigeria and Mali were quickly contained and infections were imported into Senegal, the United States, Great Britain, Spain, and Italy. The World Health Organization declared it a public health emergency of international concern.

Almost two years after the infection of the Guinean youngster, the WHO reported 25,575 cases of EVD and 11,313 fatalities (WHO 2015c). The week of October 4, 2015, was the first since March 2014 in which no new cases were reported worldwide, and the last patient under care was released from a treatment center in Sierra Leone. The outbreak seemed to be under control. Nevertheless, late in October, new cases were reported in Guinea. The unprecedented scale of the epidemic, with widespread and intense communication of the disease in Sierra Leone, Liberia, and Guinea, was startling. As the epidemic progressed, the cumulative incidence of cases in Sierra Leone and Guinea showed signs of slowing as control measures took effect. The greatest danger that an uncontrolled epidemic of catastrophic magnitude might ensue appeared to be from Liberia where for some time the epidemic followed a trajectory of exponential growth, indicative of unchecked disease transmission (Fisman, Khoo, and Tuite 2014). Second only to the shocking magnitude of the local humanitarian disaster was the concern that the epidemic might become a global pandemic. From an evolutionary virologist's standpoint, EBOV was enjoying unprecedented access to the human organism and an opportunity to establish many independent and extended chains of transmission. This, of course, is the fuel for evolution under the distinct selective pressures of its new human host species. The question was posed: Would adaptive evolution result in genotypic changes in the virus with consequences for the outcome of the epidemic? Could evolution transform EBOV into a successful global pathogen like the AIDS virus that emerged fifty years earlier?

What We Were Afraid to Say about Ebola

The worst-case scenarios for the EBOV epidemic have certainly not played out. That is not to say they are implausible and need not be considered, or that future outbreaks will not have different and perhaps more consequential outcomes. Ebola viruses share the capabilities of all RNA viruses that render them particularly adept at traversing genetic barriers between species. It is their extraordinary adaptive capacity and evolutionary invention that allow them to often spread from one host ecological niche to exploit a new host species. It is therefore imprudent to assign a zero probability to the potential for Ebola virus to adapt to humans at some point in the future and become an endemic pathogen. Let's consider the principal preoccupations of the international medical and scientific community during the epidemic and see whether they were disproportionate or entirely appropriate. We will then look at them in light of what we know about RNA virus evolution and the teachings that have emerged from the analysis of hundreds of EBOV isolates collected during the progression of what we can now refer to as the 2013–2015 EBOV outbreak.

A successful zoonotic virus must achieve a basic reproductive number exceeding unity, that is an $R_0 > 1$, indicating that each primary infection results in more than a single secondary infection. Two characteristics of EBOV (and indeed other filoviruses) are fundamental barriers to a successfully zoonosis in the human species. First, it is well established that transmission of EBOV requires relatively intimate contact with the bodily fluids of an infected person; the virus is certainly not transmitted freely by coughing or sneezing like influenza or rhinovirus, and it cannot pass across a crowded room like the measles virus that exhibits truly airborne transmission. Second, only symptomatic patients appear to be infectious, and those patients become severely and precipitously ill; their illness is readily apparent to those around them, and they are in no fit state to travel extensively (CDC). These are the limitations that Ebola virus would have to overcome if it were to become a successful disease-at-large in functioning human society.

In March of 2015, Michael Osterholm, an epidemiologist at the University of Minnesota, published a well-reasoned opinion article in the journal *mBio*, aptly quoting Donald Rumsfeld in its title: "Transmission of Ebola Viruses, What We Know and What We Do Not Know." The authors examined some pertinent themes relating to EBOV transmission including the potential for transmission from mildly symptomatic or asymptomatic patients, whether serial passage (transmission among hosts) in outbreak settings can impact virus transmission, and of course whether the virus may be transmitted by the airborne route. Sadly, his *mBio* article was preceded to press by his September 12, 2014, op-ed piece in the *New York Times* entitled "What We're Afraid to Say about Ebola." The article struck a more emotive and alarmist tone and was directed to the layperson, stating, "The current Ebola virus's hyper-evolution is unprecedented" and venturing that "if certain mutations occurred, it would mean that just breathing would put one at risk of contracting Ebola. Infections could spread quickly to every part of the globe, as the H1N1 influenza virus did in 2009, after its birth in Mexico." The first statement is certainly false; it was a premature assertion and has not withstood the scientific scrutiny of the epidemic as a whole (I will discuss this in the next sections). While the second statement is true at face value, it does not acknowledge the genetic distance that so often intervenes between two different phenotypes. A high genetic barrier has denied the avian H5N1 virus the capacity to evolve transmission by the respiratory route in man. Moreover, when artificially achieved in the laboratory, it lost virulence. Indeed, despite the remarkable success of viruses to adapt to new host species, changes in the route of transmission and tissue tropism are rare in the extreme. This is true even for RNA viruses that are endowed with such evolutionary dexterity. Yes, it does happen, but not often. Here are two notable occurrences: the transition of avian influenza virus from a highly adapted enteric pathogen to a virulent and highly contagious respiratory virus in its new human host is one. Notably, the change in mode of transmission and tropism required associated compensatory changes. The successful human influenza virus exhibits greater pathogenicity in its new host, allowing it to be transmitted via the symptomology of the disease (in respiratory droplets). A second unusual case is that of porcine enteric coronavirus, in which a discrete genetic change in a viral coat protein can turn it into a respiratory pathogen, porcine respiratory coronavirus (Zhang et al. 2006). Although this new virus is readily transmitted between pigs by the respiratory route, it is far less pathogenic than the ancestor from which it diverged. An outcome reminiscent of the experimentally created variants of avian H5N1. The success of a new viral lineage with altered mode of transmission is therefore multifactorial and typically must be associated with other phenotypic changes. More often than not the new lineage must make a trade-off in virulence!

Before delving into what we have learned empirically about the evolution of EBOV during its recent extended sojourn in the human host, we should review some possible scenarios of virus adaptation in the human population. The specter of increased virulence is, of course, a most worrisome outcome of host adaptation (should it occur). Viruses of higher virulence can replicate to higher titers in their hosts and as a consequence can be transmitted more efficiently. The viruses that are shed at higher titers will likely have a selective advantage over their less virulent siblings. This notwithstanding, increased virulence is often detrimental to the lineage resulting in a lower reproductive number; it can lead to the earlier death of the host and a shorter infectious period (in which virus shedding occurs), limiting the opportunity for transmission. Selective pressures must strike a balance, and we should not assume that they will favor increased viral virulence. A variety of adaptive changes could lead to a more successful viral lineage and affect the outcome of an epidemic. Attenuated virulence that permits infected individuals to experience extended viremia and perhaps remain infectious but mildly symptomatic for extended periods may increase the reproductive number. It may be a more successful infectious agent with the potential to spread to multiples of the number of individuals that EBOV has infected. If even a small percentage of such patients were to go on to develop severe or fatal disease complications, the magnitude of the medical challenge might still be amplified considerably. Equally, although today the CDC places the incubation period of EBOV at twenty-one days, if a virus with an incubation period of 100 days emerges, we would expect very different epidemiologic outcomes. Virus dispersal over much greater geographic distances would be possible and the challenge of containing the disease substantially more daunting.

It is known that some individuals are infected by EBOV but do not show signs of overt disease (Leroy et al. 2000); perhaps they have a genotype that allows them to resist viral virulence. It is quite feasible that if the virus "samples" enough human hosts (and individual genotypes) it may infect individuals who are "superspreaders"; these might be asymptomatic shedders or patients that resist the pathology of the disease while not suppressing virus replication. Such individuals may catalyze epidemic spread and emerge as important sources of new viral lineages that can be transmitted more efficiently from person to person.

Collectively these speculations on potential evolutionary trajectories of EBOV fall into the category of known unknowns. We can simply observe that they are more or less probable outcomes of a zoonotic transmission and that rare events will be more likely to occur in a larger epidemic with more viral transmissions. Even what we might consider to be known knowns of Ebola virus biology and disease are based on only a few studies and a few patients. Just months into the 2013–2015 outbreak, the number of infected individuals exceeded all of those infected by Ebola viruses since their first recognition in 1976.

Evolution or Adaptive Change

Many groups of researchers contributed to the understanding of EBOV evolution during the 2013–2015 epidemic. Some previously tacit assumptions about Ebola virus outbreaks have solidified and new epidemiologic insights into the epidemic gained by observing genetic changes in diverging EBOV phylogenies. Never before have scientists had the opportunity to sequence longitudinal isolates of Ebola virus that has undergone sustained human-to-human transmission. EBOV has a negative-sense strand RNA genome replicated by an RNA polymerase, and we expect misincorporation of one or two nucleotides per progeny genome will create a diverse intra-host population of viruses composing the quasispecies. This quasispecies is subject to purifying or diversifying selective pressures within the host and of course during transmission to a new host, when it experiences a bottleneck. The size of the bottleneck is defined by the minimum infective dose. The first analysis of available genome sequences from viruses isolated in Guinea and Sierra Leone emerged in the fall of 2014 (Gire et al. 2014; Baize et al. 2014). Several observations spoke to the nature of the outbreak. It had arisen from a single cross-species transfer into the human population, all isolates being rooted to a virus that emerged around the beginning of that year in Guinea. The virus was a distinct clade that had diverged from the lineages circulating in central Africa approximately a decade before. Each of the recorded EBOV outbreaks appears to have been caused by independent zoonotic events from a common pool of genetic diversity in a standing animal reservoir of viruses. These observations instruct us that in all likelihood EBOV was circulating in an animal reservoir across broad regions of Africa long before its first emergence in 1976. This is contrary to the proposition of some earlier researchers who posited that the pattern of sequential outbreaks reflected the advancing front of an epidemic emergent Ebola virus lineage spreading across Africa (Wittmann et al. 2007).

Study of the EBOV genomes in seventy-eight patient isolates from Sierra Leone (Gire et al. 2014) confirmed that two separate isolates transmitted from Guinea had founded that epidemic. These researchers also documented a variety of mutations in the isolates, some of which were nonsynonymous and could, therefore, be indicative of the potential for adaptive evolutionary change. Moreover, the nucleotide substitution rate was calculated to be higher within the epidemic than between outbreaks. This was the observation that prompted Osterholm to cry wolf and suggest that the virus was mutating faster than normal and was in danger of evolving airborne transmission. It will become clear as I discuss later research by the same and other groups that this was not the case. To be absolutely fair to the authors, it is necessary to add that their observations of higher within-epidemic substitution rates came with appropriate caveats. It was likely an overestimate due to incomplete purifying selection, a phenomenon that allows some deleterious mutations to persist in the mixed population before they are purged by selection. In this regard the scientific content of the paper was correct and balanced, if read and understood thoroughly.

The Sierra Leonean epidemic was the subject of more comprehensive later studies from many of the same authors (Park et al. 2015), who by now had access to EBOV sequences from more than 300 patients and were also able to characterize minor intra-host variants of the virus. One more layer of the evolutionary puzzle could be explored: the emergence of polymorphisms, mutants constituting a significant minority of the viral quasispecies that circulate in a single patient. The outbreak in Sierra Leone was indeed initiated by the transmission of two EBOV lineages, SL1 and SL2, from Guinea. SL2 appeared to have evolved from SL1 earlier and had accumulated four additional mutational changes. In turn, a third lineage emerged soon after the virus invaded Sierra Leone. First identified as a minor variant in a blood sample from a single patient, it soon emerged as the dominant viral lineage driving the epidemic in Sierra Leone. The ability to detect minor variants differing in frequency in different patients speaks to an intriguing aspect of EBOV transmission, which may illustrate a major barrier to the evolution of increased transmissibility or airborne transmission. The authors noted that some patient isolates had minor sequence polymorphisms, variants differing at one nucleotide position from the dominant genotype. The different sequences were often cotransmitted to the new host, consistent with the notion that the infectious viral inoculum is constituted of a substantial population of viruses that represent the diversity of the bloodborne variants of the donor host. This is indicative of a very wide bottleneck and suggests that successful infection requires a large number of virus particles to be transferred to the new host. The empirical observation that direct contact with infected body fluids is required for successful human-to-human transmission is consistent with the need for a large virus inoculum for successful virus transmission. This may be an important impediment to the evolution of EBOV with potential for transmission in smaller aerosol droplets or droplet nuclei that are typical vehicles of respiratory transmission and which will carry only a small payload of virus particles.

The search for mutational changes indicative of adaptive evolution during the epidemic extended to a genome-wide analysis of nonsynonymous and synonymous nucleotide changes. These studies provided strong indicators that positive selection is taking place in many EBOV gene sequences that are enriched in nonsynonymous mutations. As expected, the observed frequency of nonsynonymous nucleotide substitutions declines with the duration of the outbreak, consistent with an ongoing process of purifying selection as predicted by Gire and colleagues when they studied the earlier genome isolates (Gire et al. 2014). Positive or diversifying selection is less constrained within the 2013–2015 outbreak than it was between outbreaks, a feature suggestive of a higher degree of selective pressure for adaptation of the virus to its new host. Particularly evident is that positive selection in the mucin domain of the viral glycoprotein occurs at higher levels than in other viral genes both within the outbreak and between outbreaks. It is very likely that the recognition of exterior virion glycoproteins by host antibodies results in strong selection for escape variants during infection; if this is the case, it follows that the reservoir host also raises a substantial antibody response in the face of EBOV infection. The researchers were able to substantiate these suspicions. In elegant work, they used knowledge of the human antibody response to deduce which peptide sequences in the EBOV proteins would most likely be recognized by human B cells. These B-cell epitopes were enriched in nonsynonymous mutations, which are escape mutations that allow the virus to evade the ongoing antibody response. This phenomenon of immune escape recalls the adaptive evolution of HIV and influenza virus that each fluidly access genotypic variation in their antigenic proteins to avoid immune clearance by the host.

Knowledge gained to date on the genomic variation in the EBOV epidemic certainly substantiates that nucleotide substitutions can be fixed in the population by positive selection, and that variant lineages can diverge and evolve independently (Gire et al. 2014; Park et al. 2015; Olabode et al. 2015; Simon-Loriere et al. 2015). Nevertheless, the observed mutational changes have not been linked to any change in disease-causing potential or transmission, and close examination of the changes in EBOV protein sequences that have evolved during the epidemic betrays no hints of functional changes (Olabode et al. 2015). Our inability to perceive the evolution of adaptive phenotypic change in EBOV lineages could be misleading. No close examination of the viruses in model systems has yet been pursued. Such studies might reveal subtle phenotypic differences and fitness advantages of the different virus genotypes. It remains uncertain then whether the adaptive change that can evolve within the course of an Ebola virus epidemic, even as extensive as the one we have just witnessed, can be substantial enough to affect the outcome of future epidemics. Doubtless, like other RNA viruses,

Ebolaviruses have the capacity for remarkable genetic plasticity and rapid adaptive evolutionary change that can permit cross-species transmission. There are examples of Ebola virus species with radically different virulence phenotypes in human and also perhaps the potential to be transmitted as respiratory viruses. The Ebola Reston virus is the signal example; it is savagely pathogenic in primates but causes no overt disease in human. It has also been shown to spread epidemically among monkeys and to be transmitted between pigs and monkeys by the respiratory route. Such observations and our ignorance of the genetic differences that underpin the distinct phenotype of the Reston ebolavirus compared to EBOV emphasize that we should not underestimate the potential of Ebola virus to surprise us, should it have adequate opportunity to spread in human populations or undergo an adaptive change in intermediate hosts such as primates. It is perhaps most germane to point out that EBOV has quite evidently created and tested more genetic diversity than ever before in human populations. Osterholm may have been unwarranted in some of his alarmist claims, but he was right to point out that "each new infection represents trillions of throws of the genetic dice." While exploring this new genetic territory, mutational changes became fixed in virus lineages, and it successfully retained its fitness and virulence for humans. It did, however, show no hint of becoming a *more* threatening pathogen. Conventional wisdom suggests now that this outbreak is over, the genetic diversity that the virus has explored will be lost, and the next zoonotic infection will originate from a distinct virus lineage currently circulating in animals and naive to the human host. This will be the case unless the virus has found serendipitous sanctuary as a persistent infection in some hosts or if it infects an animal reservoir from which it could in the future reemerge.

EBOV Persistence

The 2013–2015 EBOV outbreak certainly broke the mold; the prior collective experience of zoonotic filovirus diseases was limited to just a few thousand infections in all, caused by a variety of genetically different filovirus lineages. An unprecedented number of individuals survived the disease in this outbreak. These Ebola survivors, some 13,000 or more, are the lucky ones, but it now appears that they have not come through unscathed. They are experiencing a bewildering assortment of postinfectious consequences of their disease, including joint pain, fatigue, and arthralgia syndrome. Others have vision problems or have suffered attacks of meningitis. Take the case of American physician Ian Crozier. He recovered from the disease he contracted as a WHO volunteer in Sierra Leone, but many months later he had to return to the hospital to be treated for severe inflammation and pressure in one eye. The virus had taken refuge in his eye, an organ whose interior is shielded from the full powers of our immune system; scientists were able to recover live EBOV from fluid within his eye where it had persisted since his illness. These sequelae are not at all rare, and there is now abundant evidence that the virus can be recovered from breast milk and from semen of convalescent patients for extended periods of time after the apparent resolution of the disease (Bausch et al. 2007; Rodriguez et al. 1999; Deen et al. 2015). At least one example of EBOV transmission between sexual partners has been documented (Mate et al. 2015). These are unsettling observations; like a forest fire that appears to have been extinguished, the epidemic continues to smolder in the peat substrata, threatening to reignite and demanding continued vigilance. The extraordinary tenacity of the virus to persist in the human body over long periods of time may have grave consequences for our ability to truly extinguish Ebola. We may have revealed "accidental" persistence only because on the occasion of this outbreak the virus was permitted to infect such a diverse sample of the human population. It may be a rare event, but it may allow the virus to reemerge at a later time (and in a different place). Should I dare to speculate that the virus has accidentally stumbled upon a mechanism that will permit the genetic diversity that it has explored during the epidemic to be "archived" in these recovered victims? Might reemerging EBOV Makona lineages, evolved during the epidemic, pick up where they left off and continue their march of adaptive evolution toward endemicity in the human population?

· 13 ·

VIRAL ZOONOSES *and* ANIMAL RESERVOIRS

AMONG *FILOVIRIDAE,* all the constituent species of the Ebolavirus and Marburgvirus genera have been associated with outbreaks of hemorrhagic fever, either in humans or primates. Although human infections are relatively rare, the last fifteen years has seen an accumulation of evidence that they circulate widely in primate species across large areas of central African forests. The death of substantial numbers of great apes is associated with outbreaks, and indeed Ebola virus infections are threatening the survival of gorillas and common chimpanzees whose last bastion habitat is in equatorial Africa's Gabon and Congo. The disease vies with habitat destruction and hunting as the cause of a decline in great ape populations of the more than 50 percent between 1983 and 2000 (Walsh et al. 2003). In 2004 scientists broadly surveyed nonhuman primate populations for the presence of antibodies that react with the virus. One in eight had antibodies to EBOV, indicative of ongoing or prior infection, revealing that the virus circulates abundantly in great ape populations (Leroy, Telfer, et al. 2004). They also found evidence of infection in other primates: drills, baboons, mandrills, and *Ceripithecus*, as well as one duiker, a species of antelope. Furthermore, EBOV antibodies are readily found in primates sampled in many other African countries that have never experienced a human outbreak. The same investigators extended their analysis to the phylogenetic identity of the EBOV lineages in wild apes (Wittmann et al. 2007; Leroy, Rouquet, et al. 2004). Here they found evidence of substantial genetic diversity in the circulating viruses. They also revealed that particular virus lineages were efficiently spread within family units but there were also multiple independent transmission events from an animal reservoir to the primates, just as we have seen for human Ebola disease outbreaks that are caused by genetically distinct lineages. In an especially striking observation, the genomes of EBOVs isolated from two individual chimpanzee carcasses that lay side by side were sequenced. The investigators naturally expected that they would find the virus in both animals to be very closely related and from the same chain of transmission (the individuals likely being from the same family or community group). This was not the case; they discovered that the chimpanzees were infected by viruses of different lineages. The related chimpanzees had been independently infected by genetically divergent viruses.

It appears then that multiple variant EBOV strains co-circulate in African forests at unexpectedly high levels and are epidemically spread through nonhuman primate communities. Most introductions of filoviruses into human populations are from infected primates, but the relatively high mortality that the viruses cause in primates is not consistent with them being the true reservoir species for the virus. It would be unusual for a maintenance reservoir host to exhibit such high levels of overt disease and mortality. Another culprit was therefore sought, and quite recently bats were implicated as the principal natural reservoir of filoviruses. Over the past decade or so there has been an accumulation of evidence supporting this notion: first serological data, evidence of antibodies directed against the viruses, and then filovirus RNA sequences isolated from healthy bats (Leroy et al. 2005; Olival and Hayman 2014). Typical of viruses circulating in their natural reservoir, the genetic diversity of the viruses in wild bat populations was far greater than that seen in virus isolated during outbreaks. Even more striking was that multiple diverse virus genotypes were represented in bat populations from a single geographic location. Opportunity is key to zoonosis and the geographic range of the reservoir host should include the areas where disease breaks out. This is indeed the case in Africa where evidence of Ebola virus infection could be found in three fruit bat species, *Hypsignathus monstrosus*, *Epomops franqueti*, and *Myonycteris torquata*, that share the forest with nonhuman primates in regions of Africa where outbreaks have occurred (Leroy et al. 2005). It is likely that many bat species (and, of course, other animals cannot be ruled out) can carry filoviruses, and one can imagine that in each species a different virus-host cell interface will exist such that subtly different selective pressures will be at work on the circulating viruses. This may contribute to a continuous process of balancing selection and the maintenance of genetic diversity that is available for cross-species transmission.

Most human infections result from contact with infected primates, but there are examples of direct transmission of the virus from bat to human populations. In somewhat anecdotal accounts of the beginning of the 2013–2015 Ebola Makona outbreak, it was reported that the first infections were of children playing in a hollow tree just outside the village of Meliandou. The tree was said to be home to a colony of the *lolibelo*, a local name referring to "mice that can fly" (actually an insectivorous bat species, *Mops condyluru*). Children would disturb the bats by poking sticks inside the tree and capturing them to roast on skewers. It seems that those innocent meals of bushmeat may have ignited what became an outbreak of global proportions.

The Usual Suspects

Bats in Africa harbor Ebola viruses; Marburg virus infects bats in Egypt. These bat populations infrequently, but with some reliability, transmit the viruses to human populations either directly or via intermediate hosts. Based on the accumulation of evidence that bats are major reservoir species for filoviruses and a source of zoonoses scientists are scouring wild bat populations elsewhere to uncover traces of other filoviruses. They have not been disappointed. When scientists investigated unusually high mortality in an insectivorous bat population in Spain, they discovered a filovirus that represented a new and distinct genus (Negredo et al. 2011). The Ebola-like virus was christened Lloviu virus. It remains unclear whether Lloviu virus was responsible for the mortality in *Miniopterus schreibersii* bats and whether the bats are the natural host of the virus. Nevertheless, the broad distribution of this bat species across Western Europe gives cause for concern. Fruit bats in Bangladesh carry antibodies reactive to EBOV and Reston ebolavirus (which itself circulates in the Philippines), and Chinese scientists have detected and sequenced RNA from bats in China that is closely related to filoviruses (Olival et al. 2013; He et al. 2015). All signs are that filoviruses are broadly distributed in natural reservoir species on many continents, certainly across Europe to Africa and mainland Asia. The facility with which Ebolavirus species circulating in African fruit bat populations spills into human populations is quite exceptional, however; there are no records of human infections by Lloviu virus in Europe or suspected filovirus-attributable hemorrhagic fever outbreaks in China. It seems likely that anthropogenic epidemiologic factors are influencing the likelihood of zoonoses; the hunting and eating of bushmeat in Africa and the intrusion of humans into wildlife habitats must play a significant role. Another influential factor that should not be ignored is the availability of prevalent intermediate hosts. In the forests of Africa, it is the Old World primates that serve as the primary vector of the disease into human populations. Lacking such suitable intermediates, filoviruses in the Americas and Europe may have considerably less opportunity to move into human hosts, in spite of their ubiquitous presence in bat populations.

It is an unavoidable that anthropogenic factors will govern the likelihood of future viral zoonoses. The human population continues to expand, extending its footprint and becoming more globally interconnected. Social structures and behaviors are continually evolving, intensive farming practices are becoming the norm, and the trafficking of exotic wildlife is more commonplace. These factors and the inexorable march of climate change are all man-made changes that we bring on ourselves. It is our own behavior that is providing ever increasing opportunities for emergence of new diseases. It is difficult to believe that there has been a significant increase in the number of potential zoonotic pathogenic viruses in recent years. They have always been circulating in animal reservoirs. The increase in viral zoonoses we are observing is simply a consequence of human behavior, which is creating more opportunities for cross-species transmissions. Viruses with the means and opportunity will expand into new environmental niches and explore new genetic space.

Filovirus Origins

Filoviruses were only recently recognized as zoonotic pathogens. Is this indicative of a recent evolutionary history? Over the past fifty years, scientists have accumulated complete genome sequences of Ebolaviruses and Marburgviruses associated with successive outbreaks of distinct lineages. Now Lloviu virus is available as an outlier genome representing a third genus of the family. Phylogenetic analysis of almost 100 genomes now permits evolutionary biologists to project the date of the most recent common ancestor of modern filoviruses to at least 10,000 years ago (Carroll et al. 2013). As discussed in the previous chapters, such estimates, based on genome sequences sampled from the tips of the evolutionary tree—those terminal branches that represent the last 50 years of growth of what is a 10,000-year-old tree—are limited in their reliability (they are very likely on the low side). Filoviruses are clearly not a recent evolutionary invention, but rather contemporary descendants of ancient precursor viruses, which themselves may be long extinct.

Bats and Viral Zoonoses

Many other zoonotic viruses that garner a great deal of attention and notoriety have their origins in bat populations. Bats appear to be repeat offenders when it comes to acting as natural reservoirs for human zoonotic infections. These often have severe pathological consequences and potential global impact. SARS-CoV and MERS-CoV (coronaviruses) caused highly contagious and severe respiratory illnesses in humans after transmission to intermediate hosts from infected bats. Virulent paramyxoviruses such as Nipah and Hendra viruses, as well as rabies virus and related lyssaviruses, all have bat reservoir species. More than fifteen different families of viruses infect over 200 distinct families of bats (O'Shea et al. 2014). It is not only the diversity of virus species calling bats their natural host that is remarkable, but that for the most part infected bats show no overt signs of disease. It seems that bats and their viruses have often coevolved to equipoise. There are ancient roots to the relationships of bats with many virus phylogenies; bats are the most ancestral taxon infected by paramyxoviruses, coronaviruses, and lyssaviruses, to name just a few (Brook and Dobson 2015). The fragments of filovirus-like DNA sequences found in bat genomes are also a reminder that their association stretches deep into evolutionary time (Taylor, Leach, and Bruenn 2010).

It is not clear whether bats evolved to be more tolerant of viral infections or if they are able to contain them more effectively, or in a different manner, than other mammals. Despite the commensal relationships evolved between bats and their viruses, cross-species transmission to human often manifests clinically severe disease. That bats are sources of many zoonoses is not entirely surprising: bats are the most abundant mammals and bat species make up about 20 percent of all mammalian species. Among mammals, only *Rodenta* has more species (although being less numerous) and they are also major sources of zoonotic infections; nevertheless, bats host more viruses per species than rodents (Dobson 2005). Strikingly, sympatry of bat species, facilitating host switches between closely related species occupying the same geographic range, was found to be a highly significant risk factor for zoonotic transmission from bats (Luis et al. 2013). If we briefly expand our consideration beyond viruses to all zoonotic human pathogens (viruses make up half of all human zoonotic pathogens), it is evident we are at greatest risk of zoonotic infection by pathogens that infect multiple species of hosts (Woolhouse and Gowtage-Sequeria 2005).

A Special Relationship

The facility with which bats host multiple zoonotic viruses and the dangerous potential of spillover infections of humans represent a serious public health concern and a scientific curiosity. Concrete explanations of the mechanistic basis for what appears to be a unique host-virus relationship remain elusive (Calisher et al. 2006). The origin of bats at least 50 million years ago is evident in the fossil record and the evolutionary divergence of conserved gene sequences. Paramyxovirus and lyssavirus phylogenies are of a similar age, suggesting that they codiverged with bat host species. It is therefore quite possible that the viruses of bats evolved to interact with cellular biochemical pathways that were inherited from the mammalian common ancestor and conserved through evolution. Might they therefore be generalists and have a preserved capacity for cross-species transmission to other mammals?

Bats have the ability to fly and in this regard are unique among mammals. Despite their small size, high heart rates, and the remarkably high metabolic rates needed for sustained flight, bats have unexpectedly long life spans. They also enter a daily state of torpor and hibernate during the cold season in temperate climates. Some bat species have unusually large population sizes; they can number in the hundreds of thousands of individuals.

Such populations are certainly large enough to support the persistence of epidemic pathogens, even if they are not persistent infections but are cleared, creating lifelong immunity. Measles and smallpox viruses could persist in human populations only once population centers exceeded 500,000 individuals (refer to Black's work discussed in Chapter 6).

All of these features of bat species have captured the attention of scientists as the potential basis for their ability to host virus infections with relative impunity. No single unifying hypothesis has yet won out. "Flight as fever" has been posited as one explanation (O'Shea et al. 2014). This model contemplates that the high body temperatures sustained by bats in flight mimic the beneficial effect of the fever response to infections. Fevers stimulate the immune response while raising the temperature of the host to levels that are not optimal for viral replication. Such a strategy might improve the ability of bats to control virus infections and perhaps reduce the severity of viral-induced pathology. Other groups favor alternative explanations indirectly related to the capacity for flight. They implicate particularities of the innate antiviral signaling pathways of bats and their unique resilience to oxidative stresses inherent in their high metabolic rate (Brook and Dobson 2015). The energy expenditure required for muscle-powered flight is considerable, and a side effect of the necessarily high metabolic rates is the generation by mitochondria of potentially damaging reactive oxygen species (ROS). It has been proposed that bats may be better adapted than flightless mammals to avoid the negative effects of ROS. This may partly explain the unexpected longevity of bats since cellular programs that respond to ROS-mediated mitochondrial DNA damage have been associated with aging. Since the innate cellular antiviral response leads to cytokine release and recruitment of immune cells that are also a source of ROS, it is thought bats may be able to manage the adverse effects of viral infection with the same mechanisms they have evolved to survive metabolic stresses associated with flight. These investigators also made note that mammalian genes involved in DNA damage and repair (genes that encode functions that mitigate ROS-mediated damage to DNA) were subject to positive evolution concomitant with the acquisition of the ability to fly. Furthermore, the mitochondrion has an erstwhile and underappreciated role in antiviral innate immune signaling. It could all be based in flight and mitochondria.

A uniquely adapted physiology and evolved innate immune system may therefore be central to the ability of bats to be ideal reservoir hosts. They can mobilize a protective antiviral response without unduly causing collateral immunopathalogic damage to their tissues. Mandl and coworkers reviewed immune responses to emerging zoonotic viruses, reasoning that understanding the mechanisms underlying the dangerous pathology of these viruses in human hosts may shed light on how viral pathology is managed by their natural reservoir hosts (Mandl et al. 2015). They noted that the severe disease manifestation of emerging zoonotic infections of humans are most often associated with aberrant innate immunologic responses. The severe disease caused by SARS-CoV had the hallmarks of an uncontrolled and overexuberant innate immune response. Poor outcomes were associated with elevated and chronic interferon production (Mandl et al. 2015). The progressively fatal pathology of HIV-1 infection leading to AIDS in untreated individuals can be contrasted to the lack of pathology in SIV-infected monkeys. The decline of the human immune system in the face of HIV-1 infection is now recognized to result from chronic immune activation, dysfunction, and apoptosis of many immune effector cell populations. The continuous viral replication and ongoing stimulation of the innate immune response in infected cells leads to excess release of interferon and other proinflammatory signaling molecules. A greater, rather than lesser, immune response (or at least its dysregulation) is responsible for the devastating cascade that progressively erodes immune competence. On the other hand, in SIV-infected African green monkeys there is no evidence of aberrant immune activation and there is no observed dysfunction of T cells or chronic interferon production (Mandl et al. 2008; Silvestri et al. 2003). Moderation of immune responses and reduction of the immunopathalogic consequences of fighting viral zoonotic diseases appears to be one important component in adaptive evolution toward the minimal pathology suffered by natural reservoir hosts.

Tolerance and Resistance

The arms race between a zoonotic virus and its new host is a result of selective pressures acting on each genome, that of the virus and that of the host. Take a moment to consider the selective pressures operating on hosts during coadaptation. Maladapted hosts can actually suffer fatal self-inflicted damage as a result of the uncontrolled immunopathalogic consequences of attempting to contain the virus, but unbridled viral replication may equally have serious consequences. There is abundant evidence that the arms race between virus and host results in positive selection on innate antiviral response genes but in contrast, reservoir species often exhibit chronic persistent asymptomatic virus infection. It appears that the genetic arms race between host and virus can result in selection for mechanisms of tolerance and mechanisms of resistance.

Population geneticists have considered these propositions in detail (Schneider and Ayres 2008). It is interesting to note that mutations in the host genes which render the individual host more resistant to the disease will provide a high competitive advantage to their genome. In the early stages after the emergence of this variant, there will be a strong selective pressure for the possession of the beneficial trait, which reduces virus infection and pathology. Counterintuitively, however, as the allele spreads through the population and becomes more prevalent, virus infections will be correspondingly reduced and the relative competitive advantage of those possessing the allele will diminish. Moreover, when 80 percent or so of the population carry the allele for disease resistance, the impact of the herd effect will result in the epidemic stalling—inadequate susceptible hosts are available. Such traits therefore have diminishing benefit to the host, as their frequency in the population increases, and in principle should not become fixed in the population, save for chance phenomena. Contrary to this situation, genes that promote tolerance to the viral infection can rapidly prevail and be driven to fixation in the population by selective pressure. An individual expressing the tolerance trait will have a greater viral burden and will be more infectious; as more individuals in the population become tolerant of infection, the selective pressure on the population will only become stronger, until all individuals are tolerators.

There are a few hints that tolerance genes may have evolved in host cells. In an earlier chapter, I alluded to the filovirus-like sequences that can be found in bat genomes today. Those researchers pondered whether these fragments of viral information, exapted into the host genome, may be remnants of an evolved mechanism of tolerance (or defense) against virus infection (see Chapter 14 for further details). If each cell in an organism expresses proteins that are antigenically identical to those expressed by predator virus pathogens, then the host immune system will recognize that particular antigen as *self*, the individual will have *immunological* tolerance, and no immune response will be mounted. I also examined how exaptation of viral sequences into the genome can provide restriction factors that limit viral infection of host cells. The number of examples of these phenomena that can be confidently recognized in contemporary genome sequences are few, but it is quite feasible that these strategies commonly evolved, though in deep evolutionary time. Their relevance for genome fitness may have since declined, leading to lack of conservation of the sequences such that they are now obscured by mutational drift.

In an effort to highlight the contrast between two alternative strategies of adaptation I presume to simplify the possible trajectories that host adaptation can take in response to zoonotic viral invasion. There will be many polymorphic alleles at play. Moreover, humans sexually reproduce and are diploid, therefore homo- and heterozygotic phenotypes must also be considered together with the powers of recombination that shuffle the genetic information passed into the male and female gametes. Tolerance is certainly not the only mechanism of virus-host coevolution, otherwise there would be no arms race, no selective pressure on the virus, and we know that is not the case. Viral adaptive evolution to new host species is evidence that they are under selective pressure to evade both innate and adaptive immune responses (we discussed them at length in previous chapters). The process by which a host species must select from its genetic diversity in the face of a variety of complex selective pressures is long and tortuous; it will involve assembly of constellations of alleles (the equivalent of the flu dream team), some of which may remain polymorphic and under balancing selection in the population. Elements of tolerance and increased control (with the trade-off of some immune pathology) may be incorporated into the complex equation of natural selection. If the coevolutionary détente is achieved, as it appears to have been between African green monkeys and SIVagm for instance, it will necessarily take a very long time. On the other hand, we can speculate that sometimes the evolutionary agility of viruses (or the plodding rate of mammalian coevolution) could not prevent host species extinction, and consequent failure of the virus to establish a lineage with extended host range.

So bats are special. But they are far from unique in their capacity to act as natural reservoirs of viruses. Birds, rodents, and primates are all perfectly coevolved with some viral species and are sources of human zoonoses. Birds are the worrisome reservoir host from which pandemic influenzas may emerge, and have been the source of the spread of West Nile virus across the continental United States in recent years. Rats host hantaviruses, the cause of hantavirus pulmonary syndromes, which evermore frequently emerge, and of course primates have been the source of one of humankind's gravest challenges in recent millennia, the HIV pandemic. Bats, however, seem to be particularly well suited and have adapted more readily than other species to become successful reservoir hosts for multiple species of viruses with potential as the source of human zoonoses. Some scientists have speculated that selective pressures acting on bat-virus coevolution are particularly acute. This "flight as fitness barometer" thesis suggests that the high level of sustained energy expenditure demanded by flight places a premium on the fitness of bats. In other words, bats suffering even modestly reduced fitness will be rapidly weeded out of the population. Equally, viruses that reduce bat fitness sufficiently to preclude sustained flight will also suffer negative selective pressures. The same principle could apply to birds that have similar demands for flight (and indeed it could be extended to small rodents that have very high metabolic rates that demand sustained activity to maintain their calorific intakes). Should small reductions in host fitness translate into an acute selective pressure it will create a stringent selective filter that might be expected to maximize the rate at which maximal fitness in the population is achieved. The selection will operate both on the bat and the virus, reciprocally fueling an accelerated arms race. The flight as fitness barometer hypothesis is likely true as far as it goes, but it is unlikely to offer a complete explanation of bat specialness. It can only operate successfully and result in perfect virus-host coadaptation if the necessary genetic space is accessible and can be explored on the viable fitness landscape. The genetic distance to adaptive co-fitness of the genotypes must be narrower and more readily bridged for bats and their viruses.

Perhaps it is not necessary to invoke an accelerated arms race between bats and their viruses to explain why bats live with numerous highly coevolved viruses. It may simply be based on long-standing coevolution with ancestral viruses. After all, simian species are very well adapted to their respective naturalized lentiviruses, and even humans have highly coevolved apparently commensal relationships with a variety of viruses in our viromes, ranging from polyomaviruses to herpesviruses. I think it is reasonable to speculate, however, that particular evolved characteristics of bat immune physiology may allow them to more readily adapt to a variety of different viral pathogens. They may have evolved a common toolset that can be easily adapted for different viruses. In evolutionary terms they can readily coevolve to exist with many different viruses because the genetic distance they must bridge and the genetic space they must explore to maximize tolerance and virus control is smaller than for other mammals.

Ironically, it may be our very immunological maladaptation to viruses such as filoviruses and coronaviruses that places us at lower risk of these viruses becoming endemic across the population. Paul Ewald is an evolutionary biologist at the University of Louisville and an advocate of the concept in pathogen host coevolution relating to the trade-off that occurs between virulence and transmissibility. On the potential of the EBOV Makona to become pandemic he told *Esquire* magazine in 2014, "But the silver lining is that the more severe the individual case the less likely it is to be a problem for the population as a whole. The more severe it is for the individual, the less likely it is to be transmitted" (Junod 2014). Zoonotic viruses that are highly pathogenic in human populations are maladapted and their genotypes may be a substantial genetic distance removed from a genotype with the true potential to form a successful lineage in humans. We are not, however, in a position to state with certainty that virus evolution cannot bridge that gap in unforeseen ways.

Looking toward the future then, special vigilance is certainly required as we anticipate a continued flow of diverse viral genetic lineages from the virus populations circulating in bats. Fruit-eating bats and insectivorous bats figure strongly in our story. As expanding human endeavors bring our activities into juxtaposition with bat populations, the opportunities for zoonotic transmissions increase. Species of large fruit-eating bats called flying foxes are the natural hosts of the Nipah virus, a paramyxovirus which infected pigs in Malaysia in 1999. The pigs were an intermediate host and over a million were culled to staunch zoonotic spread of the epidemic to humans. More than 250 cases of the viral encephalitis were reported and more than 100 were fatal. The WHO continues to report sporadic Nipah outbreaks in humans in Bangladesh and

India, where more than half of Indian flying foxes, members of *Pteropus* genus, were seropositive for Nipah virus antibody (Calisher et al. 2006). WHO implicates the consumption of raw fruits or date palm juice with the zoonosis (WHO 2015a). The bats forage around the palm trees and drink from the open buckets in which the juice collects. It is believed that bat saliva, urine, or feces contaminate the juice that is sold in local villages. Pressures on flying fox populations are resulting in their redistribution to more urban areas, where they live cheek by jowl with livestock and their human farmers. We return to the same principle: the accelerating pace of zoonotic transmission of novel viruses into humans is attributable to anthropogenic epidemiologic factors. Only behavior modification or medical management of this future health burden will mitigate the risks of future zoonoses for human populations.

· 14 ·

ENDOGENOUS RETROVIRUSES: OUR VIRAL HERITAGE

Darwin and Mendel knew nothing of DNA as such, yet they divined the principles of genetics on which the modern synthesis of evolutionary theory was later founded by twentieth-century evolutionists such as Huxley, Dobzhansky, Haldane, Wright, and others. Nevertheless, I dare say that even those luminaries would have been astonished to learn that our DNA differs by a mere 1–2 percent from that of a chimpanzee (Olson and Varki 2003; CSA 2005). In the early 1970s, the brilliant virologist Howard Temin championed the *provirus hypothesis*. It advanced a new concept: that of retroviral reverse transcription and the integration of a DNA copy of the viral genome, the *provirus*, into the host chromosome as an obligate step in retrovirus replication. In 1975 he would share the Nobel Prize in Physiology or Medicine for the discovery of *reverse transcriptase* (RT), the retroviral protein responsible for synthesis of the proviral DNA. I believe even he would have been surprised that 8 percent of our DNA is composed of his retroviral DNA proviruses (Lander et al. 2001).

The retroviruses are one among many classes of retrotransposable genetic elements that encode a reverse transcriptase. These *RT-retroelements* include retroviruses and a variety of other mobile genetic elements that can amplify themselves via reverse transcription of an RNA intermediate and reintegration into the host chromosome. They are widely distributed in all cellular organisms and phylogenetic analysis of the RT gene places its origin in group II self-splicing introns found in *Bacteria* and *Achaea* (Koonin, Dolja, and Krupovic 2015a). RT-retroelements with an extracellular phase include retroviruses and are found only in eukaryotes, suggesting they likely descended from the more ancient and ubiquitous RT-retroelements. In a manner of evolution we see shared by all viruses, retroviruses (and related *env*-like gene-containing elements) acquired their *env*-like genes, and potential for extracellular transmission, by horizontal gene transfers from other viruses (Malik, Henikoff, and Eickbush 2000). The origin of retrovirus *env* is uncertain but it is likely assembled from the envelope of other eukaryotic RNA viruses, with which it shows some structural and functional similarity (Koonin, Dolja, and Krupovic 2015a). The chimeric nature of retroviruses is further underscored by phylogenetic analysis of the RNAase H gene, an ancient component of RT-retroelements that appears to have been replaced by a eukaryotic counterpart in recent evolution (Malik and Eickbush 2001).

Retrovirus-related retrotransposable elements in our genomes include LINE elements (long interspersed nuclear elements) and SINE elements (short interspersed nuclear elements). They are organized differently from retroviruses and notably never acquired an extracellular phase in their replication cycle and do not undergo transmission between hosts. In this regard they escape being categorized as viruses. Collectively, the veritable zoo of mobile genetic elements comprises a remarkable 40 percent of our genome. The retroviral sequences littered across all vertebrate genomes are termed *endogenous retroviral elements* (ERVs). Endogenous retroviral sequences themselves are so plentiful that they take up more space in our genomes than genes encoding human proteins. Their existence is evidence of waves of retroviral infection and germline infiltration throughout vertebrate evolution. We have only recently begun to realize the powerful influence that retroviruses wielded over the evolution of vertebrate genomes and the identity of our species. They were catalysts of genetic instability that fueled evolutionary change; today they are vestiges of their former selves, fossils of viruses that once preyed on vertebrate hosts. Their remains are evidence of pyrrhic victories of sorts in many wars and arms races that have taken place between retroviruses and hosts (Stoye 2012). Most endogenous retroviruses have long been silenced by host cell *restriction mechanisms*. Associated with no phenotype, and under no selective pressure they become nonviable after millions of years of mutational drift, resulting in the accumulation of mutations or deletions in their coding sequences and control elements. How did these viruses end up in our heritable genetic material and how have they influenced vertebrate evolution? To answer these questions, we must return to the beginning and the arrival of the earliest proviral colonists of the vertebrate genome. It is an exercise in viral paleontology; much can be deduced from the fossil record of vertebrate genome sequences. The time in evolutionary history when a particular retrovirus became endogenous to a species is recorded in the phylogenetic tree of descendants that share similar proviruses. Some of them occupy identical locations in the genomes of long-diverged species; their integration sites are flanked by homologous host chromosomal sequences in each descendant species. They are unique only in their patterns of nucleotide substitutions; after speciation the proviruses evolve independently in each genome.

Genome Invasion by Retroviruses

All retroviruses share a core ensemble of essential genes (*gag*, *pol*, and *env*) and replicate by copying their RNA genome into a double strand of DNA flanked by long terminal repeats, which integrates permanently into the genome of the host cell. Gene expression and viral genome synthesis are accomplished by making copies from the integrated provirus. As of today, HIV-1 is an exclusively *exogenous* retrovirus. Its proviruses endure in the genome throughout the lifetime of each infected cell. They are not, however, heritable and not transmitted vertically to the host's offspring. The reason for this is that HIV-1 infects and establishes proviruses in the somatic cells of our body. In order to become endogenized and then vertically transmitted to our offspring, HIV-1 must invade our germline cells, which give rise to gametes (spermatozoa in males and ova in females). If indeed HIV-1 does infect human germline cells with any regularity, it appears that it precludes successful fertilization and formation of a viable embryo. Invasion of the germline by a retrovirus is clearly a long-odds bet; the infected gamete must survive, it must by chance successfully form a zygote that is viable, and survive to adulthood to reproduce. The ERVs in the human genome have been our passengers for millions of years. They began as single provirus insertions in one chromosome of the germline of a single individual. In effect, they were rare *polymorphisms* that spread from a single individual and ultimately prevailed in the population, becoming fixed alleles. Genome infiltration and population-wide fixation of the endogenous provirus is therefore likely the result of a very rare sequence of events. Nevertheless, more than thirty separate genetic lineages of endogenous retroviruses are found in primate genomes, so we can safely conclude that it has occurred at least that many times.

Each lineage of ERV in vertebrates exists as multiple copies widely dispersed across the genome. Following genome invasion, a wave of provirus proliferation ensues. At this time, the newly endogenous proviruses are actively expressed and make new infectious virus particles that can subsequently reinfect the germline cells. Newly integrated proviruses are thus formed, resulting in the proliferation of the ERV across the genome. In some instances, the usual infectious cycle can be "short-circuited." Infectious genomes synthesized within the germline cell may be directly reverse transcribed and reintegrated into the genome to create a new provirus in a different location in the genome. This may not occur or may be very rare for ERVs in general; however, this strategy has evolved in some mouse endogenous retroviruses that continue to be mobile today. For example, IAP (Intracisternal A particle) has lost its envelope gene and evolved mechanisms to assemble and bud virus particles within the cell. Such selfish genetic elements no longer have an extracellular phase. Since such transpositions do not entail the production of extracellular virus and reinfection, it would be a more efficient mechanism for selfish replication of the genomic parasite within a single genome.

The pivotal period in the relationship of the virus with its host is during retroviral invasion of the germline. At this time the disease-causing retrovirus will continue to circulate among the population as an infectious exogenous virus, causing new infections. The virus will also be produced from active endogenous proviruses in the members of the population that inherited ERV loci in every cell of their bodies. Indeed, the activity of these proviruses may also create the manifestations of the disease by replicating in somatic cells. Over the course of generations, the activity and proliferation of ERV sequences in the germline may be detrimental to host fitness and has the potential to engender dangerous levels of genomic disruption. This will create strong selective pressures that favor host individuals whose genomes evolve the necessary defensive responses to suppress endogenous virus expression and withstand infection by the exogenous virus. At first blush, it seems that germline invasion and endogenization of a retrovirus must have negative consequences for the host. Why then have ERV loci become fixed alleles in vertebrate populations? What advantage can they possibly confer upon the host genome?

We can readily envisage one likely advantage conferred by ERVs during the early stages of invasion of the population by the endogenous retrovirus allele. It is a common laboratory observation that retroviral infection of cells in tissue culture leads to *superinfection resistance* (it is also referred to as *infection interference*): the infected cells become immune to further infection. Viral envelope glycoproteins expressed from the provirus in infected cells migrate to the cell membrane. Here they are readied for incorporation into virus particles that bud from the cell surface. These env proteins on the surface of the cell can also engage the same cell surface receptor molecules that an infecting virus must target for attachment and penetration of the cell. Infected cells producing env on their surface are thus rendered immune to infection: there are no available receptors for viral attachment. A host organism with an endogenous retrovirus is thus granted some immunity to infection by the exogenous disease causing retrovirus. In another example of this phenomenon (discussed later in this chapter in the section on endogenous viral elements), vertebrate hosts benefit by exploiting retroviral genes as components of host restriction that are protective against retroviral infection. In each case the ERV genes become a beneficial Mendelian locus that can be subject to positive selection in populations preyed upon by exogenous retroviruses. The germline acquisition of a retrovirus has other potential advantages to the host in that the antigens of the ERV will now be recognized by the immune system as "self." Replication of the ERV or infection with the exogenous virus will no longer trigger an immune response and the attending immunopathalogic consequences so often seen in virus infections.

These factors can only be part of the story; the viruses that invaded our germline and became our ERVs are no longer our predators and no longer pose any danger to the endogenized host. Nevertheless, they can be envisaged to be very influential in early stages of retroviral colonization when the ERV allele is positively selected and proliferates in the host gene pool. An ERV locus may have some short-term benefit, but its long-term expression and potential to mediate genomic instability is dangerous to the host. Absent a particular beneficial phenotype of the ERV, the host genome must ensure its moderation or face the consequences: unacceptably high rates of retrotransposition and genomic instability.

Today we appear to peacefully coexist with our complement of ERVs. This is testament that the vertebrate genome has prevailed in subjugating the successive waves of genomic invasion by retroviruses. The number of ERVs in the genome of vertebrates is highly variable, ranging from one to thousands of proviruses and perhaps ten times more solitary retrovirus virus long terminal repeats (LTRs) for a given lineage. Today most appear to have no function, confer no important phenotype, and evolve in a neutral fashion, with no consequence for genome fitness. They accumulate synonymous and nonsynonymous nucleotide substitutions with no bias and undergo deletion, rearrangement, and truncations, rendering them simply retroviral remnants with no function and no coding capacity. Many exist simply as a single solitary LTR sequence, an atoll of viral DNA devoid of it coding sequences, the result of homologous recombination between the LTRs.

The cell mobilizes several arms of its regulatory apparatus to lock down and silence its endogenous retrovirus invaders. *Host restriction factors* such as TRIM5α and APOBEC3 are employed to restrict the spread of the viral infection in the host somatic tissues (Malim and Bieniasz 2012). There is accumulating evidence that restriction factor genes are key weapons in the arms race between virus and host. They can be shown to have evolved in lockstep with waves of invading retroviruses each evolving and counterevolving under selection to maintain the upper hand in the relationship. The emergence of mutations in the natural receptor for the virus can also render the host species immune to infection. Most importantly the cell can leverage an elaborate regulatory apparatus that places a stranglehold on integrated proviruses. It can modify the fundamental structural architecture of the invading DNA, silencing its expression. Most ERVs are silenced by methylation of cytosine residues at GC sequences in the DNA strand; others are smothered by modification of the chromatin organization that packages the genomic material in the nucleus and controls the ability of genes and whole regions of chromosomes to be actively expressed (Maksakova, Mager, and Reiss 2008).

All of these mechanisms come into play during the uneasy period of retroviral endogenization. As we will see, the introduction of ERV genetic material into the germ line of vertebrates is analogous to the integration of phages into the prokaryotic genome. Each is a double-edged sword, catalyzing both dangerous genetic instability and evolutionary possibilities for the recipient genome. The establishment of the "end state" in which ERVs become stable elements in the genome and no longer proliferate is a protracted affair. The vast majority of the almost 100,000 ERVs in our genome today were fixed in the genome before hominids emerged; each ERV genome can be traced in the genome of every human individual and our closest relatives, the Old World primates. This dates the invasion of the primate germline to an earlier common ancestor of chimps and humans (Stoye 2012). Only one group of human ERVs, termed HERV-K, has loci not fixed in the genome: they remain as polymorphisms in human populations. HERV-K first integrated into a common ancestor of Old World primates and humans more than 30 million years ago. It has remained active for most of this period, and there are at least twelve new elements in the human genome that cannot be found in chimpanzees, from which we diverged only 6 million years ago (Belshaw et al. 2004). The existence of polymorphic loci in human populations betrays their relatively recent activity.

Belshaw and colleagues (Belshaw et al. 2004) published their analysis of the proliferation of this group of ERVs after its first acquisition in the primate genome. Their conclusions paint a complex picture of retroviral endogenization and proliferation across the genome by multiple cycles of reinfection by viable HERV-K viruses. Somewhat to their surprise, purifying selection operated on those viruses such that they retained the ability to generate new descendant ERVs by cycles of reinfection of the primate genome over 30 million years. A small minority of active, unfixed HERV-K viruses has continually seeded a dynamic, growing population of HERV-K elements. Despite being part of the host genome and governed by the nucleotide substitution rates inherent to vertebrate genomes, the HERV-K virus has continued to evolve independently of the primate genome and has continued to be under selection for replicative viability. The vast majority of HERV-K elements do become fixed and inactivated by mutational decay, and indeed the prototypical human genome sequence (Lander et al. 2001) contains no fully intact and functional HERV-K proviruses. Nevertheless, the continued growth of the HERV-K population evident in human subpopulations with different complements of HERV-K insertions in their genomes suggests that functional viruses have existed until very recently and might perhaps still be found in pockets of the human population.

Endogenization in Progress

The koala, an iconic marsupial of Australia, has a reputation as a sickly species. In the wild, almost one in twenty koalas succumb to lymphoma or leukemia; the number is thought to be 60–80 percent in captive animals. They also appear to be highly susceptible to chronic infections of a sexually transmitted chlamydial disease. Each of these observations is the calling card of an immunodeficiency, and by 2004 the press was touting "koala immune deficiency syndrome" (KAIDS). Some twenty years earlier, researchers at the University of Sydney had found retrovirus-like virus particles in koala leukemia samples. The virus was subsequently detected in koala genomic DNA by Jon Hanger and colleagues (2000) at the University of Queensland who were studying hematopoietic cell cancers in koalas. It could be found in healthy and diseased animals, and they concluded that it was an endogenous retrovirus. The researchers found it strange, however, that in contrast to most ERVs, the chromosomal provirus was fully functional. It retained intact all of its coding and control sequences and produced virus particles when blood lymphocytes were cultured in the laboratory. What was more, the koala retrovirus (KoRV) was genetically a close relative of gibbon ape leukemia virus (GALV), a retrovirus that infects the Gibbon ape and causes a similar disease to that of the koalas (Johnsen et al. 1971).

Following up these discoveries, scientists from the University of Queensland led by Paul Young (Tarlinton, Meers, and Young 2006) showed that different koalas had different numbers of proviruses in their DNA. Although some proviruses were shared between individuals, none were common to all animals. They also observed that unlike healthy koalas, diseased individuals had high levels of virus in their blood. Taken together this was more consistent with KAIDS being the result of exogenous retrovirus infection rather than an endogenous virus. Nevertheless, the scientists went on to show that the virus was integrated into germline DNA and that the individual proviruses followed Mendelian patterns of inheritance. Offspring inherited the specific provirus insertions from their parents. However, the number and pattern of KoERVs differed between individuals; they acted as polymorphic alleles in the population, a fact that convinced the scientists that they were witnessing the endogenization of KoRV into the koala population in real time. This would be the first opportunity for scientists to take a ringside seat and observe retroviral invasion of the germline of a wild species. We can expect that the intense evolutionary arms race that is to ensue between the virus and the host genome will be readily observable for the first time at the molecular and genome level.

The dynamics of KoRV invasion is readily apparent in epidemiological studies of geographically distinct koala populations in South Eastern Australia (Simmons et al. 2012; Tarlinton, Meers, and Young 2006). Prevalence of KoRV in northernmost areas approached 100 percent, and the average number of proviruses in the koala genome was 150, while more southerly populations had mixed prevalence. Here the copy number of virus genomes in each cell ranged from 1.5 to a mere 0.0001 KoRV proviruses per genome, a number only consistent with an exogenous retroviral infection. Koalas on Kangaroo Island, located off the southwest coast of Australia, are an isolated population. In the early part of the twentieth century the island was stocked from koalas, probably from French Island in Victoria. There was no evidence of KoRV infections on the island; the founding population must have been free of KoRV. Furthermore, in contrast to the populations of koalas in southeastern Queensland where one in twenty koalas surveyed over an eighteen-month study period succumbed to leukemia, there was no evidence of hematologic disease in Kangaroo Island koalas. Overall the epidemiology of koala immunodeficiency syndrome disease prevalence and Koala ERVs in koala genomes suggests a disease front is spreading from north to south in the koala's territory. The virus is gradually spreading through the koala population, both as an epidemic exogenous viral disease and as inherited Mendelian gene loci.

Many questions remain to be answered after the first revelation of the ongoing epidemic and koala germline invasion by KoRV. Ranking high among these are: When did the virus enter the koala population and what are its genetic origins? What will be the consequences for the embattled koala species? The first of these questions can be partially addressed. The proposal of Tarlinton and colleagues that KoRV entered the Australian koala population within the last century, spurred a group led by Alex Greenwood at the Institute for Zoo and Wildlife Research in Berlin to go looking for the virus in museum samples around the world. They used specialized facilities and techniques for extracting and studying ancient DNA (Avila-Arcos et al. 2013). These samples are often unstable and degraded, and the amount of DNA obtained for sequence analysis can be minuscule. Nevertheless, from twenty-eight museum koala skins collected as early as the late 1800s, they succeeded in sequencing mitochondrial DNA from eighteen koalas. This was the control signal that confirmed the technique was working. Of these eighteen only three were negative for KoRV and of these, two were from South Australian samples. Fifteen of sixteen northern koala specimens had KoRV DNA in their cells. More than a century ago, KoRV was already ubiquitous in the koala population. It is remarkable that its spread through the population has progressed so slowly. The authors suggested that this might be due to the relatively solitary behavior of the koala or through the geographic isolation of subpopulations by habitat fragmentation.

The origin of KoRV is a much greater puzzle. Together, GALV and KoRV represent a distinct monophyletic clade within the gammaretroviruses: they have surely descended from a common ancestor. The origin of GALV remains somewhat obscure; it first emerged as a pathogen that caused an epidemic of leukemias and lymphomas in captive gibbons in a Thai primate colony in 1972. Subsequently, Asian rodent cells were found to have an endogenous retrovirus antigenically similar to GALV that could infect primate cells. Hence, it is possible that an endogenous murine retrovirus caused a cross-species infection in gibbon apes (Tarlinton, Meers, and Young 2008). Apparently, the ERV of one host species can serve as a reservoir of viruses that can emerge later as pathogen in another. The question remains as to where KoRV came from; it does not share a geographic range with either Asian rodents or gibbons. Only recently has a possible solution emerged. Scientists from Queensland University succeeded in identifying a novel endogenous retrovirus in a native Australian rodent, *Melomys burtoni* (Simmons et al. 2014), that appears to be the missing evolutionary link between GALV and KoRV. It is a seductive hypothesis that a cross-species jump of *Melomys burtoni* retrovirus from the grassland rodents to koala started the epidemic. But we remain confounded: the Asian genus of rodent is not found in Australia. A unifying theory of the origins of GALV and KoRV may await the identification of a third host.

The genetic relatedness of GALV and KoRV is indisputable: they are by far each other's closest genetic relatives. Several laboratories have attempted to probe the evolutionary changes that might be ongoing during the current KoRV epidemic and its progressive endogenization into the koala genome. It is notable that while KoRV has relatively low virulence, its gibbon counterpart is an extremely aggressive pathogen. Two sequence motifs, one in the *gag* gene and one in the *env* gene, are associated with virulence in related gammaretroviruses. In KoRV, each of these motifs is mutated (this was also the case in all of the museum samples of the virus collected in the 1800s). Substitution of these motifs in GALV with those of the koala retrovirus results in an attenuation of GALV virulence (Oliveira et al. 2007). Researchers ponder whether this divergence of GALV and KoRV occurred after the KoRV epidemic began or whether it was a necessary prerequisite that enabled the virus to infect koalas. Is this attenuated phenotype of KoRV the secret to its capability to establish germline infection of the koala? It is entirely possible that GALV, an exclusively exogenous infection, is just too pathogenic to successfully colonize the primate germline. A similar hypothesis has been extended to explain the rarity of endogenous lentiviruses in vertebrate genomes (only two have been described to date); the pathogenicity of these viruses may generally preclude the formation of a viable zygote.

The outcome of the KoRV epidemic remains a great concern (Stoye 2006). Today there are fewer than 50,000 wild koalas living in fragmented habitats. The limitations of such a restricted population on the ability of the species to survive the current epidemic are clear. Perhaps of equal concern to ecologists is that genomic surveys of koala populations reveal them to have a remarkably limited genetic diversity. The population appears to have passed through bottlenecks, perhaps as a result of infectious disease or depopulation by hunting. In any case, today the genetic background of koalas is relatively homogeneous, a factor that places them at higher risk of extinction as a result of KoRV. The population may be unable to muster the necessary genetic variation to compete in the upcoming arms race. Germline colonization will be a strong selective pressure on the koala genome to evolve viral restriction mechanisms that can mitigate KoRV pathogenicity. The arms race is already escalating, as the first variant isolates of the KoRV with altered receptor utilization have been isolated. We can speculate that such isolates have an evolutionary advantage. The possession of endogenous KoRV may provide some superinfection immunity against wild type KoRV and perhaps mitigate disease manifestations by reducing viremia. Virus variants that can gain entry to the cell through an alternative portal may be able to elude such protection and retain pathogenicity in the host. The exogenous KoRV, an RNA virus, displays the evolutionary agility of its brethren and is already adapting to surmount host challenges to its infectivity. Will the koala population respond with similar alacrity and genetic innovation, given its small size and limited diversity, or will it face extinction?

Change Agents

A provirus that takes up residence in the germline of a host strikes out on an evolutionary pathway independent of its exogenous viral siblings. Exogenous retroviruses live by the same rules as all other selfish genetic parasites that must move from host to host to perpetuate their genetic lineage. Governing all aspects of their evolution are selective pressures operating on the relative success with which they amplify their genomes in a host and are transmitted to successive susceptible hosts. The ERV has crossed a line; it is no longer an independently evolving selfish element but, in the parlance of Richard Dawkins, a mere selfish gene. It is now in service to the host genome, and it must compete for its place in the host population gene pool like any other vertically inherited Mendelian gene locus. ERVs that affect the fitness of their host detrimentally will be at a disadvantage to those that are neutral or beneficial. This is analogous to bacteria that have profligate loads of prophages in their genomes. The vast majority of those prophages are defective; over time their replicative capacity and coding sequences have been inactivated by mutational decay because they offered their host genome no benefit or were too often harmful. Stable prophages in bacterial lineages may have introduced useful gene functions to the bacterium by phage conversion, or confer on the bacterial lineage protection against infection by related phages. Each of these scenarios plays out by the ERVs of vertebrates as they find their place in the host germline. On the one hand, transcriptionally active ERVs, particularly if they have the capacity to copy themselves and proliferate across the genome, represent the noisy and disruptive babble of ill-behaved pupils in a schoolroom. The teacher must ensure their obedience and silence if the class is to proceed and the lesson learned. On the other hand, the proviral DNA sequences are the class clowns, a gift of supplementary genetic diversity for the host cell whose expression may offer benefits. They comprise genetic control elements whose integration in the region of a particular gene can influence entire programs of cellular gene expression. ERV protein coding sequences can be molded by evolution, without jeopardizing existing functions. We have discussed gene duplication and its role in the creation of new genes in both viral and cellular genomes; the manufacture of a redundant and nonessential copy of a gene leaves it free to evolve divergently and take on new functions. The litter of ERVs across vertebrate genomes, mainly in the form of defective and deleted proviruses, mutationally inactivated protein coding sequences, and isolated LTRs, confirms that the vast majority of endogenized retroviral genomes decay to nonfunctional sequences. If they have been silenced by cellular regulatory mechanisms or if their gene products are not under purifying or positive selective pressures, their DNA sequences will be under no selective pressure to retain their functionality. Over time, they accumulate random mutations and their sequences drift without consequence to the host. That ERVs mostly become such inconsequential DNA is a topic of hot debate; it is, however, an easy matter to pick several examples that illustrate how hosts can benefit from their ERVs. As for all matters of evolution, we bear witness only to the successful events that are now fixed in genomes. Evolutionary failures, no matter that they far outnumber the successes, go unrecorded, rapidly purged from the gene pool.

There is abundant evidence that ERV integration at particular loci can profoundly potentiate the evolution of the host. Human ERVs are believed to have entered the genome between 10 and 50 million years ago and, with the possible exception of HERV-K elements, they are no longer active. Mice and other rodents, however, possess some ERVs that remain viable for retrotransposition to this day, and in some cases exogenous retroviral counterparts still circulate and cause infections. Intracisternal A particles in the mouse genome are one such example. They remain able to make copies of themselves that can reintegrate into the germline chromosomes. Such retrotransposition events are documented to cause at least 10 percent of spontaneous mutations in the mouse genome that give rise to observable phenotypes (Stocking and Kozak 2008). Given that mutations which are lethal to the gamete, the egg, or embryo will go unseen, the magnitude of the jeopardy to the mouse genome of active ERVs is quite clear. Hosts must exert some control over the expression of ERVs and suppress their mutagenic potential, particularly in embryonic and germline cells. Nevertheless, just as deleterious consequences of ERV activity are likely commonplace, some examples of beneficial outcomes exist. In early primates, we can pinpoint a particular ERV integration event in the locus of the pancreatic amylase gene that conferred upon our ancestors the ability to express their amylase genes in the salivary gland (Samuelson, Phillips, and Swanberg 1996; Meisler and Ting 1993). This heritable change provided for tissue-specific expression of the gene and gave us our sweet tooth. Here, the introduction of new gene regulatory DNA sequences close to the transcriptional start site of the amylase gene allowed salivary secretion of amylase. The resulting phenotype must have offered advantages to primates as they developed a diet containing more complex carbohydrates.

Homologous sequences in any genome are well known to align and be the nexus of homologous recombination between two pieces of DNA. The proliferation of ERVs of identical or highly homologous sequences has created genomes with more than 1,000 copies of such sequences. Genetic researchers therefore had reason to believe these sequences dispersed in the genome would allow for recombination events leading to rearrangement, deletion, or duplication of genome sequences. In 2001 researchers in one of the leading laboratories studying retroviral genetics went in search of direct evidence that such events are a generality and could contribute to evolution (Hughes and Coffin 2001). They chose to focus on HERV-K elements since there is strong evidence that they have been active in the primate genome for much of the last 30 million years. They analyzed the DNA sequences and deduced the phylogenetic relationships of thirty-five full-length HERV-K proviruses in human DNA and compared their distributions in other primates. Some were shared with multiple other primates; others were specific to humans or were shared only with gorillas and chimpanzees. This is clear evidence that they resulted from evolutionarily recent integrations, some indeed after hominid species emerged. Their research provided further insights. They exploited fundamental knowledge about proviral integration: LTRs that flank each provirus are identical in sequence at the moment the provirus is integrated into the genome. Furthermore, each LTR sequence is linked to a unique piece of flanking cellular DNA, allowing its location to be mapped definitively in the genome sequence. It is expected that in ERVs each of the provirus LTRs will accumulate nucleotide substitutions at the same low rate as cellular DNA. LTRs of the same provirus are created by one integration event at the same time and will consequently cluster most closely together in a phylogenetic tree. Hughes and Coffin's work (2001) revealed that the LTRs of some HERV-K loci did not cluster together on the phylogenetic tree, and the researchers concluded that their flanking LTRs originated from different viruses, integrated at different times. Inter-element recombination had taken place between two individual proviruses, creating HERV-K elements with "mismatched" LTRs. In some instances, there was evidence that the tell-tale host flanking sequences were also mismatched. In all, almost one in five full-length HERV-K elements had undergone a recombination event that must have resulted in substantial reshuffling and rearrangement or deletion of cellular sequences. This analysis placed no more than thirty-five or so HERV-K elements under the magnifying glass, but our genome has almost 100,000 ERV sequences. If this phenomenon can be generalized (and why not?), the genetic innovation resulting from this genomic "burden" of ERVs and the consequences in our evolution cannot be underestimated.

Genome rearrangements mediated by ERV loci must have vigorously stirred the evolutionary pot, most often creating genomic anomalies that were denied inheritance. Every recombination event results in two reciprocal chromosomal products. Notably no evidence for the existence of the reciprocal products of these HERV-K recombination events was found. There is clearly too little tolerance for most genomic rearrangements, and they are never propagated but consigned to the genetic scrap heap, allowing only the healthy to flourish. Such events have obviously been productive for our evolution; ERV loci have been implicated in catalyzing genome duplication and diversification of the histocompatibility gene locus and thus may have assisted in the expansion and development of our adaptive immune repertoires (Kambhu, Falldorf, and Lee 1990). As comparative genomics unfolds in future years, we are sure to uncover more instances of genomic disruption mediated by ERVs which have been and will continue to be a factor in genetic variation of vertebrates, for the better and for the worse.

Domestication of ERV Genes

The physical remodeling of the host genome is one potential fallout from integrated ERV sequences. Recombination events driven by sequence homology among multiple ERVs distributed across the genome are just one aspect of disruptive innovation that can influence the evolution of their host genome. Again, by analogy with the relationship of bacteriophages with their hosts, the potential for gene conversion, in which the host cell acquires useful genes from its parasite, exists also for eukaryotic cells. Prophages often provide "ready made" solutions, in the form of useful gene functions that offer a selective advantage to the host cell (refer to Chapters 2 and 3). This is also the case for retroviral proviruses, although the diversity of gene functionality that they can provide the host is more limited. Notably, retrovirus *env* genes have proven to be most useful to their vertebrate hosts, and their functionality has been repeatedly co-opted during vertebrate evolution. That some ERV genes have been advantageous to evolving vertebrates is betrayed by their conservation over many millions of years as functional genes. Rather than decaying by random mutational drift, natural selection on the host ensured that their protein coding sequences remained intact and capable of expressing functional proteins. They display a higher rate of synonymous than non-synonymous mutations, indicative of purifying selection to preserve a particular function. They must have provided a fitness advantage to the genetic lineage of their host and been adopted. The host domesticated the viral gene and repurposed it. The *env* gene product is a viral glycoprotein whose principal role in viral infection is to serve as the entry machinery of the virus. All *env* gene products are fusogenic proteins, mediating the fusion of two membranes, those of the virus envelope and the cell. They vary in other functionalities and different retrovirus env proteins recognize different and distinct cellular receptor proteins; murine and primate retrovirus *env* genes products are also known to be immunosuppressive. All of these functionalities have been found useful to vertebrates, leading to the evolution and conservation in vertebrate genomes of many genes with their origins in retroviral *env* genes.

The receptor specificity of ERV envelope proteins has proved to be a useful function for cells to co-opt and retain. A mouse gene termed *Fv4* (Friend virus susceptibility gene-4) is one such example. It originates from a defective murine ERV that retains the ability to express its cognate env protein on the cell surface. Mice that possess the active (resistance) allele of *Fv4* have been shown to be resistant to the pathologic effects of an exogenous murine retrovirus known as Friend virus (Odaka et al. 1981). In wild mouse populations in California, the same gene was found to render them resistant to a retrovirus-induced lymphoma. The *Fv4* gene product on the cell surface engages the cellular receptor of particular virus strains, causing receptor down regulation on the cell surface and leaving no attachment sites for exogenous viruses to dock and initiate infection. This superinfection interference is likely a common selective advantage of *env* adoption by hosts who benefit from the resistance to infection and disease. In a similar fashion, also in the mouse, the evolution of particular intracellular retroviral restriction factors has resulted from exploitation of retroviral *gag* gene products. The mouse *Fv1* gene is a homolog of a retroviral *gag* gene. Several *Fv1* residues show evidence of strong positive selection (Yap et al. 2014), indicating evolution molded the retroviral gene for the specific benefit of the host. *Fv1* appears to function in a manner similar to TRIM5α, interfering with infection by binding retroviral capsid structures and disrupting their ordered disassembly in the host cell cytoplasm (Hilditch et al. 2011). Not only do *Fv1* alleles disrupt murine retrovirus infection, they also appear to interfere with infection by other types of retroviruses, suggesting that they have evolved to be critical elements of the host antiretroviral machinery in the mouse.

The immunosuppressive qualities of retroviral env proteins have been attributed to the possession of certain peptide sequences, the immunosuppressive domain (ISD), in the transmembrane subunit of retroviral env proteins. Its existence first emerged when, in 1985, scientists began efforts to unravel immunosuppression caused by HIV-1. It was discovered that a subdomain of twenty amino acids, conserved in the env proteins of primate and murine retroviruses, was associated with inhibition of lymphocyte proliferation (Cianciolo et al. 1985). A French research team lead by Thierry Heidmann performed seminal studies in which retroviral env was experimentally expressed on the surface of allogeneic tumor cells engrafted in mice. An allogeneic transplanted cell is one that is immunologically foreign to the mouse into which it is engrafted. Under normal circumstances, when such tissue is transplanted into an immunologically competent mouse, a graft-versus-host response occurs, and the mouse immune system rejects the graft. This is the same response that must be calmed in patients who receive organ grafts of nonidentical tissue types. When the engrafted foreign tumor cells expressed the retroviral protein on their surface, however, they were protected from immunological rejection (Mangeney and Heidmann 1998). Identifying the protein motif that mediated this immune suppressive property of env provided a key to unlocking its function in retroviral infections. The Heidmann team created mutant retroviruses with their *env* gene immunological suppression domain "switched off." Although the mutant virus infected cells grown in culture with the same efficiency as the wild type, it was defective for infection of animals. It could successfully infect and grow in mice that were devoid of a functional immune system after X-irradiation, but it was rapidly cleared and could not successfully infect normal mice (Schlecht-Louf et al., 2010). Normal retroviral virulence required the immunosuppressive functions of the envelope protein, which were shown to counteract both the innate and adaptive arms of the host immune system.

By the early 1990s, researchers had begun to characterize ERV genomes that had conserved protein coding open reading frames despite being part of the human genome for more than 30 million years. In 1993 a team led by Dr. Robin Weiss at the Institute of Cancer Research Chester Beatty Laboratories was first to bat. They described the preferential expression of the endogenous retrovirus ERV-3 in the placenta in syncytiotrophoblasts. Here were the first clues that ERV env proteins might play a role in forming the placental immunoprotective barrier that is constituted from a fused cell layer of syncytiotrophoblasts (Boyd et al. 1993). As it turned out, a case could not be made that ERV3 plays a critical role in the formation of the human placenta; 1 in 100 people have a polymorphism that renders it nonfunctional. Nevertheless, the field was galvanized to pursue this line of investigation and soon other HERV *env* genes were under scrutiny as candidates. These were env proteins from two different ERV families, HERV- W, and HERV-FRD and they were termed *syncytin-1* and *syncytin-2*. They are highly conserved among related species, showing strong signs of purifying selective pressure. They are both expressed in the placenta and bind receptors that are also expressed there. What was more, when expressed in cells in culture, they readily caused them to fuse to form large multinucleated syncytial cells, a feature typical of syncytiotrophoblast formation. These *syncytins* are retroviral proteins that primates exploit to catalyze placenta formation.

ERV envelope genes possess unique properties that make them suitable for use in forming the placenta: they are fusogenic proteins and they have immunosuppressive properties. Eutherian (placental) mammals distinguish themselves from nonplacental animals in the ability of the female to nurture the fertilized ovum and growing embryo within the body. The placenta is a transient tissue of embryonic origin whose evolution made it unnecessary to partition the embryo into a protective egg, which matured outside the mother's body. It serves two purposes for the maturing embryo: it is a conduit for respiratory gasses and nourishment supplied by the mother, and it provides an environment of immune tolerance. The fetus is necessarily half-foreign tissue, an allograft within the mother. It draws half of its genetic, and hence antigenic, identity from maternal and half from paternal genes. If the fetus is to mature within the mother, it must be isolated from the maternal immune system such that a graft-versus-host response does not reject it. The placenta forms early after implantation of the embryo. Syncytins mediate the formation of a continuous fused layer of cells around the embryo, isolating it from the mother, yet allowing essential nutrients to traverse from the mother's system. Although the observations on human *syncytin-1* and *-2* were compelling, it was left to scientists to definitively link syncytins to placental formation by studying mice. Here two syncytins (dubbed A and B) from murine ERVs were implicated, and genetic experiments with mice defective in these genes confirmed that their dysfunction disrupted placental formation. Notably, however, *syncytin-A* and *-B* were not syntenic with the human syncytins. That is, the human and mouse genes are not descended from common ancestral syncytins; they have arisen by separate ERV gene capture events from different families of ERV in human and mouse ancestors. Regardless, all of the syncytins recognized today in various mammalian species retain both their fusogenic properties as well as their immunosuppressive domains. That different placental mammals use distinct syncytins was not unexpected. Developmental biologists had long recognized that the placenta is one of the most structurally and functionally diverse organs of mammals. Neither did it escape their attention that, while the emergence of eutherian mammals can be dated quite accurately to at least 150 million years ago, the syncytins that are currently responsible for placenta in extant species of eutherian mammals are less than 50 million years old (Lavialle et al. 2013). The explanation for this discrepancy must lie in fact that placental eutherians emerged by domestication of ERV *env* genes, but as waves of subsequent retroviral invasion of the genome occurred, the first syncytins were replaced on multiple occasions by newly co-opted ERV env proteins. In evolutionary terms, it seems that placental mammals continually "traded up" their syncytins when a new model offering a selective advantage over its precursor was available. The original syncytin genes, no longer under selective pressure for functional activity would then be predicted to undergo mutational decay, perhaps remaining only as vestigial nonfunctional fossils in our genomes today.

Endogenous Viral Elements

If the abundance of retroviral DNA in our genome came as a surprise to scientists, the paucity of signature remains from other virus families was not a shock. Obligatory to retroviral replication is the establishment of an integrated provirus from which progeny genomes are copied. The virus must only infect germline cells for endogenization to become a possibility. On the other hand, other classes of viruses do not naturally integrate into host DNA during their replicative cycle. As you will now fully appreciate, virus lifestyles are diverse indeed. Their genomes are double strands of DNA or RNA, single strands of DNA or RNA of positive or negative polarity; they are circular, linear, or segmented; some use the cell nucleus, while others find no reason to enter the nucleus at all and replicate exclusively in the cytoplasm. It was therefore of little surprise that nucleotide sequences from these viruses are not abundantly represented in eukaryotic host cell genomes. In recent years, however, the emergence of whole genome sequencing and more advanced bioinformatic tools, reveals that eukaryotic genomes do carry a remarkable diversity of endogenous viral elements (EVEs). If one looks exhaustively, some genetic material from the genomes of viruses in all of these virus categories are found in host genomes (Johnson 2010). This result emerged from the work of several teams of scientists who combed mountains of sequence information (Belyi, Levine, and Skalka 2010a, b). In species as diverse as vertebrates, including mammals, marsupials, birds, lampreys, and fish, as well as insects they found an abundance of diverse virus-derived, but nonretroviral, sequences. In contrast to ERVs that were first endogenized as complete viral genomes that subsequently proliferated in the host germline, this does not appear to be the case for nonretroviral EVEs. One to a few fragments of virus sequences or at most a single gene is integrated. That these rare integration events occur at all is thought to be attributable to viral messenger RNAs being accidentally reverse transcribed and integrated into the genome by the enzymatic machinery of retroviruses or retrotransposons. This is particularly likely where whole messenger RNA sequences have been incorporated into the genome. The initiation of reverse transcription of LINE element genomes undergoing a retrotransposition event is initiated at the polyadenylated terminus of the LINE element transcript. It is quite likely that abundant viral mRNA transcripts, which are also polyadenylated, might hitchhike on this mechanism and integrate in the same manner as a mobilized LINE element. Nonspecific accidental or aberrant nonhomologous recombination events may also occur and contribute viral sequences to host genomes.

It must be exceedingly rare for fragments of viral DNA with no function to survive as alleles in their new host population and ultimately, through chance events, become fixed. The existence of such fragments of ancient virally derived DNA sequences is evidence that despite having no beneficial function (or at least none that we can perceive today) some have survived in host genomes. They are useful genetic markers to evolutionary biologists because sequences without a function, they languish in host genomes, subject to neutral mutational drift and decay at a rate characteristic of the evolutionary clock of the host genome. In vertebrate genomes, it is expected that nucleotide substitution rates in such neutral DNA will result in a constant rate of evolutionary change over time. Together with knowledge of the genomes of phylogenetically related species, it is possible to deduce the date at which the viral DNA entered the genome in the most recent common ancestor. These data confirm that many of the viral fragments in host genomes were derived from quite ancient viral genomes. In the studies of forty-eight vertebrate species, Belyi, Levine, and Skalka (2010a, b) discovered eighty EVEs in nineteen vertebrate species related to negative-stranded RNA viruses that circulate today. These genome fragments were dated to more than 40 million years ago. A phylogenetic comparison of the EVE sequences with related contemporary viruses revealed that the ancient sequences cluster together but are not part of the same monophyletic clade that evolved into today's circulating viruses. In other words, today's viruses are not direct descendants of the viruses that were circulating 40 million years ago. The fragments of viral genomes in vertebrate DNA are truly fossils of long-extinct lineages of viruses that have been more recently replaced by contemporary lineages.

These are indeed compelling observations, but just as the functionless sequence fragments inform us on the evolutionary history of ancient viruses, some EVEs contain longer functional coding sequences. Some of these did not mutationally decay at the rate expected of neutral sequences and are conserved as open reading frames to this day. Evolution captures random events but as with the domestication of ERV genes, the probability of such sequences retaining their function over such long evolutionary periods of time is negligible, unless they are under strong selective constraints. They must provide some service that is relevant to the host (or at least has been important for some significant period of their evolutionary journey together) such that their function is preserved by purifying selective pressure.

The most notable of these EVEs in vertebrates are of negative-stranded RNA virus origin and, remarkably, are from just two orders of the *Mononengavirales*: the bornaviruses and Ebola/Marburg viruses. Each of these viruses causes a lethal infection in some species. Bornaviruses are not pathogenic in human but cause fatal neurological diseases in susceptible animals such as sheep, horses, and cows; Ebola virus is the feared agent of hemorrhagic fevers in humans, but appears to be less pathogenic in species of bats believed to be its natural reservoir.

Bornavirus-derived EVEs are widespread in mammalian genomes from all geographies. Belyi and colleagues found them in thirteen species in all, and most frequently they comprised sequences derived from the viral nucleocapsid gene. Sequences related to Ebola virus/Marburg virus were found in six species and originated primarily from two genes of the virus, the nucleoprotein (NP) and the polymerase complex cofactor (VP35). The authors noted that the sequences were often acquired in multiple independent events over extended periods of evolutionary time. This is surely another indicator that the integrated genetic material offered a substantial competitive advantage to the gene pool of the host population. The authors of the study serve up a plausible and attractive hypothesis to explain the conservation of these ancient viral sequences in vertebrate genomes. The expression of these proteins may confer upon the host a degree of resistance to the viral infection, in a manner analogous to the murine retroviral viral restriction factor *Fv1*. That retroviral *gag* gene homolog purloined from the virus genome protects hosts from retrovirus infections by physical interaction with the retroviral *gag*-containing capsids. The fragments of bornavirus and Ebola/Marburg virus—like proteins expressed in cells appear to compose subdomains of viral proteins that function as multimeric protein assemblies in the infected cell. It is possible then that the proteins expressed from the EVE will be similar to, but perhaps not identical to, the homolog expressed by the infecting virus. These mismatched proteins may corrupt the functioning of the native viral proteins, leading to an antiviral protective effect. If a bricklayer's hod contains a few bricks of irregular size, their incorporation into his wall will cause defects both in its structural integrity as well as its aesthetics. The EVE gene products might be such agents of interference. That some EVEs act as viral restriction proteins is an attractive explanation for their conservation for many millions of years. The concept is further strengthened by the distribution of the EVEs in question across different vertebrate species. A persuasive correlation is seen between the possession of an EVE and resistance to the pathogenic effects of the related virus. Cows and horses have no bornavirus-like EVEs and are susceptible to lethal infection, while we have them and bornavirus infections are not clinically evident in humans. Equally, primates that are highly vulnerable to Ebola/Marburg virus have no related endogenous viral sequences, but they can be found in bats. Could this be one contributor to certain bat species being resistant to the pathogenic effects of Ebola?

Indirect support for this concept recently emerged from studies of endogenous bornavirus-like elements (EBLs) in the genome of the thirteen-lined ground squirrel (Fujino et al. 2014). Kan Fujino and his collaborators focused study on a ground squirrel EBL that is transcribed into mRNA that comprises an open reading frame with 77 percent amino acid sequence with the modern Borna disease virus (BDV) nucleoprotein. Such conservation of the expressed EBL sequence led them to speculate that it is indeed a co-opted viral gene functionally conserved through evolution. They made a strong case by expressing a cloned copy of the ground squirrel EBL in BDV-infected cells, where it inhibited replication of the virus. It became incorporated into viral ribonucleoprotein complex and inhibited the viral RNA replicase. Its mode of action is most likely attributable to hetero-multimerization with the native BDV nucleoprotein, rendering the complex functionally inactive. The endogenization of this virus-like sequence may provide the ground squirrel with protection from infection by related exogenous viruses. It seems quite plausible then that the endogenous filovirus-like elements of bats are conserved factors which protect bats from related viruses and allow wild bat populations to harbor a reservoir of these hemorrhagic fever viruses that can reemerge periodically to infect susceptible species such as humans and other primates.

Just as murine retroviruses provided some species with the genetic material with which to develop immunity to infection, so too did these viruses. Development of this antiviral response may have been just one escalation in the arms race between these viruses and their natural host. As we see for other infections, reduced pathogenesis can lead to a more persistent and a less harmful infection of the host that is evolutionarily beneficial to both parties. Otherwise, strictly epidemic infections that are rapidly lethal may put the virus at risk of extinction, unless the virus lineage can persist in a reservoir species. EVEs may have contributed to the provision of such a reservoir species. The ancient viruses that left EVEs in host genomes 50 million years ago appear to have met extinction, and are not the direct ancestors of the viruses that circulate today. This may, in fact, be one force that has shaped the Ebola virus, Marburg virus, and bornavirus lineages that are circulating today.

· 15 ·

VIRUSES AS HUMAN TOOLS

Today we manage many endemic human viral diseases with vaccines. The deployment of the smallpox vaccination and the campaign to eradicate the disease culminated with the 1977 declaration that the world was free of smallpox. Highly effective childhood vaccines for viral diseases are almost universally available today; the mumps, measles, and rubella vaccine (MMR) and polio vaccine have all but eliminated these once-feared diseases in the developed world. Before, they were a serious concern to parents. In my childhood these virus infections were a rite of passage and like many children of my age during the 1960s, I acquired immunity to measles, rubella, mumps, and varicella zoster viruses the hard way. These concerns might be considered quite baroque today (even more so given that NASA's declared objective of putting men on the moon would be achieved later in the same decade), but it would not be so without the advent of viral vaccines.

We have succeeded in turning viruses on themselves and using them as tools. Their employment in vaccines to fight infectious diseases is now a well-established area of endeavor. These tools can truly be regarded as biological (viral) control agents. Live virus and live-attenuated virus vaccines are used to control viruses of humans and domestic animals. There have also been celebrated attempts to use viruses to control mammalian pests that they infect—a form of bioremediation. I will review one example of virus-mediated pest control (a salutary failure that I expect readers of this book will understand and anticipate). It is the canonical parable of virus-host coevolution. It can be understood in the context of the biology of viral evolution and is a remarkable example of the Red Queen in action.

Myxoma Virus: Biological Control

In 1896 an epidemic disease spread through a colony of European rabbits imported to the Pasteur Institute in Montevideo, Uruguay. This was a cross-species transmission of a myxoma virus, a poxvirus whose natural host was tapeti, a species of South American rabbits. Myxoma virus caused relatively benign fibrotic lesions in tapeti, but was highly pathogenic in European rabbits. It caused a severe generalized disease (myxomatosis) resulting in almost 100 percent mortality. Mosquitoes transmitted the disease. About a hundred years earlier, the rabbits themselves caused an epidemic. European settlers of the Australian continent brought rabbits with them to their new home. What started out as a favorite food of the settlers, perhaps a source of fur, and possibly an animal for recreational hunting, spread continent-wide during the eighteenth century. By 1920 Australia was home to 10 billion rabbits. As early as 1919, a Brazilian scientist, Dr. H. de Beaurepaire Aragao, believed he saw a utility for the myxoma virus. He wrote to the Australian government to advocate the use of myxomatosis to deal with their rabbit infestation (Fenner 1983). Despite this initiative, little came of it until after the Second World War, when a special section of the Commonwealth Scientific and Industrial Research (C.S.R.O) was established with the mandate to control the pest.

Efforts to control rabbit populations by spreading myxoma virus infections among them started in earnest in 1950. They were noticed by Dr. Frank Fenner, an up-and-coming, and later eminent, Australian virologist, and author of the classic textbook *The Biology of Animal Viruses*, published in 1974. We are fortunate for his prescient interest in what turned out to be a real-world experiment in virus-host evolution. His diligent epidemiologic and laboratory studies of myxoma virus field isolates from wild rabbit populations illuminated the natural history and evolution of the emerging disease in its new host. Fenner documented population changes, infection rates, and the virulence of myxoma viruses isolated from rabbit populations as the epidemic proceeded. Quite recently, analysis of the same virus isolates at the level of their complete genome sequence has completed the picture. A molecular understanding of the process of myxoma virus-rabbit coevolution is now at hand.

Following the earliest introductions of the myxoma virus into rabbit populations, Fenner and colleagues began their studies of the disease in rabbits in the Murray Valley. Inoculation and release of 100 infected rabbits into a population of 5,000 reduced the rabbit population to 50, a 99.8 percent case fatality rate. Notably, however, when some uninfected rabbits entered the population from a remote region and breeding occurred, the population swelled to more than 500. An outbreak of the disease then occurred reducing the population to 60. Remarkably, the surviving animals all had antibodies to the myxoma virus. They had not avoided infection; they had survived it. The case fatality rate was now 90 percent, more than tenfold lower than the previous season (Fenner 1983). Fenner took the isolates from these infections to his laboratory and used them to infect laboratory rabbits. They were able to demonstrate that the strains circulating after one year were of lower virulence than the virus initially released. In work spanning thirty years, Fenner and colleagues carefully monitored the virulence of circulating myxoma virus field isolates. It is a remarkably detailed accounting of the evolution of virulence phenotype as the virus adapted to a new host, and provides elegant data to illustrate how viral virulence and transmission trade-off against one another to achieve a balance that is optimal for fitness within the new host population (Marshall and Fenner 1960; Fenner 1983). The myxoma virus isolates were classified by virulence grade from I (highest virulence) to V (lowest virulence grade) based on different case fatality rates and survival times after infection. The most virulent viruses, classified grade I, killed 99 percent of infected animals within thirteen days, while the viruses scored with grade V virulence caused no more than 50 percent mortality with survival exceeding fifty days. Fenner and his colleagues assembled their vast collection of data from years spanning 1950 to 1981. For each year of testing, viruses were collected from hundreds of rabbits and each isolate classified by virulence grade; they were thus able to picture the frequency distribution of virulence in the circulating virus population (i.e., the number of viruses of each different grade of virulence). When the virus was first introduced, 100 percent of the virus isolates were of grade I virulence. Within five years only 13 percent of virus isolates were so highly virulent; viruses ranged in virulence from grade I to grade IV. Within eight years, fewer than 1 percent were of grade I virulence and 15 percent had grade V virulence. Each and every year, however, after the first year of introduction, the most prevalent viruses, and always more than half, had grade III virulence. The virus did not continue to evolve to lower virulence; it seems that an equilibrium was being established with an optimal virulence phenotype predominating. So how can this be explained? I think that now after our extensive discussions on viral evolution and the selective pressures at work on the virus population, this should be quite clear—it is as one would predict.

Viral lineage success is always based in the parameters of virulence and transmission; biting mosquitoes transmit myxomatosis and sufficient viral titers must be achieved in the circulation of the rabbit for mechanical transmission. The infectious period is that time during which sufficient viral titers are in the blood to mediate transmission. A rabbit that develops high virus titers in the blood, but dies quickly after infection, will have a short infectious period; it follows that a less virulent virus may provide a more extended infectious period. On the other hand, viruses with further reduced virulence may have lower titers in the blood, be cleared by the host quickly, and cause little disease thus also have a short infectious period. "Strains of grade III virulence were highly infectious for the lifetime of the rabbits that died and for a much longer period in those that survived," observed Fenner, in his Florey Lecture of 1983 at the Royal Society (Fenner 1983).

So what is happening to the rabbits? Are they evolving too? The answer is, of course, yes! In carefully designed experiments they studied rabbits from surviving populations after successive epizootics (outbreaks) of myxomatosis. They identified animals that were seronegative for myxoma virus antibodies and hence had not been infected by the virus. The animals from the progressively more virus-experienced populations were challenged by infection with an identical grade III virulent virus. It was immediately evident that populations which had been exposed repeatedly to myxoma virus epizootics had become more resistant to the pathogenesis of the virus. Infected rabbits from populations that had experienced seven epizootics of the virus exhibited a case fatality rate of just 30 percent, while the baseline population had a 90 percent fatality rate. These results were complemented by experimental breeding experiments (Sobey 1969) that demonstrated similar levels of resistance to a grade III virus could be achieved over six generations of breeding survivor animals. Both of these results are definitive, regarding the evolution of phenotype, and the rapidity of host change was quite surprising. Under natural selective pressures these changes can occur quickly, but at the cost of substantial mortality in the host population of animals. The expected arms race was clearly under way between myxoma virus and European rabbits. Just as virulence diminished in the virus, finding an optimum at grade III, the host also evolved a degree of resistance to the virus. Fenner asked whether the two phenotypes were interacting. If host resistance increases, would selection on the virus then favor more virulent viruses in response, as the arms race continues? The answer was affirmative: as the rabbit populations became more resistant to the virus, the frequency at which more virulent virus isolates were recovered began to rise.

The plan to control rabbits in Australia using myxoma virus was ultimately a failure, a casualty of the Red Queen. It did, however, provide us with a powerful example of evolution in process on both a virus and a vertebrate host occurring over a strikingly short time period. The experiment is not unique and has been reproduced. A French landowner, bothered by rabbits on his land, imprudently introduced myxomatosis in order to reduce their numbers. Three years afterward, it had spread naturally across Europe and to Britain where it was estimated that 90 percent of the rabbit population died. Today the populations have rebounded; the evolution of viral attenuation and population rebound exactly mimicked that seen in Australia (Kerr et al. 2012).

Genomics of an Attenuated Poxvirus

What is possible today, but was impossible in the laboratories of Fenner and his contemporaries, is the ability to examine the field isolates of myxoma virus at the level of their whole genome nucleotide sequence. Colleagues from C.S.I.R.O in Australia and from the United States have done just that in an attempt to elaborate the genetic changes underlying the attenuation of virulence that the circulating virus lineages experienced (Kerr et al. 2012). Their findings were striking, and in part a reflection of the complexity of poxvirus genomes and the very strong selective pressure applied by a new host. Kerr and colleagues reported the whole genome sequences of the first myxoma virus used to infect rabbits in Australia, together with those of viruses of various levels of virulence as defined by Fenner and isolated over more than forty years of the epidemic. The foremost observation was that during the period studied, the virus experienced a nucleotide substitution rate of 10^{-5} substitutions per site per year, an evolutionary rate higher than ever recorded for a DNA virus; nonsynonymous changes were abundant, indicative of strong positive selective pressures. Perhaps more consequential for our understanding of poxvirus evolution is the observation that mutations in the attenuated viruses occurred in multiple different genes and were often associated with loss of gene function. The myxoma virus genome, as is typical for poxviruses, possesses a centrally disposed core of relatively conserved genes, flanked by more highly variable genes with roles in host range and immune evasion. Myxoma virus is estimated to have twenty to forty genes with these functions (Cameron et al. 1999; Stanford, Werden, and McFadden 2007). A preponderance of the mutations observed in the forty-nine years of myxoma virus evolution occurred in these flanking region genes as expected. Even so attenuation of grade I virulent virus could not be associated with a shared mutational change. While the tendency of evolution was toward a shared phenotype of similar but moderate virulence (grade III), the genetic pathways to achieve it were extraordinarily diverse.

Evolutionary rates higher than other DNA viruses have previously been tentatively associated with variola virus, and may be one feature of poxviruses endowing them with host switching capabilities that approach those of RNA viruses. The results of Kerr and colleagues do, however, stress a key difference between poxviruses with their genomic flexibility in size and composition and RNA viruses whose genomes are restricted in size and densely packed with coding information. The complexity of the myxoma virus genome evidently provides a multitude of possible pathways to improved fitness in the new host. On the other hand, as we saw with zoonotic HIV, a single adaptive mutation appeared to provide the "key" initial genetic change to unlock successful chimpanzee infection. RNA viruses generate genetic diversity at much higher rates, permitting them to explore the more restricted genetic space that provides for their adaptive evolution to a new species.

Orthopoxviruses: Past Solutions and Future Problems

Vaccination to prevent viral diseases is one of the highly celebrated successes of modern medicine, and it has been written about extensively. Vaccines are sophisticated tools used to control viruses, invented and deployed to protect populations of otherwise susceptible human hosts from particular viral infections. Today these tools come in a variety of models with different pros and cons associated with different technologies that often utilize complex biotechnological processes. I will restrict myself to two different models: live heterologous virus vaccines and live-attenuated virus vaccines. For our purpose, I will limit discussion to a few observations on why these tools have been successful and how their effectiveness relates to virus evolution and speciation.

The topic of smallpox has arisen several times in these pages. Its global eradication was a signal success for society. The variola virus was driven to extinction, unable to maintain its basic reproductive number greater than unity. Comprehensive vaccine distribution programs, together with ingenious vaccination strategies were employed. A commonly used and effective strategy was *ring vaccination* of susceptible subjects in a ring around the location of disease cases. This creates a buffer zone of vaccinated and hence immune individuals around the outbreak, locally stemming the epidemic spread of infection. The virus was ultimately snuffed out because there were inadequate numbers of susceptible human hosts for the virus to infect and no natural nonhuman reservoir for the virus exists.

Edward Jenner had the seminal insight that milkmaids who contracted cowpox did not catch smallpox; as country doctor he administered the first vaccine to Joseph Phipps in 1796. Before this time the only available recourse for prevention of smallpox was *variolation*, a procedure in which a person's skin was scarified with an infected needle. It had been recognized in Asia as early as the tenth century that the pockmarked survivors of smallpox never caught the disease again. Today we recognize this as protective immunity. Lady Montagu, who observed the procedure successfully carried out on her son in Constantinople (now Istanbul), Turkey, imported the practice of variolation to Britain, where it was first used on her daughter in 1721. When skillfully performed, the procedure resulted in only mild disease in most individuals. It is speculated that the novel route of administration places the virus at a disadvantage, resulting in moderated pathology, perhaps by provoking a superior and more protective immune response. Nevertheless, the results were erratic and 2 to 3 percent of those inoculated died. Considering that the case fatality rate of smallpox often exceeded 30 percent, this was a risk that people were ready to take. The advent of Jenner's vaccine, however, put a rapid end to variolation.

Both variola and cowpox viruses are in the genus *Orthopoxvirus*, but while the host range of variola is restricted to humans where it is especially virulent, cowpox can infect a broad variety of mammals including (but are certainly not restricted to) humans, rats, carnivores, cows, and even elephants—a broad host range. It is generally less virulent than variola virus and causes only mild disease in humans. The narrow host range of the smallpox virus is likely due to a lack of host range genes necessary for poxviruses to exploit multiple hosts. As a consequence of its restricted host range, it has no reservoir host species. As you will see, viruses for which vaccination campaigns have been highly successful are all strictly human viruses. It would have been impossible to eradicate smallpox if a reservoir of the virus existed in wild animal populations.

Jenner began to vaccinate against smallpox with cowpox isolated from bovine lesions (recall that Twort was harvesting lesions for vaccine when he discovered bacteriophages of *Staphylococcus* that contaminated his samples) but today various strains of vaccinia virus are used. This is an orthopoxvirus that is closely related phylogenetically to both variola and cowpox, and like cowpox, it has a broad host range and causes a relatively mild disease. It is the close evolutionary relationship of the orthopoxvirus genomes that results in antigenic cross-reactivity between the viruses of this genus. Our immunity to smallpox pursuant to vaccination with vaccinia virus is actually a result of cross-protection by an immune response directed to vaccinia virus antigens. It is the low pathogenicity of the broad spectrum vaccinia and cowpox viruses that suits them for use in a vaccine. Smallpox vaccination results in relatively broad immunity to all orthopoxviruses. Today the virus has long since been declared extinct, so we no longer vaccinate against the disease. Only small groups of individuals, scientists that work in the laboratory with orthopoxviruses, routinely receive the vaccine as a precaution against accidental infection. Smallpox vaccination was stopped in 1972 in the

United States, and in 1979 the WHO advised a worldwide cessation of vaccination. As this last generation of vaccinees ages and dies, they will be replaced by a vaccine naive population. It has been suggested that in the absence of this immunity, the population will become more susceptible to the zoonotic emergence of virulent poxviruses (Shchelkunov 2013). Some of the broad host range poxviruses have a far larger gene complement than the virulent variola, leading to speculation that evolution by mutation, gene loss, or recombination may have the potential to recreate new virulent zoonotic orthopoxviruses. Recurrent zoonoses of monkeypox infections are a serious concern. The natural hosts of the virus are African rodents, and since the cessation of vaccination in the Republic of Congo, human monkeypox incidence has increased thirty-fold (Rimoin et al. 2010), a startling statistic. This should put us on guard that orthopoxviruses certainly have the potential for serious disease outbreaks given the right constellation of genetic change and anthropogenic social epidemiologic change.

Live-Attenuated Viruses

Live-attenuated viruses have become a go-to for vaccine development. The MMR vaccine is composed of live-attenuated vaccine strains of measles, mumps, and rubella viruses. Poliovirus, varicella zoster, and now recently influenza virus vaccines are available in live-attenuated virus form. How are live-attenuated vaccine strains created? The answer is that they are evolved. A well-characterized virus isolate, representative of the strains that cause the disease, is grown in cells from a different species of animal. The measles virus was inoculated in partially permissive embryonated chicken eggs or chick embryo fibroblasts in culture. The virus growing in these cultures is passaged serially in the same nonhuman cells. The chick cells in which the virus is propagated represent a new and unnatural species for growth of what has naturally evolved to be a human-adapted virus. The normal purifying selective pressures exerted on the virus by the human host are removed. Selective pressures favoring high levels of replication in chicken cells replace them. Under normal circumstances the human measles virus lineage is under selective pressure to maintain optimal virulence and transmission *in vivo* in human populations. Serial passage *in vitro* relaxes the selective pressures on a virus; it need not be pathogenic in an organism or battle its adaptive immune ecosystem, nor does it need to be transmitted in the normal sense. Measles transmission between humans is airborne and highly contagious, a quality that must be maintained by rigorous purifying selective pressure; in tissue culture the virus is under no such selection, it relies on laboratory workers to manually transfer it to a fresh culture of cells. In fact, it is a common observation that viruses maintained over long periods in tissue culture lose some or all of their virulence for pathogenicity *in vivo*.

Passaging a virus repeatedly in a foreign cell type *in vitro* therefore relaxes selective pressures of natural host that maintain its pathogenic potential *in vivo*. They are supplanted by selective pressures of the foreign host cell type. The resulting viruses reproducibly lose their potential for pathogenesis in their native species; it is as if they have begun the process of speciation, but have only adapted to grow in the cells of the new species and are not able to cause disease in the organisms of the new species. The attenuation process is one of genetic divergence from the canonical wild-type lineage sequence. Of five different measles vaccine strain genomes sequenced, all were found to share a select set of nucleotide substitutions, while other mutations were restricted to one or a subset of the vaccine strains. Common pathways to attenuation appear to be favored and the numbers of mutational steps to attenuation are remarkably few. None of the five vaccine strains differed from the wild-type low-passage seed strain of the Edmondston strain of measles virus by more than 0.3 percent and contained no more than twenty-five nonsynonymous mutational changes (Parks et al. 2001).

Such an attenuation regimen was successfully used with other viruses, notably mumps, rubella, and poliovirus. All are RNA viruses that create genetic diversity at a high rate during replication, and they are therefore highly susceptible to altered selective pressures which can quickly fix new mutations in the lineage that prevails in the artificial system *in vitro*. The use of these strains as vaccines is highly effective. No disease manifestations are evident but the restricted replication of the attenuated virus in humans is able to trigger a highly robust immune response that provides protection against the wild-type circulating isolates of the viruses. In effect, vaccine producers have created a virus that is no longer equipped for infection of the human host; it can only cause dead-end infections. The inoculation of the virus results in an abortive infection which efficiently provokes and is cleared by an immune response similar to that caused by natural infection.

Vaccination with a live-attenuated virus vaccine can be compared with the cross-species transmission of viruses that cause zoonotic infection. If the vaccine strain is an RNA virus, it can be expected that the vaccine inoculum will be made up of a population of genotypes representing the quasispecies which is created during replication and amplification of the virus in host cells. Furthermore, virus replication in the vaccine recipient creates genetic diversity, providing an opportunity for reversion of the attenuated phenotype. This is a real but rarely realized disadvantage of live-attenuated virus vaccines.

The orally delivered live-attenuated poliovirus vaccine developed by Albert Sabin and introduced in 1950s is the key weapon in the ongoing campaign to globally eradicate poliovirus. It is hoped that it will be the second virus (after smallpox) to be forced into extinction by man-made vaccine technology. Already polio rates worldwide have been reduced by 99 percent but the virus remains endemic in some countries where vaccination rates remain low (WHO 2015b). The vaccine is highly effective because the virus infects and replicates to high titers in our gut cells. It does not cause disease but it provokes the development of a potent and lifelong immunity to the virus. It is this replication, however, that necessarily results in the creation of genotypic variants with the potential for greater pathogenicity and most importantly, neurovirulence. In undervaccinated populations the evolution of these *vaccine-derived polioviruses* (VDPV) has resulted in outbreaks of polio that jeopardize the success of the polio eradication program (Burns et al. 2014). The failure of the poliovirus RNA polymerase to proofread its product leads to the misincorporation of nucleotides in each and every cycle of genome replication. The Sabin 1 vaccine strain of poliovirus serotype 1 differs from the parental virulent wild type P1/Mahoney strain by fifty-six discrete mutations scattered across the 7,600 nucleotide RNA genome (Christodoulou et al. 1990). Nevertheless, some of the mutations have a greater influence on the attenuated phenotype than others, and during replication in the gut the reversion of just one or a few of these attenuating mutations can generate VDPV strains with the potential for transmission and the attendant risks of poliomyelitis (Burns et al. 2014).

Since the polio vaccination is designed to protect against infection by each of the three circulating serotypes of poliovirus, it incorporates three distinct attenuated viral strains (Sabin 1–3), each with its own unique signature of attenuating mutations. There exists then a possibility of genetic exchange or recombination between different vaccine strains if they infect the same cell in the gut. In some territories, closely related non-poliovirus human enteroviruses also circulate at high levels, and interspecies recombinants can contribute to the emergence of novel VDPVs (Arita et al. 2005; Joffret et al. 2012). Although VDPV can trigger outbreaks of poliovirus infections in populations where vaccination rates are low, they have little consequence in the highly vaccinated populations of the developed world. Despite the emergence of revertant viruses in individual vaccine recipients, the risk that such viruses will cause neurologic manifestations is less than one in a half million. Nevertheless, recourse to the use of the inactivated poliovirus vaccine may be necessary in localities where pockets of undervaccinated people persist despite vaccination campaigns (Grassly 2013).

Attenuation by Design

The ease with which attenuation of RNA viruses can be achieved by serial passage in a modified environment highlights the fact that despite their highly error-prone replicative processes, they have genome sequences that are finely tuned for fitness and rigorously curated by purifying selective pressures. Their evolved and optimized genotypes are balanced on a razor's edge. The phenotype of a virus is obviously dependent on the amino acid sequences of its proteins; it can also be profoundly influenced by the RNA sequence itself. Some RNA sequences operate as control elements in and of themselves or fold into complex structures that manifest functionality. To appreciate this point we need only examine the nature of attenuating mutations in Sabin-1 poliovirus. One of the most important mutations for attenuation is in the 5' noncoding region of the genome and results in destabilization of base pairing that is important for maintaining complex secondary structure of the genome (Minor et al. 1993). Other constraints on the genome sequence are also evident and were revealed by investigators from Stony Brook University and from the Centers for Disease Control and Prevention (Burns et al. 2006; Coleman et al. 2008; Mueller, Papamichail, and Coleman 2006). This constraint is that of codon bias, which is evidently maintained through selective pressures on genomes. Our own genome and that of different bacterial species betray codon bias just as viral genomes do. Noting that a 300 amino acid protein sequence can be coded by 10^{151} distinct combinations of 300 codons selected from our redundant genetic code, it begged the question as to whether the actual sequence represented a fitness optima (Coleman et al. 2008). Indeed, it does: it is not just what viral genes encode but *how* they encode it that contributes to fitness. The investigators had access to computational tools that allowed them to recode the poliovirus capsid protein P1, changing the codon bias, yet maintaining the ability of the RNA chain to fold in an authentic fashion like its parental wild type sequence.

The results of the studies opened up a whole new avenue of research into live-attenuated vaccines: synthetic attenuated virus engineering (SAVE). A virus was created with 631 synonymous mutations in its P1 coding sequence, designed to bias it toward the use of codons that are rarely preferred in human cells. The result was a highly attenuated virus that caused no disease in an animal model of virus infection, and like the naturally evolved live-attenuated polioviruses developed by Sabin, it proved to be a highly effective vaccine. Unlike Sabin's strains, however, the multiplicity of genetic changes contributing to attenuation is expected to render the phenotype far more stable and resilient to reversion *in vivo*. This technology could prove extremely useful in the development of safe and stable attenuated viruses that raise an immune response almost identical to that against the natural infection. There are now many examples of the genetic engineering of synthetic attenuated virus vaccines; most notably it has been employed to create a live-attenuated vaccine against a strain of human influenza, a virus that, unlike poliovirus or smallpox virus, we cannot hope to eradicate and for which vaccination remains the lynchpin of disease management.

Virus Therapeutics

Today it may seem rather intuitive that viruses themselves can be adapted and used as tools to generate immunity to the very diseases that their natural counterparts cause. It is also obvious that viruses can be used to control populations of host organisms susceptible to their pathogenic effects. This approach was a failure in Australia, where the control of European rabbit pests using a poxvirus was stymied by coevolution of the virus and host. Surprisingly, and to this author against all odds, it continues to be the strategy pursued in this very endeavor today. Calciviruses are being developed and used in Australia as a part of their long-term project to control rabbit proliferation and the destruction of habitat and vegetation that they wreak across the continent (CSIRO 2015). Less intuitively obvious is that newly developed virus-based technologies may in the future prove to be invaluable tools for the treatment of a variety of medical disorders unrelated to virus infections, ranging from rare inherited genetic disorders to cancer. Could this be the redemption of viruses?

Doctor's Little Helpers

The creation of medically useful designer viruses is a relatively young enterprise that has been going on in research laboratories and clinics for just a few decades. Reminiscent of gene transduction between prokaryotes by bacteriophages, eukaryotic virus genomes can be engineered to incorporate therapeutic genes and introduce them into cells by infection. Many different viruses have been explored for the purpose of delivering a genetic payload with therapeutic utility into human cells. The scientific endeavor has not, however, been without setbacks. Early work used modified Moloney murine leukemia virus, a gammaretrovirus, and targeted the restoration of immunity in children with a genetic deficiency that causes X-linked severe combined immunodeficiency (X-SCID). Researchers replaced the envelope glycoprotein gene of the virus with the growth factor receptor gene that is defective in X-SCID patients. Infection of the patient's bone marrow cells with the designer virus resulted in effective transduction of the gene. When the cells were retransfused into the patients, their immune deficiency was successfully corrected. Sadly, however, one in four of the successfully treated patients succumbed to T cell acute lymphoblastic leukemia as a consequence of the treatment regimen (Hacein-Bey-Abina et al. 2008). In these patients the murine retrovirus, with its beneficial genetic payload, had integrated into the cell genome in close proximity to cellular *oncogenes*, genes whose upregulation is associated with cancer. The hematologic malignancies of T cells in the treated patients were a sober reminder of the remaining cancer-causing potential of the virus, which is associated with leukemias in mice. Later clinical studies have mitigated this safety concern while providing similar effectiveness. The scientists inactivated the retrovirus transcriptional control enhancer elements, rendering them incapable of activating cellular oncogenes (Hacein-Bey-Abina et al. 2014). After all, retroviruses have provided us with a useful tool for treating diseases of gene deficiencies in immune cells (Naldini 2015; Jacobson et al. 2012).

Other promising work employs genetically engineered adenovirus associated virus (AAV), a parvovirus, as a nonintegrating gene therapy vector that poses no risk of activating cancer-related genes. This type of gene therapy vector was used to introduce genes to rectify gene deficiencies in Leber congenital amaurosis (caused by gene RPE65 deficiency in the eye), lipoprotein lipase deficiency (a severe lipid disorder), and aromatic L-amino acid decarboxylase (AADC) deficiency (a disease that results in severe neurologic development disorders), to name a few (Jacobson et al. 2012; Hwu et al. 2012; Gaudet et al. 2013). The therapy is delivered locally to the tissue that manifests the gene deficiency phenotype; for example, into ocular tissue (RPE65 deficiency) or brain tissue (AADC deficiency). In the case of lipoprotein lipase deficiency, the gene is delivered into muscle cells, which act as a cellular factory manufacturing the missing protein. Although many such approaches are showing some promise of efficacy, there have been few home runs to date and the list of approved gene therapies is very short. Nevertheless, it is still early in the employment of viruses in gene therapy and too soon to predict how great their impact will be on these disorders and other challenges in human health.

Oncolytic Viruses

A field of explosive progress in recent years has become that of *oncolytic virotherapy*. In November 2014 the U.S. Food and Drug Administration licensed *talimogene laherparepvec* (T-vec), the first *oncolytic virus*, a genetically modified herpes simplex virus for the treatment of melanoma (FDA 2015). Viruses that we normally think of as disease-causing pathogens are now under exploration as cancer therapies. Genetically tailored variants of viruses are being worked up in laboratories: the measles virus, Newcastle disease virus, rhabdo-, herpes-, adeno-, and poxviruses among them. Remarkably, these comprise RNA and DNA viruses from multiple and of diverse families. Ever since cancer was first recognized and described in the 1800s, it has been tacitly understood that some patients showed improvement as a consequence of contracting an infectious disease. In particular, it was recognized that patients with hematologic malignancies benefited from a bout with the flu (Kelly and Russell 2007). The extraordinary potential of virolytic therapy in cancer was first demonstrated only in the 1990s, the same time period that gene therapy was being explored for the treatment of X-SCID. Scientists working in the laboratory of Dr. Don Coen at Harvard University exploited the unique property of certain herpes simplex virus type 1 mutant viruses. Viruses with mutations in their thymidine kinase (TK) gene grow normally in rapidly dividing cells but slowly or not at all in quiescent cells. The TK gene is normally required to supplement the provision of nucleotide precursors to support the virus's rapid viral DNA synthesis required for replication in host cells that are not actively dividing. Cells that divide at high rates, such as cancer cells, are metabolically very active and can readily provide abundant DNA precursors in the absence of the viral TK enzyme. Coen and his team hypothesized that a TK-negative HSV-1 virus could be a selective anticancer agent and selectively kill cancer cells, while sparing normal tissue. In seminal experiments published in 1991, they realized this potential by infecting and curing mice that had been implanted with human tumor cells (Martuza et al. 1991). Virolytic therapy grounded in scientific design and pursued as evidence-based medicine had truly arrived.

The basis for the potential of all virolytic therapeutics resides in the exquisite selectivity they exhibit for infecting and killing cancer cells. The very nature of cancer cells makes them extremely susceptible to virus infection: they divide in an uncontrolled fashion and are metabolically hyperactive, thus they exhibit greatly diminished capacity for apoptosis and innate immune defense against virus infection. While normal cells reduce metabolic activity, activate apoptotic signaling pathways, and block cell cycle progression in response to virus infection, cancer cells remain oblivious (Pikor, Bell, and Diallo 2015). These are perfect conditions for the growth of viruses, particularly those that are attenuated for growth in normal cells. Consequently, oncolytic viruses are specific reagents that target cancer cells and spread from cell to cell within tumors. It has become apparent that the direct lytic effects of viruses on cancer cells is just one element of their therapeutic effects; the cytolysis of infected cells releases viral and cellular antigens that can provoke antitumor immune responses, and some cancer therapeutic viruses are engineered to deliver additional genes such as immune activators to augment these effects (Lichty et al. 2014).

Let's take T-vec as an example. It is, like Martuza's virus, based on herpes simplex virus type 1, but it has been attenuated, not by mutation in the TK gene but by deletion of a neurovirulence gene that encodes a protein called γ34.5. The mutant virus causes little or no disease in normally permissive animal models and fails to establish latency in neurons, a feature that is typical and necessary in the pathogenesis of HSV. Those treated with the virus are thus in no danger and will not suffer subsequent reactivations of herpes after treatment. T-vec was further modified: first, a viral gene responsible for immune evasion was removed, and then the human gene encoding GM-CSF (granulocyte macrophage—colony stimulating factor), an immunostimulatory cytokine, was introduced. Both of these modifications increase the immune-stimulating properties of the virus. In animal tests, inoculation of the virus into tumors results in their eradication (Liu et al. 2003). The injected tumors are productively infected by the virus, which spreads among tumor cells, killing them in succession and causing the tumor to shrink and often regress. Surprisingly, tumors at remote sites that were not directly injected with the virus were also seen to shrink. This was shown to be due to the provocation of a systemic immune reaction against the cancer cells. Cancer therapy with such a vector therefore combines virus infection induced tumor lysis with imunotherapeutic benefits.

The many different viruses available for use in these applications has opened up a plenitude of possibilities for oncolytic tumor therapies. Myxoma virus, the poxvirus of South American tapeti (and now endemic virus of rabbits), will not infect normal human cells due to the species barrier, but it replicates efficiently and lyses human tumor cells that fail to mount an appropriate antiviral response. Perhaps even more promising results have been achieved with the Edmondston strain of measles virus (MV-Edm), the virus that is normally associated with measles vaccination. Despite being attenuated, it has shown substantial efficacy and high specificity for killing tumors in several preclinical animal models of human tumor types (Bell 2014), and scientists at the Mayo clinic have pioneered clinical studies of an engineered MV-Edm in the treatment of multiple myeloma (Russell et al. 2014). It is one of the most prevalent blood cell cancers, and patients with the disease have a five-year survival rate of 40 percent. The dissemination of cancer cells in the bone marrow and focal lesions of cancer cells in the bone characterize the course of multiple myeloma. Russell and his colleagues treated two patients with the virus delivered at high doses by intravenous infusion; both responded favorably, with one undergoing complete remission of disease. The scientific team went on to elegantly demonstrate that the therapeutic virus was highly selective, infecting only tumor cells. The therapeutic virus genome was engineered to incorporate a human gene encoding a sodium iodide symporter protein (NIS). Infected cells express NIS and actively accumulate iodine in their interior. In clinical tomographic scans, the infected cells are revealed using a radioactive iodine tracer molecule. Only plasmacytoma cells concentrated the iodine123, indicating that the systemic infusion of the infectious agent exclusively targeted the cancer cells. In future studies the investigators may contemplate infusing the patients undergoing MV-Edm-NIS virolytic therapy with a strong beta-emitting iodine isotope such as ^{131}I. In this way the virolytic effects of the therapy may be intensified by localized radiotherapy.

We are certainly in the early days of oncolytic virotherapy and in the use of viruses in gene therapy applications. The use of human viruses such as MV-Edm comes with some disadvantages, among them the fact most of us are vaccinated against the virus. It follows that its utility will thus be restricted to only those cancer patients that are substantially immunocompromised. Nevertheless, the revelation that viruses normally considered to be serious human pathogens can, with human ingenuity, be tailored to become exquisitely refined therapeutic tools with the power to selectively eliminate human cancers is a remarkable reversal. To what degree these new agents will replace the surgeon's scalpel, radiation, and conventional cancer chemotherapies in the future remains to be seen. The realization that these virus treatments can directly kill cancer cells as well as deliver indirect immunotherapeutic benefits is an area of exciting future opportunity. New immune-mediated therapies are showing great promise in the treatment of many different cancers and there is surely an opportunity to garner a synergism of these treatments with oncolytic virus therapy (Ascierto, Marincola, and Atkins 2015).

· 16 ·

CONCLUSION: HUMANITY *and* VIRUSES

We have reviewed numerous examples of how viruses have been "of service" to their hosts. In no instances, be they the provision of toxin genes for pathogenic *V. cholerae* strains or syncytins for placental development in mammals, can purpose be evoked. They are the outcome of natural selection acting on the phenotype of serendipitous genetic variants; evolution captured and built on random rare events—it has neither intent nor prospective planning. The development of viruses as medical tools was, on the other hand, a prospectively calculated endeavor, undertaken knowingly. This is self-evident, but a closer examination of the concept emphasizes the unique relationship of humans with viruses.

Humans have altered the rules of the game: the Red Queen works differently for us. We do not allow evolution to determine our relationship with a virus. Humans do not rely on genotypic change and selection of the fittest to compete in the escalating arms race with viral pathogens. The development of antiretroviral drugs to combat HIV-1 (Chapter 9) was one such example. As Joseph Henrich puts it so eloquently in the title of his 2015 book, "the secret of our success" resides in our human intelligence, our cognitive capacity, and our evolved human culture. We communicate, collaborate, and accumulate knowledge that is preserved from generation to generation. These are the products of *cultural evolution*, which distinguish us from all other species, including our great ape cousins. Virus-host relationships are defined by the interaction of the virus with cells and populations of individual hosts. Viruses that infect humans must also contend with the product of the collective of human individuals, human culture: a much more vexing challenge. When a zoonotic virus spills over into humans, it is not only infecting us individual by individual, it is infecting our society. If the virus lineage is to successfully infect the human host population it must contend not only with innate and adaptive immune defenses but also with nongenetic adaptations that are mobilized by our sophisticated human culture.

Humankind has collectively responded to the challenge of the HIV-1 pandemic with nongenetic adaptive changes. Our capability in this regard is itself the product of natural selection of ancestral humans. Koalas and gorillas must rely on evolutionary change to respond to the retrovirus epidemics they currently face. Without human intervention KoRV and EBOV may lead to the demise of these animal species. To survive they must coevolve with their viral predators, but this is a process that takes millennia (Chapter 10). Humans are currently enduring the pandemic zoonosis that became HIV-1. It is perhaps too early to proclaim victory in this particular war, which will in any case leave indelible scars on human populations in the developing world and sub-Saharan Africa. Nevertheless, the arms race has been taken up on both sides (Chapter 9); we have prevailed in some battles but the war is not over; the epidemic is ongoing. It is tempting to think of society's adaptive response to HIV infection, in which we strove toward a resilient human "phenotype" over a mere generation, as almost Lamarckian in concept. It is not. It is rooted in Darwinian selection that favored genetic changes in the human lineage promoting the capacity for cultural evolution and the possession of these unique survival skills.

The Human Future and Viruses

Viruses play an inextricable role in the evolution of all life. They are not themselves life-forms, a fact that is difficult to reconcile with the complexity and vitality of the events they trigger in their hosts. Like invading armies, viruses depend on their living hosts for matériel needed to support their campaign to replicate their genetic information. Their fundamental need is energy. The cell is a source of energetically rich components and structures that viruses must tap to fuel their propagation. The genius of viruses is to incorporate the energy of living systems into their energy-rich and highly "ordered" virus particles, returning "disorder" to their environment. Their encoded information proliferates and evolves in the slipstream and at the expense of energy from living organisms. Viruses may only be sophisticated reagents in this reaction cocktail, obeying the laws of thermodynamics, but their evolution is governed by the same laws of Darwinian evolution that rule the living world.

The viral metagenome is the greatest repository of novel extant genetic information in the biosphere. The creation of this genetic diversity is a feat unequalled by any of the three domains of life, and much of it remains dark matter. As a consequence of its promiscuity and continued diversification it will continue to be the dominant source of genetic innovation in the biosphere. The viruses that we know about are a significant minority; the oceanic virome, the viromes of rodents, bats, and primates, and our own viromes are certainly more complex than yet documented and will inevitably be sources of future evolutionary innovation. The virus metagenome will continue to fuel evolution, particularly in response to change. Change, interpreted in its broadest sense, will be the catalyst that unlocks the evolutionary invention of the viral metagenome and the capacity of viruses for rapid and opportunistic evolutionary change.

The accelerating pace at which viruses are emerging to cause human disease is bewildering (Jones et al. 2008), but has the ascendency of the cultured human species placed us beyond the threat of emerging viruses? Are we insulated, as a species at least, from risk of extinction or severe population decline resulting from pandemic viral disease? Consider a future pandemic flu. It will doubtless be a global health crisis, but despite global travel, the consequences are unlikely to compare with those of the 1918 pandemic flu, which killed one in every forty individuals on the planet in the space of two years. Global health care will leverage superior preventive strategies, antiviral drugs, and effective antibiotics; superior supportive care will be available and a vaccine will be fast-tracked. All these factors make a worst-nightmare scenario implausible. Nevertheless, as individuals we should not ignore the grave potential of such a pandemic or other emergent virus infections. Severe epidemics of local scale, but global consequence, should be anticipated as a consequence of different access to health care infrastructure in geographically and economically separate groups and nations.

We will be tested, ever more frequently, by new viral diseases that emerge *de novo* from their reservoir hosts or that become geographically more widespread or virulent. These challenges to human health are catalyzed by social and economic epidemiologic factors and, of course, by global climate change. While new zoonoses can be predicted to be more frequent, they remain *known unknowns*. On the other hand, the relentless expanding geographic reach of serious insect- or tick-vectored viruses promoted by climate change, global commerce, and travel is a *known known*. In recent years, we have witnessed the devastating effects of HIV and then EBOV on the human potential and social and economic well-being of whole African nations. There is no quick fix to humanity's growing liability for emerging viral diseases; our only recourse is to be prepared for the worst.

In 2015 as part of their Emergency Preparedness Program, the WHO identified priority viral pathogens for R&D: Crimean Congo hemorrhagic fever, Ebola virus disease and Marburg, Lassa fever, MERS and SARS coronavirus diseases, Nipah, and Rift Valley fever (WHO 2016). This is only a short list, and it was noted that chikungunya, among other viruses, is also considered a threat that needs imminent attention. Of course, the WHO cannot predict or prioritize viruses that have yet to emerge; MERS would not have been on this list in 2013 because it was an *unknown unknown*. What will need to be added to this seemingly ever growing list in the future? At the height of the HIV-1 pandemic Anthony Fauci, a scientist and head of NIH's National Institute of Allergy and Infectious Diseases, noted in an interview that "back in 1918, when influenza wiped out twenty to fifty million people worldwide and hundreds and thousands of people in the United States, the people who lived through that, I think, had a good idea of what an emerging microbe might do. But then as the decades went by, they forgot it. Here we are with AIDS and people still have not had the foresight to understand that this can happen again. We are not even half over with this yet." He was reminding us to never be complacent and sanguine about the threat of another emerging pandemic virus. In the future, we could face a crisis of global proportions as we struggle to prevent and manage viral diseases. Our species may not be at risk but our quality of life certainly is.

Our globally connected human society has never been at greater risk yet better equipped to respond to such challenges as they emerge. We should expect that medical science will continue to develop breakthrough technologies. We are reliant on the social structures, human ingenuity, and the scientific know-how of our species to rapidly respond and deploy treatments (like antiretroviral drugs) or vaccines for emerging viruses. But the dependence of the human species on our complex and integrated social structures is also its greatest vulnerability. Fractured social structures and social upheaval can be directly linked to dramatically increased incidences of infectious diseases; "war and infectious disease are deadly comrades" (Connolly and Heymann 2002). Viruses are remarkable evolutionary opportunists and volatile and changing global realities mandate they command our diligence and respect if we are to prevent them from evolving into an untenable health burden on human societies. In our new world we must be aware of the potential global consequence of local social changes and political upheavals in remote parts of the world. We cannot be detached from emerging zoonoses and new patterns of viral disease on distant continents; today all continents are our backyard.

Beauty in Design

The trends speak to an unavoidable truth. Society's future will be challenged by zoonotic viruses, a quite natural prediction, not least because humanity is a potent agent of change, which is the essential fuel of evolution. Notwithstanding these assertions, I began with the intention of leaving the reader with a broader appreciation of viruses: they are not simply life's pathogens. They are life's obligate partners and a formidable force in nature on our planet. As you contemplate the ocean under a setting sun, consider the multitude of virus particles in each milliliter of seawater; flying over wilderness forestry, consider the collective viromes of its living inhabitants. The stunning number and diversity of viruses in our environment should engender in us greater awe that we are safe among these multitudes than fear that they will harm us.

Personalized medicine will soon become a reality and medical practice will routinely catalogue and weigh a patient's genome sequence. Not long thereafter one might expect this data to be joined by the patient's viral and bacterial metagenomes; the patient's collective genetic identity will be recorded in one printout. We will doubtless discover some of our viral passengers are harmful to our health, while others are protective.

But the appreciation of viruses that I hope you have gained from these pages is not about an exercise in accounting. The balancing of benefit versus threat to humanity is a fruitless task. The viral metagenome will contain new and useful gene functionalities for biomedicine; viruses may become essential biomedical tools and phages will continue to optimize the health of our oceans, ensuring optimal primary production. Viruses may also accelerate the development of antibiotic drug resistance in the post-antibiotic era and emerging viruses may threaten our complacency and challenge our society economically and socially. Simply comparing these pros and cons, however, does not do justice to viruses and acknowledge their rightful place in nature.

In humility, we should acknowledge that we are one and the same with viruses, products of Darwinian evolution. Jonathan Swift, the Irish poet (1667–1745) who would have been a skeptic of evolution, wrote: "That the universe was formed by a fortuitous concourse of atoms, I will no more believe than that the accidental jumbling of the alphabet would fall into a most ingenious treatise of philosophy." Viruses and life are, however, just such a "treatise of philosophy," born in random events and selected under nature's universal laws of thermodynamics and natural selection.

Life and viruses are inseparable. Viruses are life's complement, sometimes dangerous but always beautiful in design. All autonomous self-sustaining replicating systems that generate their own energy will foster parasites. Viruses are the inescapable by-products of life's success on the planet. We owe our own evolution to them; the fossils of many are recognizable in ERVs and EVEs that were certainly powerful influences in the evolution of our ancestors. Like viruses and prokaryotes, we are also a patchwork of genes, acquired by inheritance and horizontal gene transfer during our evolution from the primitive RNA-based world.

It is a common saying that "beauty is in the eye of the beholder." It is a natural response to a visual queue: a sunset, the drape of a designer dress, or the pattern of a silk tie, but it can also be found in a line of poetry, a particularly effective kitchen implement, or even the ruthless efficiency of a firearm. The latter are uniquely human acknowledgments of beauty in design. It is humanity that allows us to recognize the beauty in the evolutionary design of viruses. They are unique products of evolution, the inevitable consequence of life, infectious egotistical genetic information that taps into life and the laws of nature to fuel evolutionary invention.

REFERENCES

ACKNOWLEDGMENTS

INDEX

REFERENCES

Adriaenssens, E. M., L. Van Zyl, P. De Maayer, E. Rubagotti, E. Rybicki, M. Tuffin, and D. A. Cowan. 2014. "Metagenomic analysis of the viral community in Namib Desert hypoliths." *Environ Microbiol.* doi:10.1111/1462–2920.12528.

Aherfi, S., B. La Scola, I. Pagnier, D. Raoult, and P. Colson. 2014. "The expanding family Marseilleviridae." *Virology* 466–467:27–37. doi:10.1016/j.virol.2014.07.014.

Alizon, S., V. von Wyl, T. Stadler, R. Kouyos, D. S. Yerly, B. Hirschel, J. Böni, et al. 2010. "Phylogenetic approach reveals that virus genotype largely determines HIV set-point viral load." *PLOS Pathog* 6 (9). doi:10.1371/journal.ppat.1001123.

Altfeld, M., and M. Gale, Jr. 2015. "Innate immunity against HIV-1 infection." *Nat Immunol* 16 (6):554–62. doi:10.1038/ni.3157.

An, P., and C. A. Winkler. 2010. "Host genes associated with HIV/AIDS: Advances in gene discovery." *Trends Genet* 26 (3):119–31. doi:10.1016/j.tig.2010.01.002.

Angly, F. E., B. Felts, M. Breitbart, P. Salamon, R. A. Edwards, C. Carlson, A. M. Chan, et al. 2006. "The marine viromes of four oceanic regions." *PLOS Bio* 4 (11). doi:10.1371/journal.pbio.0040368.

Arita, M., S. L. Zhu, H. Yoshida, T. Yoneyama, T. Miyamura, and H. Shimizu. 2005. "A Sabin 3-derived poliovirus recombinant contained a sequence homologous with indigenous human enterovirus species C in the viral polymerase coding region." *J Virol* 79 (20):12650–57. doi:10.1128/JVI.79.20.12650–12657.2005.

Arrildt, K., S. Joseph, and R. Swanstrom. 2012. "The HIV-1 env protein: A coat of many colors." *Curr HIV/AIDS Rep* 9 (1):52–63. doi:10.1007/s11904–0114–0107–3.

Ascierto, P. A., F. M. Marincola, and M. B. Atkins. 2015. "What's new in melanoma? Combination!" *J Transl Med* 13:213. doi:10.1186/s12967–015–0582–1.

Avila-Arcos, M. C., S. Y. Ho, Y. Ishida, N. Nikolaidis, K. Tsangaras, K. Honig, R. Medina, et al. 2013. "One hundred twenty years of koala retrovirus evolution determined from museum skins." *Mol Biol Evol* 30 (2):299–304. doi:10.1093/molbev/mss223.

Baba, T., F. Takeuchi, M. Kuroda, H. Yuzawa, K. Aoki, A. Oguchi, Y. Nagai, et al. 2002. "Genome and virulence determinants of high virulence community-acquired MRSA." *Lancet* 359 (9320):1819–27.

Baize, S., D. Pannetier, L. Oestereich, T. Rieger, L. Koivogui, N. Magassouba, B. Soropogui, et al. 2014. "Emergence of Zaire Ebola virus disease in Guinea." *New Engl J Med* 371 (15):1418–25. doi:10.1056/NEJMoa1404505.

Baldo, A. M., and M. A. McClure. 1999. "Evolution and horizontal transfer of dUTPase-encoding genes in viruses and their hosts." *J Virol* 73 (9):7710–21.

Banks, D. J., S. B. Beres, and S. J. Musser. 2002. "The fundamental contribution of phages to GAS evolution, genome diversification and strain emergence." *Trends Microbio* 10 (11):515521. doi:10.1016/S0966–842X(02)02461–7.

Bar, K. J., C. Y. Tsao, S. S. Iyer, J. M. Decker, Y. Yang, M. Bonsignori, X. Chen, et al. 2012. "Early low-titer neutralizing antibodies impede HIV-1 replication and select for virus escape." *PLOS Pathog* 8 (5). doi:10.1371/journal.ppat.1002721.

Barr, J. J., M. Youle, and F. Rohwer. 2013. "Innate and acquired bacteriophage-mediated immunity." *Bacteriophage* 3 (3). doi:10.4161/bact.25857.

Bausch, D. G., J. S. Towner, S. F. Dowell, F. Kaducu, M. Lukwiya, A. Sanchez, S. T. Nichol, T. G. Ksiazek, and P. E. Rollin. 2007. "Assessment of the risk of Ebola virus transmission from bodily fluids and fomites." *J Infect Dis* 196 Suppl 2:S142–7. doi:10.1086/520545.

Bell, J. C. 2014. "Taming measles virus to create an effective cancer therapeutic." *Mayo Clin Proc* 89 (7):863–865. doi:10.1016/j.mayocp. 2014.04.009.

Belshaw, R., V. Pereira, A. Katzourakis, G. Talbot, J. Paces, A. Burt, and M. Tristem. 2004. "Long-term reinfection of the human genome by endogenous retroviruses." *P Natl Acad Sci USA* 101 (14):4894–9. doi:10.1073/pnas.0307800101.

Belyi, V. A., A. J. Levine, and A. M. Skalka. 2010a. "Sequences from ancestral single-stranded DNA viruses in vertebrate genomes: The parvoviridae and circoviridae are more than 40 to 50 million years old." *J Virol* 84 (23):12458–62. doi:10.1128/JVI.01789–10.

Belyi, V. A., A. J. Levine, and A. M. Skalka. 2010b. "Unexpected inheritance: multiple integrations of ancient bornavirus and ebolavirus/marburgvirus sequences in vertebrate genomes." *PLOS Pathog* 6 (7). doi:10.1371/journal.ppat.1001030.

Bergh, Ø., K. BØrsheim, G. Bratbak, and M. Heldal. 1989. "High abundance of viruses found in aquatic environments." *Nature* 340 (6233):467–68. doi:10.1038/340467a0.

Biebricher, C. K., and M. Eigen. 2005. "The error threshold." *Virus Res* 107 (2):117–27. doi:10.1016/j.virusres.2004.11.002.

Black, F. L. 1966. "Measles endemicity in insular populations: Critical community size and its evolutionary implication." *J Theor Biol* 11 (2):207–11.

Black, F. L. 1975. "Infectious diseases in primitive societies." *Science* 187 (4176):515–18.

Bogaert, D., A. van Belkum, M. Sluijter, A. Luijendijk, R. de Groot, H. C. Rumke, H. A. Verbrugh, and P. W. Hermans. 2004. "Colonisation by Streptococcus pneumoniae and Staphylococcus aureus in healthy children." *Lancet* 363 (9424):1871–72. doi:10.1016/S0140–6736(04)16357–5.

Bohlman, M. C., S. P. Morzunov, J. Meissner, M. B. Taylor, K. Ishibashi, J. Rowe, S. Levis, D. Enria, and S. C. St. Jeor. 2002. "Analysis of hantavirus genetic diversity in Argentina: S segment-derived phylogeny." *J Virol* 76 (8):3765–73.

Bos, L. 1999. "Beijerinck's work on tobacco mosaic virus: Historical context and legacy." *Philos T Roy Soc B*. doi:10.1098/rstb.1999.0420.

Bour, S., and K. Strebel. 1996. "The human immunodeficiency virus (HIV) type 2 envelope protein is a functional complement to HIV type 1 Vpu that enhances particle release of heterologous retroviruses." *J Virol* 70 (12):8285–300.

Boyd, E. F., A. J. Heilpern, and M. K. Waldor. 2000. "Molecular analyses of a putative CTXphi precursor and evidence for independent acquisition of distinct CTX(phi)s by toxigenic Vibrio cholerae." *J Bacteriol* 182 (19):5530–38.

Boyd, M. T., C. M. Bax, B. E. Bax, D. L. Bloxam, and R. A. Weiss. 1993. "The human endogenous retrovirus ERV-3 is upregulated in differentiating placental trophoblast cells." *Virology* 196 (2):905–9. doi:10.1006/viro.1993.1556.

Boyer, M., N. Yutin, I. Pagnier, L. Barrassi, G. Fournous, L. Espinosa, Catherine Robert, et al. 2009. "Giant Marseillevirus highlights the role of amoebae as a melting pot in emergence of chimeric microorganisms." *P Natl Acad Sci USA* 106 (51):21848–53. doi:10.1073/pnas.0911354106.

Brault, A. C., A. M. Powers, D. Ortiz, J. G. Estrada-Franco, R. Navarro-Lopez, and S. C. Weaver. 2004. "Venezuelan equine encephalitis emergence: Enhanced vector infection from a single amino acid substitution in the envelope glycoprotein." *P Natl Acad Sci USA* 101 (31):11344–49. doi:10.1073/pnas.0402905101.

Brault, A. C., A. M. Powers, and S. C. Weaver. 2002. "Vector infection determinants of Venezuelan equine encephalitis virus reside within the E2 envelope glycoprotein." *J Virol* 76 (12):6387–92. doi:10.1128/JVI.76.12.6387–6392.2002.

Breitbart, M., and F. Rohwer. 2005. "Here a virus, there a virus, everywhere the same virus?" *Trends Microbiol* 13 (6). doi:10.1016/j.tim.2005.04.003.

Breitbart, M., P. Salamon, B. Andresen, J. M. Mahaffy, A. M. Segall, D. Mead, F. Azam, and F. Rohwer. 2002. "Genomic analysis of uncultured marine viral communities." *P Natl Acad Sci US A* 99 (22):14250–55. doi:10.1073/pnas.202488399.

Brook, C. E., and A. P. Dobson. 2015. "Bats as 'special' reservoirs for emerging zoonotic pathogens." *Trends Microbiol.* doi:10.1016/j.tim.2014.12.004.

Brüssow, H., and R. W. Hendrix. 2002. "Phage genomics: Small is beautiful." *Cell* 108 (1):13–16. doi:10.1016/S0092-8674(01)00637-7.

Burns, C. C., O. M. Diop, R. W. Sutter, and O. M. Kew. 2014. "Vaccine-derived polioviruses." *J Infect Dis* 210 Suppl 1:S283–93. doi:10.1093/infdis/jiu295.

Burns, C. C., J. Shaw, R. Campagnoli, J. Jorba, J. Vincent, A. Quay, and O. Kew. 2006. "Modulation of poliovirus replicative fitness in HeLa cells by deoptimization of synonymous codon usage in the capsid region." *J Virol* 80 (7): 3259–72. doi:10.1128/JVI.80.7.3259–3272.2006.

Bushman, F. D., N. Malani, J. Fernandes, I. D'Orso, G. Cagney, T. L. Diamond, H. Zhou, et al. 2009. "Host cell factors in HIV replication: Meta-analysis of genome-wide studies." *PLOS Pathog* 5 (5):e1000437. doi:10.1371/journal.ppat.1000437.

Calisher, C. H., J. E. Childs, H. E. Field, K. V. Holmes, and T. Schountz. 2006. "Bats: Important reservoir hosts of emerging viruses." *Clin Microbiol Rev* 19 (3):531–45. doi:10.1128/CMR.00017–06.

Cameron, C., S. Hota-Mitchell, L. Chen, J. Barrett, J. X. Cao, C. Macaulay, D. Willer, D. Evans, and G. McFadden. 1999. "The complete DNA sequence of myxoma virus." *Virology* 264 (2):298–318. doi:10.1006/viro.1999.0001.

Carrington, J. C., and V. Ambros. 2003. "Role of microRNAs in plant and animal development." *Science* 301 (5631):336–38. doi:10.1126/science.1085242.

Carroll, L. 1871. *Through the Looking Glass.* London: Macmillan.

Carroll, S. A., J. S. Towner, T. K. Sealy, L. K. McMullan, M. L. Khristova, F. J. Burt, R. Swanepoel, P. E. Rollin, and S. T. Nichol. 2013. "Molecular evolution of viruses of the family Filoviridae based on 97 whole-genome sequences." *J Virol* 87 (5):2608–16. doi:10.1128/JVI.03118–12.

Casjens, S. 2003. "Prophages and bacterial genomics: What have we learned so far?" *Mol Microbiol.* doi:10.1046/j.1365–2958.2003.03580.x.

Cauldwell, A. V., J. S. Long, O. Moncorge, and W. S. Barclay. 2014. "Viral determinants of influenza A virus host range." *J Gen Virol* 95 (Pt 6):1193–210. doi:10.1099/vir.0.062836–0.

CDC. "Ebola (Ebola virus disease)." Atlanta, GA: Centers for Disease Control and Prevention. http://www.cdc.gov/vhf/ebola/index.html.

CDC. 2009. "Outbreak of swine-origin Influenza A (H1N1) Virus Infection—Mexico, March—April 2009." Atlanta, GA: Centers for Disease Control and Prevention. http://www.cdc.gov/mmwr/preview/mmwrhtml/mm58d0430a2.htm.

Cech, T. R. 1986a. "A model for the RNA-catalyzed replication of RNA." *P Natl Acad Sci* 83 (12):4360–63. doi:10.1073/pnas.83.12.4360.

Cech, T. R. 1986b. "RNA as an enzyme." *Sci Am* 255 (5):64–75.

Cech, T. R. 1993. "Catalytic RNA: Structure and mechanism." *Biochem Soc T* 21 (2):229–34.

Cech, T. R. 2000. "The Ribosome Is a Ribozyme." *Science* 289 (5481):878–79. doi:10.1126/science.289.5481.878.

Chaipan, C., D. Kobasa, S. Bertram, I. Glowacka, I. Steffen, T. Tsegaye, M. Takeda, et al. 2009. "Proteolytic activation of the 1918 influenza virus hemagglutinin." *J Virol* 83 (7):3200–3211. doi:10.1128/JVI.02205-08.

Charleston, M. A., and D. L. Robertson. 2002. "Preferential host switching by primate lentiviruses can account for phylogenetic similarity with the primate phylogeny." *Syst Biol* 51 (3):528–35. doi:10.1080/10635150290069940.

Chirico, N., A. Vianelli, and R. Belshaw. 2010. "Why genes overlap in viruses." *P Roy Soc B-Biol Sci* 277 (1701):3809–17. doi:10.1098/rspb.2010.1052.

Chopera, D. R., Z. Woodman, K. Mlisana, M. Mlotshwa, D. P. Martin, C. Seoighe, F. Treurnicht, et al. 2008. "Transmission of HIV-1 CTL escape variants provides HLA-mismatched recipients with a survival advantage." *PLOS Pathog* 4 (3):e1000033. doi:10.1371/journal.ppat.1000033.

Christodoulou, C., F. Colbere-Garapin, A. Macadam, L. F. Taffs, S. Marsden, P. Minor, and F. Horaud. 1990. "Mapping of mutations associated with neurovirulence in monkeys infected with Sabin 1 poliovirus revertants selected at high temperature." *J Virol* 64 (10):4922–29.

Cianciolo, G. J., H. P. Bogerd, R. J. Kipnis, T. D. Copeland, S. Oroszlan, and R. Snyderman. 1985. "Inhibition of lymphocyte proliferation by a synthetic peptide homologous to envelope proteins of human and animal retroviruses." *T Assoc Am Physicians* 98:30–41.

Clavel, F., D. Guetard, F. Brun-Vezinet, S. Chamaret, M. A. Rey, M. O. Santos-Ferreira, A. G. Laurent, et al. 1986. "Isolation of a new human retrovirus from West African patients with AIDS." *Science* 233 (4761):343–46.

Clemente, J. C., L. K. Ursell, L. W. Parfrey, and R. Knight. 2012. "The impact of the gut microbiota on human health: An integrative view." *Cell* 148 (6):1258–70. doi:10.1016/j.cell.2012.01.035.

Clokie, M. R. J., and N. H Mann. 2006. "Marine cyanophages and light." *Environ Microbiol* 8. doi:10.1111/j.1462–2920.2006.01171.x.

Coetzer, M., R. Nedellec, T. Cilliers, T. Meyers, L. Morris, and D. E. Mosier. 2011. "Extreme genetic divergence is required for coreceptor switching in HIV-1 subtype C." *J Acq Immun Def Synd* 56 (1):9–15. doi:10.1097/QAI.0b013e3181f63906.

Coffey, L. L., N. Vasilakis, A. C. Brault, A. M. Powers, F. Tripet, and S. C. Weaver. 2008. "Arbovirus evolution in vivo is constrained by host alternation." *P Natl Acad Sci USA* 105 (19):6970–75. doi:10.1073/pnas.0712130105.

Coffin, J. M., S. H. Hughes, and H. E. Varmus. 1997. *Retroviruses.* Woodbury, NY: Cold Spring Harbor Laboratory Press.

Coleman, J. R., D. Papamichail, S. Skiena, B. Futcher, E. Wimmer, and Steffen Mueller. 2008. "Virus attenuation by genome-scale changes in codon pair bias." *Science* 320 (5884):1784–87. doi:10.1126/science.1155761.

Colomer-Lluch, M., J. Jofre, and M. Muniesa. 2011. "Antibiotic resistance genes in the bacteriophage DNA fraction of environmental samples." *PLOS ONE* 6 (3). doi:10.1371/journal.pone.0017549.

Colson, P., X. De Lamballerie, N. Yutin, S. Asgari, Y. Bigot, D. K. Bideshi, X. W. Cheng, et al. 2013. "'Megavirales,' a proposed new order for eukaryotic nucleocytoplasmic large DNA viruses." *Arch Virol* 158 (12):2517–21. doi:10.1007/s00705–013–1768–6.

Colson, P., X. de Lamballerie, G. Fournous, and D. Raoult. 2012. "Reclassification of giant viruses composing a fourth domain of life in the new order Megavirales." *Intervirology* 55 (5):321–32. doi:10.1159/000336562.

Connolly, M. A., and D. L. Heymann. 2002. "Deadly comrades: War and infectious diseases." *Lancet* 360 Suppl:s23–24.

Copenhaver, C. C., J. E. Gern, Z. Li, P. A. Shult, L. A. Rosenthal, L. D. Mikus, C. J. Kirk, et al. 2004. "Cytokine response patterns, exposure to viruses, and respiratory infections in the first year of life." *Am J Resp Crit Care Med* 170 (2):175–80. doi:10.1164/rccm.200312–1647OC.

Crotty, S., C. E. Cameron, and R. Andino. 2001. "RNA virus error catastrophe: direct molecular test by using ribavirin." *P Natl Acad Sci US A* 98 (12):6895–6900. doi:10.1073/pnas.111085598.

Crotty, S., D. Maag, J. J. Arnold, W. Zhong, J. Y. Lau, Z. Hong, R. Andino, and C. E. Cameron. 2000. "The broad-spectrum antiviral ribonucleoside ribavirin is an RNA virus mutagen." *Nat Med* 6 (12):1375–9. doi:10.1038/82191.

CSA (Chimpanzee Sequencing and Analysis Consortium). 2005. "Initial sequence of the chimpanzee genome and comparison with the human genome." *Nature* 437 (7055):69–87. doi:10.1038/nature04072.

CSIRO. 2015. "Controlling those pesky rabbits." http://www.csiro.au/en/Research/BF/Areas/Managing-the-impacts-of-invasive-species/Biological-control/Controlling-those-pesky-rabbits.

D'Arc, M., A. Ayouba, A. Esteban, G. H. Learn, V. Boue, F. Liegeois, L. Etienne, et al. 2015. "Origin of the HIV-1 group O epidemic in western lowland gorillas." *P Natl Acad Sci USA* 112 (11):1343–52. doi:10.1073/pnas.1502022112.

D'Hérelle, F. 1917. "Sur un microbe invisible antagoniste des bacilles dysentériques." *C. R. Acad Sci Paris* 165:373–75.

D' Hérelle, F., and G. H. Smith. 1930. *The Bacteriophage and Its Clinical Applications.* Springfield, IL, Baltimore, MD: C. C. Thomas.

Daniel, M. D., N. L. Letvin, N. W. King, M. Kannagi, and P. K. Sehgal. 1985. "Isolation of T-cell tropic HTLV-III-like retrovirus from macaques." *Science.* doi:10.1126/science.3159089.

Daros, J. A., S. F. Elena, and R. Flores. 2006. "Viroids: An Ariadne's thread into the RNA labyrinth." *EMBO Rep* 7 (6):593–98. doi:10.1038/sj.embor.7400706.

Davis, B. M., and M. K. Waldor. 2003. "Filamentous phages linked to virulence of Vibrio cholerae." *Curr Opin Microbiol* 6 (1):35–42. doi:10.1016/S1369-5274(02)00005–X.

Davison, A. J., and N. D. Stow. 2005. "New genes from old: Redeployment of dUTPase by Herpesviruses." *J Virol* 79 (20):12880–92. doi:10.1128/JVI.79.20.12880–12892.2005.

Davison, A. J. 2002. "Evolution of the herpesviruses." *Vet Microbiol* 86 (1–2): 6988. doi:10.1016/S0378–1135(01)00492–8.

Dawkins, R., and J. R. Krebs. 1979. "Arms races between and within species." *P Roy Soc Lond B Bio* 205 (1161):489–511.

Dawkins, R. 1976. *The selfish gene.* New York: Oxford University Press.

Dean, M., M. Carrington, C. Winkler, G. A. Huttley, M. W. Smith, R. Allikmets, J. J. Goedert, et al. 1996. "Genetic restriction of HIV-1 infection and progression to AIDS by a deletion allele of the CKR5 structural gene. Hemophilia Growth and Development Study, Multicenter AIDS Cohort Study, Multicenter Hemophilia Cohort Study, San Francisco City Cohort, ALIVE Study." *Science* 273 (5283):1856–62.

DeCaprio, J., A., and R. Garcea, L. 2013. "A cornucopia of human polyomaviruses." *Nat Rev Microbiol* 11 (4):264–76. doi:10.1038/nrmicro2992.

De Crignis, E., S. Guglietta, B. T. Foley, M. Negroni, A. F. Di Narzo, V. Waelti Da Costa, M. Cavassini, P. A. Bart, G. Pantaleo, and C. Graziosi. 2012. "Nonrandom distribution of cryptic repeating triplets of purines and pyrimidines (RNY)(n) in gp120 of HIV Type1." *AIDS Res Hum Retrov* 28 (5):493–504. doi:10.1089/AID.2011.0208.

Deen, G. F., B. Knust, N. Broutet, F. R. Sesay, P. Formenty, C. Ross, A. E. Thorson, et al 2015. "Ebola RNA persistence in semen of Ebola virus disease survivors—Preliminary report." *New Engl J Med.* doi:10.1056/NEJMoa1511410.

Del Prete, G.Q., M. F. Kearney, J. Spindler, A. Wiegand, E. Chertova, J. D. Roser, J. D. Estes, et al. 2012. "Restricted replication of xenotropic murine leukemia virus-related virus in pigtailed macaques." *J Virol* 86 (6):3152–66. doi:10.1128/JVI.06886–11.

Delviks-Frankenberry, K., O. Cingoz, J. M. Coffin, and V. K. Pathak. 2012. "Recombinant origin, contamination, and de-discovery of XMRV." *Curr Opin Virol* 2 (4):499–507. doi:10.1016/j.coviro.2012.06.009.

De Paepe, M., M. Leclerc, C. R. Tinsley, and M. A. Petit. 2014. "Bacteriophages: An underestimated role in human and animal health?" *Front Cell Infect Microbiol* 4:39. doi:10.3389/fcimb.2014.00039.

Desnues, C., B. La Scola, N. Yutin, G. Fournous, C. Robert, S. Azza, P. Jardot, et al. 2012. "Provirophages and transpovirons as the diverse mobilome of giant viruses." *P Natl Acad Sci USA* 109 (44):18078–83. doi:10.1073/pnas.1208835109.

DHHS. 2014. "U. S. government gain-of-function deliberative process and research funding pause on selected gain-of-function research involving influenza, MERS, and SARS Viruses." Washington, DC: Department of Health ad Human Services http://www.phe.gov/s3/dualuse/documents/gain-of-function.pdf.

Diener, T. O. 1989. "Circular RNAs: relics of precellular evolution?" *Proceedings of the National Academy of Sciences* 86 (23):9370–74. doi:10.1073/pnas.86.23.9370.

Diener, T. O., and W. B. Raymer. 1967. "Potato spindle tuber virus: A plant virus with properties of a free nucleic acid." *Science* 158 (3799):378–81.

Dobson, A. P. 2005. "Virology. What links bats to emerging infectious diseases?" *Science* 310 (5748):628–29. doi:10.1126/science.1120872.

Domingo, E., D. Sabo, T. Taniguchi, and C. Weissmann. 1978. "Nucleotide sequence heterogeneity of an RNA phage population." *Cell* 13 (4):735–44. doi:10.1016/0092–8674(78)90223–4.

Doorbar, J., W. Quint, L. Banks, I. G. Bravo, M. Stoler, T. R. Broker, and M. A. Stanley. 2012. "The biology and life-cycle of human papillomaviruses." *Vaccine* 30 Suppl 5:F55–70. doi:10.1016/j.vaccine.2012.06.083.

Doudna, J. A., and J. W. Szostak. 1989. "RNA-catalysed synthesis of complementary-strand RNA." *Nature* 339 (6225):519–22. doi:10.1038/339519a0.

Drake, J. W., and C. B. Hwang. 2005. "On the mutation rate of herpes simplex virus type 1." *Genetics* 170 (2):969–70. doi:10.1534/genetics.104.040410.

Duchêne, S., E. C. Holmes, and S. Y. Ho. 2014. "Analyses of evolutionary dynamics in viruses are hindered by a time-dependent bias in rate estimates." *P Roy Soc B-Biol Sci* 281 (1786). doi:10.1098/rspb.2014.0732.

Eamens, A., M. B. Wang, N. A. Smith, and P. M. Waterhouse. 2008. "RNA silencing in plants: Yesterday, today, and tomorrow." *Plant Physiol* 147 (2):456–68. doi:10.1104/pp. 108.117275.

Edwards, B. H., A. Bansal, S. Sabbaj, J. Bakari, M. J. Mulligan, and P. A. Goepfert. 2002. "Magnitude of functional CD8+ T-cell responses to the gag protein of human immunodeficiency virus type 1 correlates inversely with viral load in plasma." *J Virol* 76 (5):2298–305. doi:10.1128/jvi.76.5.2298–2305.2002.

Eigen, M. 1993. "Viral quasispecies." *Sci Am* 269 (1):42–9.

Elde, N. C., S. J. Child, M. T. Eickbush, J. O. Kitzman, K. S. Rogers, J. Shendure, A. P. Geballe, and H. S. Malik. 2012. "Poxviruses deploy genomic accordions to adapt rapidly against host antiviral defenses." *Cell* 150 (4):831–41. doi:10.1016/j.cell.2012.05.049.

Emerman, M., and H. S. Malik. 2010. "Paleovirology—modern consequences of ancient viruses." *PLOS Biol* 8 (2):e1000301. doi:10.1371/journal.pbio.1000301.

Faruque, S. M., M. J. Albert, and J. J. Mekalanos. 1998. "Epidemiology, genetics, and ecology of toxigenic Vibrio cholerae." *Microbiol Mol Biol Rev* 62 (4):1301–14.

Faruque, S. M., M. Asadulghani, M. Kamruzzaman, R. K. Nandi, A. N. Ghosh, G. B. Nair, J. J. Mekalanos, and D. A. Sack. 2002. "RS1 element of Vibrio cholerae can propagate horizontally as a filamentous phage exploiting the morphogenesis genes of CTXphi." *Infect Immun* 70 (1):163–70. doi:10.1128/IAI.70.1.163–170.2002.

FDA. 2015. "FDA approves first-of-its-kind product for the treatment of melanoma." Silver Spring, MD: U.S. Food & Drug Administration. http://www.fda.gov/NewsEvents/Newsroom/PressAnnouncements/ucm469571.htm.

Fenner, F. 1983. "The Florey Lecture, 1983: Biological control, as exemplified by smallpox eradication and myxomatosis." *P Roy Soc B-Biol Sci* 218 (1212):259–85. doi:10.1098/rspb.1983.0039.

Field, C. B., M. J. Behrenfeld, J. T. Randerson, and P. Falkowski. 1998. "Primary production of the biosphere: integrating terrestrial and oceanic components." *Science* 281 (5374):237–40.

Fisman, D., E. Khoo, and A. Tuite. 2014. "Early epidemic dynamics of the west african 2014 ebola outbreak: Estimates derived with a simple two-parameter model." *PLOS Curr* 6. doi:10.1371/currents.outbreaks.89c0d3783f36958d96ebbae97348d571.

Flint, J. S., V. R. Racaniello, G. F. V. Rall, A. M. Skalka, and L. W. Enquist. 2015. *Principles of Virology*, 4th ed. Washington, DC: American Society of Microbiology Press.

Flores, R., S. Gago-Zachert, P. Serra, R. Sanjuan, and S. F. Elena. 2014. "Viroids: Survivors from the RNA world?" *Annu Rev Microbiol* 68:395–414. doi:10.1146/annurev-micro-091313–103416.

Flores, R., S. Minoia, A. Carbonell, A. Gisel, S. Delgado, A. Lopez-Carrasco, B. Navarro, and F. Di Serio. 2015. "Viroids, the simplest RNA replicons: How they manipulate their hosts for being propagated and how their hosts react for containing the infection." *Virus Res* 209:136–45. doi:10.1016/j.virusres.2015.02.027.

Flores, R., C. Hernández, E. A. de Alba, J-A. Daròs, and F. Serio. 2005. "Viroids and viroid-host interactions." *Phytopathology* 43 (1):117–39. doi:10.1146/annurev.phyto.43.040204.140243.

Forterre, P. 2013. "The virocell concept and environmental microbiology." *ISME J* 7 (2):233–36. doi:10.1038/ismej.2012.110.

Francis, R. M., S. L. Nielsen, and P. Kryger. 2013. "Varroa-virus interaction in collapsing honey bee colonies." *PLOS ONE* 8 (3):e57540. doi:10.1371/journal.pone.0057540.

Fraser, C., T. D. Hollingsworth, R. Chapman, F. de Wolf, and W. P. Hanage. 2007. "Variation in HIV-1 set-point viral load: epidemiological analysis and an evolutionary hypothesis." *P Natl Acad Sci USA* 104 (44):17441–46. doi:10.1073/pnas.0708559104.

Frobisher, M., and J. Brown. 1927. "Transmissible Toxicogenicity of Streptococci." *B Johns Hopkins Hosp* 41:167–73.

Fujino, K., M. Horie, T. Honda, D. K. Merriman, and K. Tomonaga. 2014. "Inhibition of Borna disease virus replication by an endogenous bornavirus-like element in the ground squirrel genome." *P Natl Acad Sci USA* 111 (36):13175–80. doi:10.1073/pnas.1407046111.

Gago, S., S. F. Elena, R. Flores, and R. Sanjuán. 2009. "Extremely high mutation rate of a hammerhead viroid." *Science* 323 (5919):1308. doi:10.1126/science.1169202.

Gago-Zachert, S. 2016. "Viroids, infectious long non-coding RNAs with autonomous replication." *Virus Res* 212:12–24. doi:10.1016/j.virusres.2015.08.018.

Gamage, S. D., J. E. Strasser, and C. L. Chalk. 2003. "Nonpathogenic Escherichia coli can contribute to the production of Shiga toxin." *Infect Immun* 71 (6):3107–15. doi:10.1128/IAI.71.6.3107–3115.2003.

Gao, F., E. Bailes, D. L. Robertson, Y. Chen, C. M. Rodenburg, S. F. Michael, L. B. Cummins, et al. 1999. "Origin of HIV-1 in the chimpanzee Pan troglodytes troglodytes." *Nature* 397 (6718):436–41. doi:10.1038/17130.

Garten, R. J., T. C. Davis, C. A. Russell, B. Shu, S. Lindstrom, A. Balish, W. M. Sessions, X. Xu, E. Skepner, and V. Deyde. 2009. "Antigenic and genetic characteristics of swine-origin 2009 A (H1N1) influenza viruses circulating in humans." *Science* 325 (5937):197–201. doi:10.1126/science.1176225.

Gaudet, D., J. Methot, S. Dery, D. Brisson, C. Essiembre, G. Tremblay, K. Tremblay, et al. 2013. "Efficacy and long-term safety of alipogene tiparvovec (AAV1-LPLS447X) gene therapy for lipoprotein lipase deficiency: an open-label trial." *Gene Ther* 20 (4):361–69. doi:10.1038/gt.2012.43.

Gern, J. E., D. M. Galagan, N. N. Jarjour, E. C. Dick, and W. W. Busse. 1997. "Detection of rhinovirus RNA in lower airway cells during experimentally induced infection." *Am J Resp Crit Care Med* 155 (3):1159–61. doi:10.1164/ajrccm.155.3.9117003.

Gilbert, M. T., A. Rambaut, G. Wlasiuk, T. J. Spira, A. E. Pitchenik, and M. Worobey. 2007. "The emergence of HIV/AIDS in the Americas and beyond." *P Natl Acad Sc USA* 104 (47):18566–70. doi:10.1073/pnas.0705329104.

Gire, S. K., A. Goba, K. G. Andersen, R. S. Sealfon, D. J. Park, L. Kanneh, S. Jalloh, et al. 2014. "Genomic surveillance elucidates Ebola virus origin and transmission during the 2014 outbreak." *Science* 345 (6202):1369–72. doi:10.1126/science.1259657.

Gordon, R. J., and F. D. Lowy. 2008. "Pathogenesis of methicillin-resistant Staphylococcus aureus infection." *Clin Infect Dis* 46 (Supp 5): S350–59. doi:10.1086/533591.

Gorry, P. R., and P. Ancuta. 2011. "Coreceptors and HIV-1 pathogenesis." *Curr HIV/AIDS Rep* 8 (1):45–53. doi:10.1007/s11904–010–0069-x.

Grassly, N. C. 2013. "The final stages of the global eradication of poliomyelitis." *Philos T Roy Soc B* 368 (1623):20120140. doi:10.1098/rstb.2012.0140.

Guan, Y., K. F. Shortridge, S. Krauss, and R. G. Webster. 1999. "Molecular characterization of H9N2 influenza viruses: Were they the donors of the "internal" genes of H5N1 viruses in Hong Kong?" *P Natl Acad Sci USA* 96 (16):9363–67.

Guo, W. P., X. D. Lin, W. Wang, J. H. Tian, and M. L. Cong. 2013. "Phylogeny and origins of hantaviruses harbored by bats, insectivores, and rodents." *PLOS Pathog* 9 (2). doi:10.1371/journal.ppat.1003159.

Gupta, S., N. Ferguson, and R. Anderson. 1998. "Chaos, persistence, and evolution of strain structure in antigenically diverse infectious agents." *Science* 280 (5365):912–15.

Gwaltney, J. M., Jr., P. B. Moskalski, and J. O. Hendley. 1978. "Hand-to-hand transmission of rhinovirus colds." *Ann Intern Med* 88 (4):463–67.

Hacein-Bey-Abina, S., A. Garrigue, G. P. Wang, J. Soulier, A. Lim, E. Morillon, E. Clappier, et al. 2008. "Insertional oncogenesis in 4 patients after retrovirus-mediated gene therapy of SCID-X1." *J Clin Invest* 118 (9):3132–42. doi:10.1172/JCI35700.

Hacein-Bey-Abina, S., S-Y. Y. Pai, H. B. Gaspar, M. Armant, C. C. Berry, S. Blanche, J. Bleesing, et al. 2014. "A modified γ-retrovirus vector for X-linked severe combined immunodeficiency." *New Engl J Med* 371 (15):1407–17. doi:10.1056/NEJMoa1404588.

Hanger, J. J., L. D. Bromham, J. J. McKee, T. M. O'Brien, and W. F. Robinson. 2000. "The nucleotide sequence of koala (Phascolarctos cinereus) retrovirus: A novel type C endogenous virus related to Gibbon ape leukemia virus." *J Virol* 74 (9):4264–72.

Hare, R. 1967. *Diseases in Antiquity; A Survey of the Diseases, Injuries, and Surgery of Early Populations.* Springfield, IL: C. C. Thomas.

He, B., Y. Feng, H. Zhang, L. Xu, W. Yang, Y. Zhang, X. Li, and C. Tu. 2015. "Filovirus RNA in fruit bats, China." *Emerg Infect Dis* 21 (9):1675–77. doi:10.3201/eid2109.150260.

Hel, Z., J. McGhee, R, and J. Mestecky. 2006. "HIV infection: First battle decides the war." *Trends Immunol* 27 (6):274281. doi:10.1016/j.it.2006.04.007.

Hendrix, R. W., G. F. Hatfull, and M. C. Smith. 2003. "Bacteriophages with tails: Chasing their origins and evolution." *Res Microbiol* 154 (4):253–57. doi:10.1016/S0923–2508(03)00068–8.

Henrich, J. 2015. *The Secret of Our Success.* Princeton, NJ: Princeton University Press.

Herfst, S., E. J. Schrauwen, M. Linster, S. Chutinimitkul, E. de Wit, V. J. Munster, E. M. Sorrell, et al. 2012. "Airborne transmission of influenza A/H5N1 virus between ferrets." *Science* 336 (6088):1534–41. doi:10.1126/science.1213362.

Hicks, A. L., and S. Duffy. 2014. "Cell Tropism Predicts Long-term Nucleotide Substitution Rates of Mammalian RNA Viruses." *PLOS Pathog* 10 (1). doi:10.1371/journal.ppat.1003838.

Hilditch, L., R. Matadeen, D. C. Goldstone, P. B. Rosenthal, I. A. Taylor, and J. P. Stoye. 2011. "Ordered assembly of murine leukemia virus capsid protein on lipid nanotubes directs specific binding by the restriction factor, Fv1." *P Natl Acad Sci USA* 108 (14):5771–76. doi:10.1073/pnas.1100118108.

Hirsch, V. M., R. A. Olmsted, M. Murphey-Corb, R. H. Purcell, and P. R. Johnson. 1989. "An African primate lentivirus (SIVsm) closely related to HIV-2." *Nature* 339 (6223):389–92. doi:10.1038/339389a0.

Holmes, E. C. 2003. "Molecular clocks and the puzzle of RNA virus origins." *J Virol* 77 (7):3893–97.

Holmes, E. C., E. Ghedin, N. Miller, J. Taylor, Y. Bao, K. St. George, B. T. Grenfell, et al. 2005. "Whole-genome analysis of human influenza A virus reveals multiple persistent lineages and reassortment among recent H3N2 viruses." *PLOS Biol* 3 (9):e300. doi:10.1371/journal.pbio.0030300.

Holmes, E. C., L. Q. Zhang, P. Simmonds, C. A. Ludlam, and A. J. Brown. 1992. "Convergent and divergent sequence evolution in the surface envelope glycoprotein of human immunodeficiency virus type 1 within a single infected patient." *P Natl Acad Sci USA* 89 (11):4835–39.

Holmes, E. C., and Y. Z. Zhang. 2015. "The evolution and emergence of hantaviruses." *Curr Opin Virol* 10:27–33. doi:10.1016/j.coviro.2014.12.007.

Horimoto, T., and Y. Kawaoka. 2001. "Pandemic threat posed by avian influenza A viruses." *Clin Microbiol Rev* 14 (1):129–49. doi:10.1128/CMR.14.1.129–149.2001.

Horimoto, T., and Y. Kawaoka. 2005. "Influenza: Lessons from past pandemics, warnings from current incidents." *Nat Rev Microbiol* 3 (8):591–600. doi:10.1038/nrmicro1208.

Huff, J. L., and P. A. Barry. 2003. "B-virus (Cercopithecine herpesvirus 1) infection in humans and macaques: Potential for zoonotic disease." *Emerg Infect Dis* 9 (2):246–50. doi:10.3201/eid0902.020272.

Hughes, A. L., and R. Friedman. 2005. "Poxvirus genome evolution by gene gain and loss." *Mol Phylogenet Evol* 35 (1):186–95. doi:10.1016/j.ympev.2004.12.008.

Hughes, J. F., and J. M. Coffin. 2001. "Evidence for genomic rearrangements mediated by human endogenous retroviruses during primate evolution." *Nat Genet* 29 (4):487–89. doi:10.1038/ng775.

Hutchins, C. J., P. D. Rathjen, A. C. Forster, and R. H. Symons. 1986. "Self-cleavage of plus and minus RNA transcripts of avocado sunblotch viroid." *Nucleic Acids Res* 14 (9):3627–40.

Hwu, W-L., S. Muramatsu, S-H. Tseng, K-Y. Tzen, N-C. Lee, Y-H. Chien, R. O. Snyder, B. J. Byrne, C-H. Tai, and R-M. Wu. 2012. "Gene therapy for aromatic l-amino acid decarboxylase deficiency." *Sci Transl Med* 4 (134). doi:10.1126/scitranslmed.3003640.

Imai, M., T. Watanabe, M. Hatta, S. C. Das, M. Ozawa, K. Shinya, G. Zhong, et al. 2012. "Experimental adaptation of an influenza H5 HA confers respiratory droplet transmission to a reassortant H5 HA/H1N1 virus in ferrets." *Nature* 486 (7403):420–8. doi:10.1038/nature10831.

ISDA. 2004. *Bad Bugs, No Drugs*. Alexandria, VA: Infectious Diseases Society of America. https://www.idsociety.org/uploadedFiles/IDSA/Policy_and_Advocacy/Current_Topics_and_Issues/Advancing_Product_Research_and_Development/Bad_Bugs_No_Drugs/Statements/As%20Antibiotic%20Discovery%20Stagnates%20A%20Public%20Health%20Crisis%20Brews.pdf

Jacobs, S. E., D. M. Lamson, K. St. George, and T. J. Walsh. 2013. "Human rhinoviruses." *Clin Microbiol Rev* 26 (1):135–62. doi:10.1128/CMR.00077–12.

Jacobson, S. G., A. V. Cideciyan, R. Ratnakaram, E. Heon, S. B. Schwartz, A. J. Roman, M. C. Peden, et al. 2012. "Gene therapy for Leber congenital amaurosis caused by RPE65 mutations: Safety and efficacy in 15 children and adults followed up to 3 years." *Arch Ophthalmol* 130 (1):9–24. doi:10.1001/archophthalmol.2011.298.

Joffret, M. L., S. Jegouic, M. Bessaud, J. Balanant, C. Tran, V. Caro, B. Holmblat, R. Razafindratsimandresy, J. M. Reynes, M. Rakoto-Andrianarivelo, and F. Delpeyroux. 2012. "Common and diverse features of cocirculating type 2 and 3 recombinant vaccine-derived polioviruses isolated from patients with poliomyelitis and healthy children." *J Infect Dis* 205 (9):1363–73. doi:10.1093/infdis/jis204.

Johnsen, D. O., W. L. Wooding, P. Tanticharoenyos, and C. H. Bourgeois, Jr. 1971. "Malignant lymphoma in the gibbon." *J Am Vet Med Assoc* 159 (5):563–66.

Johnson, W. E. 2010. "Endless forms most viral." *PLOS Genet* 6 (11):e1001210. doi:10.1371/journal.pgen.1001210.

Johnston, S. L., A. Papi, P. J. Bates, J. G. Mastronarde, M. M. Monick, and G. W. Hunninghake. 1998. "Low grade rhinovirus infection induces a prolonged release of IL-8 in pulmonary epithelium." *J Immunol* 160 (12):6172–81.

Jones, K. E., N. G. Patel, M. A. Levy, A. Storeygard, D. Balk, J. L. Gittleman, and P. Daszak. 2008. "Global trends in emerging infectious diseases." *Nature* 451 (7181):990–93. doi:10.1038/nature06536.

Joyce, G. F. 1989. "RNA evolution and the origins of life." *Nature* 338 (6212):217–24. doi:10.1038/338217a0.

Juhasz, A., H. Hegyi, and F. Solymosy. 1988. "A novel aspect of the information content of viroids." *Biochim Biophys Acta* 950 (3):455–8.

Junod, T. 2014. "Why there won't be an Ebola pandemic." *Esquire*. http://www.esquire.com/news-politics/news/a30193/no-ebola-pandemic/

Kaiser, L., J. D. Aubert, J. C. Pache, C. Deffernez, T. Rochat, J. Garbino, W. Wunderli, et al. 2006. "Chronic rhinoviral infection in lung transplant recipients." *Am J Resp Crit Care Med* 174 (12):1392–99. doi:10.1164/rccm.200604–489OC.

Kambhu, S., P. Falldorf, and J. S. Lee. 1990. "Endogenous retroviral long terminal repeats within the HLA-DQ locus." *P Natl Acad Sci USA* 87 (13):4927–31.

Kaser, A., S. Zeissig, and R. S. Blumberg. 2010. "Inflammatory bowel disease." *Annu Rev Immunol* 28:573–621. doi:10.1146/annurev-immunol-030409–101225.

Kawashima, Y., K. Pfafferott, J. Frater, P. Matthews, R. Payne, M. Addo, H. Gatanaga, et al. 2009. "Adaptation of HIV-1 to human leukocyte antigen class I." *Nature* 458 (7238):641–45. doi:10.1038/nature07746.

Keele, B. F., E. E. Giorgi, J. F. Salazar-Gonzalez, J. M. Decker, K. T. Pham, M. G. Salazar, C. Sun, et al. 2008. "Identification and characterization of transmitted and early founder virus envelopes in primary HIV-1 infection." *P Natl Acad Sci USA* 105 (21):7552–57. doi:10.1073/pnas.0802203105.

Keele, B. F., J. H. Jones, K. A. Terio, J. D. Estes, R. S. Rudicell, M. L. Wilson, Y. Li, et al. 2009. "Increased mortality and AIDS-like immunopathology in wild chimpanzees infected with SIVcpz." *Nature* 460 (7254):515–19. doi:10.1038/nature08200.

Keele, B. F., F. Van Heuverswyn, Y. Li, E. Bailes, J. Takehisa, M. L. Santiago, F. Bibollet-Ruche, et al. 2006. "Chimpanzee reservoirs of pandemic and nonpandemic HIV-1." *Science* 313 (5786):523–26. doi:10.1126/science.1126531.

Kelly, E., and S. J. Russell. 2007. "History of oncolytic viruses: genesis to genetic engineering." *Mol Ther* 15 (4):651–59. doi:10.1038/sj.mt.6300108.

Kerr, P. J., E. Ghedin, J. V. DePasse, A. Fitch, I. M. Cattadori, P. J. Hudson, D. C. Tscharke, A. F. Read, and E. C. Holmes. 2012. "Evolutionary history and attenuation of myxoma virus on two continents." *PLOS Pathog* 8 (10):e1002950. doi:10.1371/journal.ppat.1002950.

Kilbourne, E. D. 2006. "Influenza pandemics of the 20th century." *Emerg Infect Dis* 12 (1):9–14. doi:10.3201/eid1201.051254.

Kim, E. Y., T. Bhattacharya, K. Kunstman, P. Swantek, F. A. Koning, M. H. Malim, and S. M. Wolinsky. 2010. "Human APOBEC3G-mediated editing can promote HIV-1 sequence diversification and accelerate adaptation to selective pressure." *J Virol* 84 (19):10402–5. doi:10.1128/JVI.01223–10.

Kistler, A. L., D. R. Webster, S. Rouskin, V. Magrini, J. J. Credle, D. P. Schnurr, H. A. Boushey, E. R. Mardis, H. Li, and J. L. DeRisi. 2007. "Genome-wide diversity and selective pressure in the human rhinovirus." *Virol J* 4:40. doi:10.1186/1743–422X-4–40.

Knipe, D., and Howley, P. 2013. *Fields Virology*, 6th ed. Philadelphia, PA: Lippincott Williams & Wilkins.

Koonin, E. V., and V. V. Dolja. 2014. "Virus world as an evolutionary network of viruses and capsidless selfish elements." *Microbiol Mol Biol Rev* 78 (2):278–303. doi:10.1128/MMBR.00049–13.

Koonin, E. V., V. V. Dolja, and M. Krupovic. 2015a. "Origins and evolution of viruses of eukaryotes: The ultimate modularity." *Virology* 479–480:2–25. doi:10.1016/j.virol.2015.02.039.

Koonin, E. V., V. V. Dolja, and M. Krupovic. 2015b. "Origins and evolution of viruses of eukaryotes: The ultimate modularity." *Virology.* doi:10.1016/j.virol.2015.02.039.

Koonin, E. V., M. Krupovic, and N. Yutin. 2015. "Evolution of double-stranded DNA viruses of eukaryotes: from bacteriophages to transposons to giant viruses." *Ann NY Acad Sci* 1341:10–24. doi:10.1111/nyas.12728.

Koonin, E. V., and W. Martin. 2005. "On the origin of genomes and cells within inorganic compartments." *Trends Genet* 21 (12):647–54. doi:10.1016/j.tig.2005.09.006.

Korber, B., M. Muldoon, J. Theiler, F. Gao, R. Gupta, A. Lapedes, B. H. Hahn, S. Wolinsky, and T. Bhattacharya. 2000. "Timing the ancestor of the HIV-1 pandemic strains." *Science* 288 (5472):1789–96.

Kruger, K., P. J. Grabowski, A. J. Zaug, J. Sands, D. E. Gottschling, and T. R. Cech. 1982. "Self-splicing RNA: autoexcision and autocyclization of the ribosomal RNA intervening sequence of Tetrahymena." *Cell* 31 (1):147–57.

Krumbholz, A., O. R. Bininda-Emonds, P. Wutzler, and R. Zell. 2009. "Phylogenetics, evolution, and medical importance of polyomaviruses." *Infect Genet Evol* 9 (5):784–99. doi:10.1016/j.meegid.2009.04.008.

Krupovic, M., and V. Cvirkaite-Krupovic. 2011. "Virophages or satellite viruses?" *Nat Rev Microbiol* 9 (11):762–3.

Krupovic, M., D. H. Bamford, and E. V. Koonin. 2014. "Conservation of major and minor jelly-roll capsid proteins in Polinton (Maverick) transposons suggests that they are bona fide viruses." *Biology Direct* 9 (1):6. doi:10.1186/1745–6150–9–6.

Krupovic, M., D. Prangishvili, R. W. Hendrix, and D. H. Bamford. 2011. "Genomics of bacterial and archaeal viruses: dynamics within the prokaryotic virosphere." *Microbiol Mol Biol Rev* 75 (4):610–35. doi:10.1128/MMBR.00011–11.

Kuiken, T., R. Fouchier, G. Rimmelzwaan, and A. Osterhaus. 2003. "Emerging viral infections in a rapidly changing world." *Curr Opin Biotechnol* 14 (6): 641–46.

Kuo, C-H. H., and H. Ochman. 2010. "The extinction dynamics of bacterial pseudogenes." *PLOS Genet* 6 (8). doi:10.1371/journal.pgen.1001050.

Kuroda, M., T. Ohta, I. Uchiyama, T. Baba, H. Yuzawa, I. Kobayashi, L. Cui, et al. 2001. "Whole genome sequencing of meticillin-resistant Staphylococcus aureus." *Lancet* 357 (9264):1225–40.

La Scola, B., C. Desnues, I. Pagnier, C. Robert, L. Barrassi, G. Fournous, M. Merchat, et al. 2008. "The virophage as a unique parasite of the giant mimivirus." *Nature* 455 (7209):100–4. doi:10.1038/nature07218.

La Scola, B., S. Audic, C. Robert, L. Jungang, X. de Lamballerie, M. Drancourt, R. Birtles, J-M. M. Claverie, and D. Raoult. 2003. "A giant virus in amoebae." *Science* 299 (5615):2033. doi:10.1126/science.1081867.

Lander, E. S., L. M. Linton, B. Birren, C. Nusbaum, M. C. Zody, J. Baldwin, K. Devon, et al. 2001. "Initial sequencing and analysis of the human genome." *Nature* 409 (6822):860–921. doi:10.1038/35057062.

Langland, J. O., and B. L. Jacobs. 2002. "The role of the PKR-inhibitory genes, E3L and K3L, in determining vaccinia virus host range." *Virology* 299 (1):133–41.

Lavialle, C., G. Cornelis, A. Dupressoir, C. Esnault, O. Heidmann, C. Vernochet, and T. Heidmann. 2013. "Paleovirology of 'syncytins,' retroviral *env* genes exapted for a role in placentation." *Philos T Roy Soc B* 368 (1626):20120507. doi:10.1098/rstb.2012.0507.

Lawrence, J. G., R. W. Hendrix, and S. Casjens. 2001. "Where are the pseudogenes in bacterial genomes?" *Trends Microbiol* 9 (11):535–40.

Lawrence, J. G., and H. Ochman. 1998. "Molecular archaeology of the Escherichia coli genome." *P Natl Acad Sci USA* 95 (16):9413–17.

Le Tortorec, A., and S. J. Neil. 2009. "Antagonism to and intracellular sequestration of human tetherin by the human immunodeficiency virus type 2 envelope glycoprotein." *J Virol* 83 (22):11966–78. doi:10.1128/JVI.01515-09.

Lee, W-M. M., C. Kiesner, T. Pappas, I. Lee, K. Grindle, T. Jartti, B. Jakiela, R. F. Lemans, P. A. Shult, and J. E. Gern. 2007. "A diverse group of previously unrecognized human rhinoviruses are common causes of respiratory illnesses in infants." *PLOS ONE* 2 (10). doi:10.1371/journal.pone.0000966.

Lemey, P., A. Rambaut, T. Bedford, N. Faria, F. Bielejec, G. Baele, C. A. Russell, et al. 2014. "Unifying viral genetics and human transportation data to predict the global transmission dynamics of human influenza H3N2." *PLOS Pathog* 10 (2):e1003932. doi:10.1371/journal.ppat.1003932.

Lemey, P., A. Rambaut, and O. G. Pybus. 2006. "HIV evolutionary dynamics within and among hosts." *AIDS Rev* 8 (3):125–40.

Leroy, E. M., S. Baize, V. E. Volchkov, S. P. Fisher-Hoch, M. C. Georges-Courbot, J. Lansoud-Soukate, M. Capron, P. Debre, J. B. McCormick, and A. J. Georges. 2000. "Human asymptomatic Ebola infection and strong inflammatory response." *Lancet* 355 (9222):2210–15.

Leroy, E. M., P. Rouquet, P. Formenty, S. Souquiere, A. Kilbourne, J. M. Froment, M. Bermejo, et al. 2004. "Multiple Ebola virus transmission events and rapid decline of central African wildlife." *Science* 303 (5656):387–90. doi:10.1126/science.1092528.

Leroy, E. M., P. Telfer, B. Kumulungui, P. Yaba, P. Rouquet, P. Roques, J. P. Gonzalez, T. G. Ksiazek, P. E. Rollin, and E. Nerrienet. 2004. "A serological survey of Ebola virus infection in central African nonhuman primates." *J Infect Dis* 190 (11):1895–99. doi:10.1086/425421.

Leroy, E. M., B. Kumulungui, X. Pourrut, P. Rouquet, A. Hassanin, P. Yaba, A. Délicat, J. T. Paweska, J-P. Gonzalez, and R. Swanepoel. 2005. "Fruit bats as reservoirs of Ebola virus." *Nature* 438 (7068):575–76. doi:10.1038/438575a.

Levy, D. N., G. M. Aldrovandi, O. Kutsch, and G. M. Shaw. 2004. "Dynamics of HIV-1 recombination in its natural target cells." *P Natl Acad Sci USA* 101 (12):4204–9. doi:10.1073/pnas.0306764101.

Li, K. S., Y. Guan, J. Wang, G. J. Smith, K. M. Xu, L. Duan, A. P. Rahardjo, P. Puthavathana, C. Buranathai, et al. 2004. "Genesis of a highly pathogenic and potentially pandemic H5N1 influenza virus in eastern Asia." *Nature* 430 (6996):209–13. doi:10.1038/nature02746.

Li, W. 1997. *Molecular Evolution*. Sunderland, MA: Sinauer Associates.

Lichty, B. D., C. J. Breitbach, D. F. Stojdl, and J. C. Bell. 2014. "Going viral with cancer immunotherapy." *Nat Rev Cancer* 14 (8):559–67. doi:10.1038/nrc3770.

Lindell, D., M. B. Sullivan, Z. I. Johnson, A. C. Tolonen, F. Rohwer, and S. W. Chisholm. 2004. "Transfer of photosynthesis genes to and from Prochlorococcus viruses." *P Natl Acad Sci USA* 101 (30):11013–18. doi:10.1073/pnas.0401526101.

Lindstrom, S. E., N. J. Cox, and A. Klimov. 2004. "Genetic analysis of human H2N2 and early H3N2 influenza viruses, 1957–1972: Evidence for genetic divergence and multiple reassortment events." *Virology* 328 (1):101–19. doi:10.1016/j.virol.2004.06.009.

Liu, B. L., M. Robinson, Z. Q. Q. Han, R. H. Branston, C. English, P. Reay, Y. McGrath, et al. 2003. "ICP34.5 deleted herpes simplex virus with enhanced oncolytic, immune stimulating, and anti-tumour properties." *Gene Ther* 10 (4):292–303. doi:10.1038/sj.gt.3301885.

Lothar, K., and K. Kiplagat. 2010. "Fluctuating Algal Food Populations and the Occurrence of Lesser Flamingos (Phoeniconaias minor) in three Kenyan Rift Valley lakes1." *J Phycology.* doi:10.1111/j.1529–8817.2010.00915.x.

Loy, T. V., K. Thys, L. Tritsmans, and L. Stuyver, J. 2012. "Quasispecies analysis of JC virus DNA present in urine of healthy subjects." *PLOS ONE* 8 (8). doi:10.1371/journal.pone.0070950.

Luis, A. D., D. T. Hayman, T. J. O'Shea, P. M. Cryan, A. T. Gilbert, J. R. Pulliam, J. N. Mills, et al. 2013. "A comparison of bats and rodents as reservoirs of zoonotic viruses: Are bats special?" *P Roy Soc B-Biol Sci* 280 (1756):20122753. doi:10.1098/rspb.2012.2753.

Lythgoe, K. A., and C. Fraser. 2012. "New insights into the evolutionary rate of HIV-1 at the within-host and epidemiological levels." *P Roy Soc B-Biol Sci* 279 (1741):3367–75. doi:10.1098/rspb.2012.0595.

Maines, T. R., L. M. Chen, Y. Matsuoka, H. Chen, T. Rowe, J. Ortin, A. Falcon, et al. 2006. "Lack of transmission of H5N1 avian-human reassortant influenza viruses in a ferret model." *P Natl Acad Sci USA* 103 (32):12121–26. doi:10.1073/pnas.0605134103.

Maines, T. R., L. M. Chen, N. Van Hoeven, T. M. Tumpey, O. Blixt, J. A. Belser, K. M. Gustin, et al. 2011. "Effect of receptor binding domain mutations on receptor binding and transmissibility of avian influenza H5N1 viruses." *Virology* 413 (1):139–47. doi:10.1016/j.virol.2011.02.015.

Maiques, E., C. Ubeda, S. Campoy, N. Salvador, I. Lasa, R. P. Novick, J. Barbé, and J. R. Penadés. 2006. "Beta-lactam antibiotics induce the SOS response and horizontal transfer of virulence factors in Staphylococcus aureus." *J Bacteriol* 188 (7):2726–29. doi:10.1128/JB.188.7.2726–2729.2006.

Maksakova, I. A., D. L. Mager, and D. Reiss. 2008. "Keeping active endogenous retroviral-like elements in check: The epigenetic perspective." *Cell Mol Life Sci* 65 (21):3329–47. doi:10.1007/s00018–008–8494–3.

Malik, H. S. 2005. "Ribonuclease H evolution in retrotransposable elements." *Cytogenet Genome Res* 110 (1–4):392–401. doi:10.1159/000084971.

Malik, H. S., and T. H. Eickbush. 2001. "Phylogenetic analysis of ribonuclease H domains suggests a late, chimeric origin of LTR retrotransposable elements and retroviruses." *Genome Res* 11 (7):1187–97. doi:10.1101/gr.185101.

Malik, H. S., S. Henikoff, and T. H. Eickbush. 2000. "Poised for contagion: evolutionary origins of the infectious abilities of invertebrate retroviruses." *Genome Res* 10 (9):1307–18.

Malim, M. H., and P. D. Bieniasz. 2012. "HIV Restriction Factors and Mechanisms of Evasion." *Cold Spring Harb Perspect Med* 2 (5):a006940. doi:10.1101/cshperspect.a006940.

Malim, M. H., and M. Emerman. 2008. "HIV-1 accessory proteins—ensuring viral survival in a hostile environment." *Cell Host Microbe* 3 (6):388–98. doi:10.1016/j.chom.2008.04.008.

Malpica, J. M., A. Fraile, I. Moreno, C. I. Obies, J. W. Drake, and F. Garcia-Arenal. 2002. "The rate and character of spontaneous mutation in an RNA virus." *Genetics* 162 (4):1505–11.

Mandl, J. N., R. Ahmed, L B. Barreiro, P. Daszak, J. H. Epstein, H. W. Virgin, and M. B. Feinberg. 2015. "Reservoir host immune responses to emerging zoonotic viruses." *Cell* 160. doi:10.1016/j.cell.2014.12.003.

Mandl, J. N., A. P. Barry, T. H. Vanderford, N. Kozyr, R. Chavan, S. Klucking, et al. 2008. "Divergent TLR7 and TLR9 signaling and type I interferon production distinguish pathogenic and nonpathogenic AIDS virus infections." *Nat Med* 14 (10):1077–87. doi:10.1038/nm.1871.

Mangeney, M., and T. Heidmann. 1998. "Tumor cells expressing a retroviral envelope escape immune rejection in vivo." *P Natl Acad Sci USA* 95 (25). doi:10.1073/pnas.95.25.14920.

Mann, N. H., A. Cook, A. Millard, S. Bailey, and M. Clokie. 2003. "Marine ecosystems: Bacterial photosynthesis genes in a virus." *Nature*. doi:10.1038/424741a.

Marsh, G. A., J. Haining, and R. Robinson. 2011. "Ebola Reston virus infection of pigs: Clinical significance and transmission potential." *J Infect Dis*. doi:10.1093/infdis/jir300.

Marshall, I. D., and F. Fenner. 1960. "Studies in the epidemiology of infectious myxomatosis of rabbits. VII. The virulence of strains of myxoma virus recovered from Australian wild rabbits between 1951 and 1959." *J Hyg (Lond)* 58:485–8.

Martinez de Alba, A. E., R. Sagesser, M. Tabler, and M. Tsagris. 2003. "A bromodomain-containing protein from tomato specifically binds potato spindle tuber viroid RNA in vitro and in vivo." *J Virol* 77 (17):9685–94.

Martinez, J. L. 2009. "The role of natural environments in the evolution of resistance traits in pathogenic bacteria." *Proc Roy Soc B-Biol Sci* 276 (1667):2521–30. doi:10.1098/rspb.2009.0320.

Martuza, R. L., A. Malick, J. M. Markert, K. L. Ruffner, and D. M. Coen. 1991. "Experimental therapy of human glioma by means of a genetically engineered virus mutant." *Science* 252 (5007):854–56.

Mate, S. E., J. R. Kugelman, T. G. Nyenswah, J. T. Ladner, M. R. Wiley, T. Cordier-Lassalle, A. Christie, et al. 2015. "Molecular evidence of sexual transmission of Ebola virus." *New Engl J Med* 373 (25):2448–54. doi:10.1056/NEJMoa1509773.

McCarthy, K. R., and W. E. Johnson. 2014. "Plastic proteins and monkey blocks: How lentiviruses evolved to replicate in the presence of primate restriction factors." *PLOS Pathog* 10 (4):e1004017. doi:10.1371/journal.ppat.1004017.

McFadden, E. R., Jr., B. M. Pichurko, H. F. Bowman, E. Ingenito, S. Burns, N. Dowling, and J. Solway. 1985. "Thermal mapping of the airways in humans." *J Appl Physiol (1985)* 58 (2):564–70.

McFadden, G. 2005. "Poxvirus tropism." *Nat Rev Microbiol* 3 (3):201–13. doi:10.1038/nrmicro1099.

McGeoch, D. J. 1990. "Protein sequence comparisons show that the 'pseudoproteases' encoded by poxviruses and certain retroviruses belong to the deoxyuridine triphosphatase family." *Nucleic Acids Res* 18 (14):4105–10.

McGeoch, D. J., A. Dolan, and A. C. Ralph. 2000. "Toward a comprehensive phylogeny for mammalian and avian herpesviruses." *J Virol* 74 (22):10401–6.

McGeoch, D. J., S. Cook, A. Dolan, F. E. Jamieson, and E. Telford. 1995. "Molecular phylogeny and evolutionary timescale for the family of mammalian herpesviruses." *J Mol Biol* 247 (3):443–58. doi:10.1006/jmbi.1995.0152.

McGeoch, D. J., F. J. Rixon, and A. J. Davison. 2006. "Topics in herpesvirus genomics and evolution." *Virus Res* 117 (1):90–104. doi:10.1016/j.virusres.2006.01.002.

McLeod, S. M., H. H. Kimsey, B. M. Davis, and M. K. Waldor. 2005. "CTXphi and Vibrio cholerae: Exploring a newly recognized type of phage-host cell relationship." *Mol Microbiol* 57 (2):347–56. doi:10.1111/j.1365–2958.2005.04676.x.

Meisler, M. H., and C. N. Ting. 1993. "The remarkable evolutionary history of the human amylase genes." *Crit Rev Oral Biol Med* 4 (3–4):503–9.

Message, S. D., and S. L. Johnston. 2001. "The immunology of virus infection in asthma." *Eur Resp J* 18 (6):1013–25.

Miller, M. A., C. Viboud, M. Balinska, and L. Simonsen. 2009. "The signature features of influenza pandemics—implications for policy." *New Engl J Med* 360 (25):2595–98. doi:10.1056/NEJMp0903906.

Mills, S., F. Shanahan, C. Stanton, C. Hill, A. Coffey, and R. P. Ross. 2013. "Movers and shakers: influence of bacteriophages in shaping the mammalian gut microbiota." *Gut Microbes* 4 (1):4–16. doi:10.4161/gmic.22371.

Minor, P. D., A. J. Macadam, D. M. Stone, and J. W. Almond. 1993. "Genetic basis of attenuation of the Sabin oral poliovirus vaccines." *Biologicals* 21 (4):357–63. doi:10.1006/biol.1993.1096.

Miranda, M. E. G., and N. L. J. Miranda. 2011. "Reston ebolavirus in humans and animals in the Philippines: A review." *J I Dis* 204 (Suppl 3):S757–60.

Modi, S. R., H. H. Lee, C. S. Spina, and J. J. Collins. 2013. "Antibiotic treatment expands the resistance reservoir and ecological network of the phage metagenome." *Nature* 499 (7457):219–22. doi:10.1038/nature12212.

Monier, A., J-M. M. Claverie, and H. Ogata. 2008. "Taxonomic distribution of large DNA viruses in the sea." *Genome Biol* 9 (7). doi:10.1186/gb-2008-9-7-r106.

Moore, L. R., G. Rocap, and S. W. Chisholm. 1998. "Physiology and molecular phylogeny of coexisting Prochlorococcus ecotypes." *Nature* 393 (6684):464–67. doi:10.1038/30965.

Morens, D. M., J. K. Taubenberger, and A. S. Fauci. 2009. "The persistent legacy of the 1918 influenza virus." *New Engl J Med* 361 (3):225–229. doi:10.1056/NEJMp0904819.

Mosser, A. G., R. Vrtis, L. Burchell, W. M. Lee, C. R. Dick, E. Weisshaar, D. Bock, et al. 2005. "Quantitative and qualitative analysis of rhinovirus infection in bronchial tissues." *Am J Resp Crit Care Med* 171 (6):645–51. doi:10.1164/rccm.200407–970OC.

Mueller, S., D. Papamichail, and J. R. Coleman. 2006. "Reduction of the rate of poliovirus protein synthesis through large-scale codon deoptimization causes attenuation of viral virulence by lowering specific infectivity." *J Virol* 80 (19):9687–96. doi:10.1128/JVI.00738-06.

Nakajima, K., U. Desselberger, and P. Palese. 1978. "Recent human influenza A (H1N1) viruses are closely related genetically to strains isolated in 1950." *Nature* 274 (5669):334–39.

Naldini, Luigi. 2015. "Gene therapy returns to centre stage." *Nature* 526 (7573):351–60. doi:10.1038/nature15818.

Negredo, A., G. Palacios, S. Vazquez-Moron, F. Gonzalez, H. Dopazo, F. Molero, J. Juste, et al. 2011. "Discovery of an ebolavirus-like filovirus in Europe." *PLOS Pathog* 7 (10):e1002304. doi:10.1371/journal.ppat.1002304.

Nelson, M. I., and E. C. Holmes. 2007. "The evolution of epidemic influenza." *Nat Rev Genet* 8 (3):196–205. doi:10.1038/nrg2053.

Nelson, M. I., C. Viboud, L. Simonsen, R. T. Bennett, S. B. Griesemer, K. St. George, J. Taylor, et al. 2008. "Multiple reassortment events in the evolutionary history of H1N1 influenza A virus since 1918." *PLOS Pathog* 4 (2):e1000012. doi:10.1371/journal.ppat.1000012.

Nelson, M. I., L. Simonsen, C. Viboud, M. A. Miller, J. Taylor, K. St. George, S. B. Griesemer, et al. 2006. "Stochastic processes are key determinants of short-term evolution in influenza a virus." *PLOS Pathog* 2 (12). doi:10.1371/journal.ppat.0020125.

Neverov, A. D., K. V. Lezhnina, A. S. Kondrashov, and G. A. Bazykin. 2014. "Intrasubtype reassortments cause adaptive amino acid replacements in H3N2 influenza genes." *PLOS Genet* 10 (1):e1004037. doi:10.1371/journal.pgen.1004037.

Nicholson, J. K., E. Holmes, J. Kinross, R. Burcelin, G. Gibson, W. Jia, and S. Pettersson. 2012. "Host-gut microbiota metabolic interactions." *Science* 336 (6086):1262–67. doi:10.1126/science.1223813.

Novick, R., P., G. Christie, E., and J. Penadés, R. 2010. "The phage-related chromosomal islands of Gram-positive bacteria." *Nat Rev Microbiol* 8 (8):541–51. doi:10.1038/nrmicro2393.

Novoa, R. R., G. Calderita, R. Arranz, J. Fontana, H. Granzow, and C. Risco. 2005. "Virus factories: Associations of cell organelles for viral replication and morphogenesis." *Biol Cell* 97 (2):147–72. doi:10.1042/BC20040058.

O'Brien, A. D., J. W. Newland, S. F. Miller, and R. K. Holmes. 1984. "Shiga-like toxin-converting phages from Escherichia coli strains that cause hemorrhagic colitis or infantile diarrhea." *Science*. doi:10.1126/science.6387911.

O'Brien, S. J., and G. W. Nelson. 2004. "Human genes that limit AIDS." *Nat Genet* 36 (6):565–74. doi:10.1038/ng1369.

O'Shea, T. J., P. M. Cryan, A. A. Cunningham, A. R. Fooks, D. T. Hayman, A. D. Luis, A. J. Peel, R. K. Plowright, and J. L. Wood. 2014. "Bat flight and zoonotic viruses." *Emer Infect Dis* 20 (5):741–45. doi:10.3201/eid2005.130539.

Ochman, H., J. G. Lawrence, and E. A. Groisman. 2000. "Lateral gene transfer and the nature of bacterial innovation." *Nature* 405 (6784):299–304. doi:10.1038/35012500.

Odaka, T., H. Ikeda, H. Yoshikura, K. Moriwaki, and S. Suzuki. 1981. "Fv-4: Gene controlling resistance to NB-tropic Friend murine leukemia virus.

Distribution in wild mice, introduction into genetic background of BALB/c mice, and mapping of chromosomes." *J Natl Cancer Inst* 67 (5):1123–27.

Olabode, A. S., X. Jiang, D. L. Robertson, and S. C. Lovell. 2015. "Ebolavirus is evolving but not changing: No evidence for functional change in EBOV from 1976 to the 2014 outbreak." *Virology* 482:202–7. doi:10.1016/j .virol.2015.03.029.

Olenec, J. P., W. K. Kim, W. M. Lee, F. Vang, T. E. Pappas, L. E. Salazar, M. D. Evans, et al. 2010. "Weekly monitoring of children with asthma for infections and illness during common cold seasons." *J Allergy Clin Immun* 125 (5):1001–6 e1. doi:10.1016/j.jaci.2010.01.059.

Olival, K. J., and D. T. Hayman. 2014. "Filoviruses in bats: Current knowledge and future directions." *Viruses* 6 (4):1759–88. doi:10.3390/v6041759.

Olival, K. J., A. Islam, M. Yu, S. J. Anthony, J. H. Epstein, S. A. Khan, S. U. Khan, et al. 2013. "Ebola virus antibodies in fruit bats, bangladesh." *Emerg Infect Dis* 19 (2):270–3. doi:10.3201/eid1902.120524.

Oliveira, N., M., H. Satija, A. Kouwenhoven, I., and M. Eiden, V. 2007. "Changes in viral protein function that accompany retroviral endogenization." *P Natl Acad Sci.* doi:10.1073/pnas.0704313104.

Olson, M. V., and A. Varki. 2003. "Sequencing the chimpanzee genome: insights into human evolution and disease." *Nat Rev Genet* 4 (1):20–28. doi:10.1038 /nrg981.

Osterholm, M. T., K. A. Moore, N. S. Kelley, L. M. Brosseau, G. Wong, F. A. Murphy, C. J. Peters, et al. 2015. "Transmission of Ebola viruses: What we know and what we do not know." *mBio* 6 (2). doi:10.1128/mBio.00137–15.

Otto, M. 2010. "Basis of virulence in community-associated methicillin-resistant Staphylococcus aureus." *Annu Rev Microbiol* 64:143–62. doi:10.1146/annurev. micro.112408.134309.

Owens, R. C., Jr., C. J. Donskey, R. P. Gaynes, V. G. Loo, and C. A. Muto. 2008. "Antimicrobial-associated risk factors for Clostridium difficile infection." *Clin Infect Dis* 46 Suppl 1:S19–31. doi:10.1086/521859.

Papi, A., and S. L. Johnston. 1999. "Rhinovirus infection induces expression of its own receptor intercellular adhesion molecule 1 (ICAM-1) via increased NF-⊠B-mediated transcription." *J Biol Chem* 274 (14):9707–20. doi:10.1074 /jbc.274.14.9707.

Park, B., V. Nizet, and G. Y. Liu. 2008. "Role of Staphylococcus aureus catalase in niche competition against Streptococcus pneumoniae." *J Bacteriol* 190 (7):2275–78. doi:10.1128/JB.00006–08.

Park, D. J., G. Dudas, S. Wohl, A. Goba, S. L. Whitmer, K. G. Andersen, R. S. Sealfon, et al. 2015. "Ebola virus epidemiology, transmission, and evolution during seven months in Sierra Leone." *Cell* 161 (7):1516–26. doi:10.1016/j.cell.2015.06.007.

Parks, C. L., R. A. Lerch, P. Walpita, H. P. Wang, M. S. Sidhu, and S. A. Udem. 2001. "Comparison of predicted amino acid sequences of measles virus strains in the Edmonston vaccine lineage." *J Virol* 75 (2):910–20. doi:10.1128/JVI.75.2.910-920.2001.

Partensky, F., W. R. Hess, and D. Vaulot. 1999. "Prochlorococcus, a marine photosynthetic prokaryote of global significance." *Microbiol Mol Biol Rev* 63 (1):106–27.

Passaes, C., C. Cardoso, D. Caetano, S. Teixeira, M. Guimarães, D. Campos, V. Veloso, et al. 2014. "Association of single nucleotide polymorphisms in the lens epithelium-derived growth factor (LEDGF/p75) with HIV-1 infection outcomes in Brazilian HIV-1+ individuals." *PLOS ONE* 9. doi:10.1371/journal.pone.0101780.

Payne, R., M. Muenchhoff, J. Mann, H. E. Roberts, P. Matthews, E. Adland, A. Hempenstall, et al. 2014. "Impact of HLA-driven HIV adaptation on virulence in populations of high HIV seroprevalence." *P Natl Acad Sci USA* 111 (50):E5393–400. doi:10.1073/pnas.1413339111.

Pedulla, M. L., M. E. Ford, J. M. Houtz, T. Karthikeyan, C. Wadsworth, J. A. Lewis, D. Jacobs-Sera, et al. 2003. "Origins of highly mosaic mycobacteriophage genomes." *Cell* 113 (2):171–82. doi:10.1016/S0092–8674(03)00233–2.

Peduzzi, P., M. Gruber, M. Gruber, and M. Schagerl. 2014. "The virus's tooth: Cyanophages affect an African flamingo population in a bottom-up cascade." *ISME J*. doi:10.1038/ismej.2013.241.

Penades, J. R., J. Chen, N. Quiles-Puchalt, N. Carpena, and R. P. Novick. 2015. "Bacteriophage-mediated spread of bacterial virulence genes." *Curr Opin Microbiol* 23:171–78. doi:10.1016/j.mib.2014.11.019.

Penfold, M. E., D. J. Dairaghi, G. M. Duke, N. Saederup, E. S. Mocarski, G. W. Kemble, and T. J. Schall. 1999. "Cytomegalovirus encodes a potent alpha chemokine." *P Natl Acad Sci USA* 96 (17):9839–44.

Pepin, J. 2011. *The Origins of AIDS* Cambridge: Cambridge University Press.

Perez-Caballero, D., T. Zang, A. Ebrahimi, M. W. McNatt, D. A. Gregory, M. C. Johnson, and P. D. Bieniasz. 2009. "Tetherin inhibits HIV-1 release by directly tethering virions to cells." *Cell* 139 (3):499–511. doi:10.1016/j.cell.2009.08.039.

Pfeiffer, J. K., and K. Kirkegaard. 2003. "A single mutation in poliovirus RNA-dependent RNA polymerase confers resistance to mutagenic nucleotide analogs via increased fidelity." *P Natl Acad Sci* 100. doi:10.1073/pnas.1232294100.

Pfeiffer, J. K., and K. Kirkegaard. 2005. "Increased fidelity reduces poliovirus fitness and virulence under selective pressure in mice." *PLOS Pathog* 1 (2). doi:10.1371/journal.ppat.0010011.

Philippe, N., M. Legendre, G. Doutre, Y. Coute, O. Poirot, M. Lescot, D. Arslan, et al. 2013. "Pandoraviruses: Amoeba viruses with genomes up to 2.5 Mb reaching that of parasitic eukaryotes." *Science* 341 (6143):281–86. doi:10.1126/science.1239181.

Pikor, L. A., J. C. Bell, and J-S. Diallo. 2015. "Oncolytic viruses: Exploiting cancer's deal with the devil." *Trends Cancer* 1 (4). doi:10.1016/j.trecan.2015.10.004.

Plotkin, J. B., and J. Dushoff. 2003. "Codon bias and frequency-dependent selection on the hemagglutinin epitopes of influenza A virus." *P Natl Acad Sci USA* 100 (12):7152–57. doi:10.1073/pnas.1132114100.

Plotkin, J. B., J. Dushoff, and S. A. Levin. 2002. "Hemagglutinin sequence clusters and the antigenic evolution of influenza A virus." *P Natl Acad Sci USA* 99 (9):6263–68. doi:10.1073/pnas.082110799.

Popovic, M., M. G. Sarngadharan, E. Read, and R. C. Gallo. 1984. "Detection, isolation, and continuous production of cytopathic retroviruses (HTLV-III) from patients with AIDS and pre-AIDS." *Science* 224 (4648):497–500.

Prince, J. L., D. T. Claiborne, J. M. Carlson, M. Schaefer, T. Yu, S. Lahki, H. A. Prentice, et al. 2012. "Role of transmitted Gag CTL polymorphisms in defining replicative capacity and early HIV-1 pathogenesis." *PLOS Pathog* 8 (11). doi:10.1371/journal.ppat.1003041.

Ptashne, M. 2004. *A Genetic Switch: Phage Lambda Revisited.* Woodbury, NY: Cold Spring Harbor Laboratory Press.

Racaniello, V., R. 2006. "One hundred years of poliovirus pathogenesis." *Virology* 344 (1). doi:10.1016/j.virol.2005.09.015.

Ram, G., J. Chen, K. Kumar, H. F. Ross, C. Ubeda, P. K. Damle, K. D. Lane, J. R. Penadés, G. E. Christie, and R. P. Novick. 2012. "Staphylococcal pathogenicity island interference with helper phage reproduction is a paradigm of molecular parasitism." *P Natl Acad Sci USA* 109 (40):16300–5. doi:10.1073/pnas.1204615109.

Rambaut, A., D. Posada, K. A. Crandall, and E. C. Holmes. 2004. "The causes and consequences of HIV evolution." *Nat Rev Genet* 5 (1):52–61. doi:10.1038/nrg1246.

Ramduth, D., P. Chetty, N. C. Mngquandaniso, N. Nene, J. D. Harlow, I. Honeyborne, N. Ntumba, et al. 2005. "Differential immunogenicity of HIV-1 clade C proteins in eliciting CD8+ and CD4+ cell responses." *J Infect Dis* 192 (9):1588–96. doi:10.1086/496894.

Raoult, D., S. Audic, C. Robert, C. Abergel, P. Renesto, H. Ogata, B. La Scola, M. Suzan, and J-M. M. Claverie. 2004. "The 1.2-megabase genome sequence of Mimivirus." *Science* 306 (5700):1344–50. doi:10.1126/science.1101485.

Raoult, D., and P. Forterre. 2008. "Redefining viruses: Lessons from Mimivirus." *Nat Rev Microbiol* 6 (4):315–19. doi:10.1038/nrmicro1858.

Redd, A. D., A. N. Collinson-Streng, N. Chatziandreou, C. E. Mullis, O. Laeyendecker, C. Martens, S. Ricklefs, et al. 2012. "Previously transmitted HIV-1 strains are preferentially selected during subsequent sexual transmissions." *J Infect Dis* 206 (9):1433–42. doi:10.1093/infdis/jis503.

Regev-Yochay, G., R. Dagan, M. Raz, Y. Carmeli, B. Shainberg, E. Derazne, G. Rahav, and E. Rubinstein. 2004. "Association between carriage of Streptococcus pneumoniae and Staphylococcus aureus in Children." *JAMA* 292 (6):716–20. doi:10.1001/jama.292.6.716.

Regev-Yochay, G., K. Trzciński, and C. M. Thompson. 2006. "Interference between Streptococcus pneumoniae and Staphylococcus aureus: in vitro hydrogen peroxide-mediated killing by Streptococcus pneumoniae." *J Biol* 188 (13). doi:10.1128/JB.00317–06.

Rho, H. M., B. Poiesz, F. W. Ruscetti, and R. C. Gallo. 1981. "Characterization of the reverse transcriptase from a new retrovirus (HTLV) produced by a human cutaneous T-cell lymphoma cell line." *Virology* 112 (1):355–60.

Riley, L. W., R. S. Remis, S. D. Helgerson, H. B. McGee, J. G. Wells, B. R. Davis, R. J. Hebert, et al. 1983. "Hemorrhagic colitis associated with a rare Escherichia coli serotype." *New Engl J Med* 308 (12):681–85. doi:10.1056 NEJM198303243081203.

Rimoin, A. W., P. M. Mulembakani, S. C. Johnston, J. O. Smith, N. K. Kisalu, T. L. Kinkela, S. Blumberg, et al. 2010. "Major increase in human monkeypox incidence 30 years after smallpox vaccination campaigns cease in the Democratic Republic of Congo." *P Natl Acad Sci* 107 (37). doi:10.1073/pnas.1005769107.

Robinson, M. C. 1955. "An epidemic of virus disease in Southern Province, Tanganyika Territory, in 1952–53. I. Clinical features." *Trans R Soc Trop Med Hyg* 49 (1):28–32.

Rodriguez, L. L., A. De Roo, Y. Guimard, S. G. Trappier, A. Sanchez, D. Bressler, A. J. Williams, et al. 1999. "Persistence and genetic stability of Ebola virus during the outbreak in Kikwit, Democratic Republic of the Congo, 1995." *J Infect Dis* 179 Suppl 1:S170–6. doi:10.1086/514291.

Rohwer, F., and R. V. Thurber. 2009. "Viruses manipulate the marine environment." *Nature* 459 (7244):207–12. doi:10.1038/nature08060.

Rohwer, F. 2003. "Global phage diversity." *Cell* 113 (2):141. doi:10.1016/S0092-8674(03)00276–9.

Roizman, B., and R. J. Whitley. 2013. "An inquiry into the molecular basis of HSV latency and reactivation." *Annu Rev Microbiol* 67:355–74. doi:10.1146/annurev-micro-092412–155654.

Rosenberg, R., M. A. Johansson, A. M. Powers, and B. R. Miller. 2013. "Search strategy has influenced the discovery rate of human viruses." *P Natl Acad Sci USA* 110 (34):13961–64. doi:10.1073/pnas.1307243110.

Rozo, M., and G. K. Gronvall. 2015. "The reemergent 1977 H1N1 strain and the gain-of-function debate." *MBio* 6 (4). doi:10.1128/mBio.01013–15.

Rudicell, R. S., J. Holland Jones, E. E. Wroblewski, G. H. Learn, Y. Li, J. D. Robertson, E. Greengrass, et al. 2010. "Impact of simian immunodeficiency virus infection on chimpanzee population dynamics." *PLOS Pathog* 6 (9):e1001116. doi:10.1371/journal.ppat.1001116.

Rusch, D. B., A. L. Halpern, G. Sutton, K. B. Heidelberg, S. Williamson, S. Yooseph, D. Wu, et al. 2007. "The Sorcerer II Global Ocean Sampling expedition: Northwest Atlantic through eastern tropical Pacific." *PLOS Biol* 5 (3):e77. doi:10.1371/journal.pbio.0050077.

Russell, C. A., T. C. Jones, I. G. Barr, N. J. Cox, R. J. Garten, V. Gregory, I. D. Gust, et al. 2008. "The global circulation of seasonal influenza A (H3N2) viruses." *Science* 320 (5874):340–46. doi:10.1126/science.1154137.

Russell, S. J., M. J. Federspiel, K-W. W. Peng, C. Tong, D. Dingli, W. G. Morice, V. Lowe, et al. 2014. "Remission of disseminated cancer after systemic oncolytic virotherapy." *Mayo Clin Proc* 89 (7):926–33. doi:10.1016/j.mayocp.2014.04.003.

Rustagi, A., and M. Gale, Jr. 2014. "Innate antiviral immune signaling, viral evasion and modulation by HIV-1." *J Mol Biol* 426 (6):1161–77. doi:10.1016/j.jmb.2013.12.003.

Sadler, H. A., M. D. Stenglein, R. S. Harris, and L. M. Mansky. 2010. "APOBEC3G contributes to HIV-1 variation through sublethal mutagenesis." *J Virol* 84 (14):7396–404. doi:10.1128/JVI.00056–10.

Salomon, R., and R. G. Webster. 2009. "The influenza virus enigma." *Cell* 136 (3). doi:10.1016/j.cell.2009.01.029.

Samuelson, L. C., R. S. Phillips, and L. J. Swanberg. 1996. "Amylase gene structures in primates: Retroposon insertions and promoter evolution." *Mol Biol Evol* 13 (6):767–79. doi:10.1093/oxfordjournals.molbev.a025637.

Schlecht-Louf, G., M. Renard, M. Mangeney, C. Letzelter, A. Richaud, B. Ducos, I. Bouallaga, and T. Heidmann. 2010. "Retroviral infection in vivo requires an immune escape virulence factor encrypted in the envelope protein of oncoretroviruses." *P Natl Acad Sci USA* 107 (8):3782–87. doi:10.1073/pnas.0913122107.

Schmid, M. F., C. W. Hecksel, R. H. Rochat, D. Bhella, W. Chiu, and F. J. Rixon. 2012. "A tail-like assembly at the portal vertex in intact herpes simplex type-1 virions." *PLOS Pathog* 8 (10). doi:10.1371/journal.ppat.1002961.

Schneider, D. S., and J. S. Ayres. 2008. "Two ways to survive infection: what resistance and tolerance can teach us about treating infectious diseases." *Nat Rev Immunol* 8 (11):889–95. doi:10.1038/nri2432.

Scholtissek, C., V. von Hoyningen, and R. Rott. 1978. "Genetic relatedness between the new 1977 epidemic strains (H1N1) of influenza and human influenza strains isolated between 1947 and 1957 (H1N1)." *Virology* 89 (2):613–17.

Schrauwen, E. J. A., T. M. Bestebroer, G. F. Rimmelzwaan, A. D. M. E. Osterhaus, R. A. M. Fouchier, and S. Herfst. 2013. "Reassortment between avian H5N1 and human influenza viruses is mainly restricted to the matrix and neuraminidase gene segments." *PLOS ONE* 8 (3). doi:10.1371/journal.pone.0059889.

Selva, L., D. Viana, G. Regev-Yochay, K. Trzcinski, J. M. Corpa, I. Lasa, R. P. Novick, and J. R. Penadés. 2009. "Killing niche competitors by remote-control bacteriophage induction." *P Natl Acad Sci USA* 106 (4):1234–38. doi:10.1073/pnas.0809600106.

Shackelton, L. A., and E. C. Holmes. 2006. "Phylogenetic evidence for the rapid evolution of human B19 erythrovirus." *J Virol* 80 (7):3666–69. doi:10.1128/JVI.80.7.3666–3669.2006.

Shackelton, L. A., A. Rambaut, O. G. Pybus, and E. C. Holmes. 2006. "JC virus evolution and its association with human populations." *J Virol* 80 (20):9928–33. doi:10.1128/JVI.00441–06.

Shackelton, L. A., and E. C. Holmes. 2004. "The evolution of large DNA viruses: Combining genomic information of viruses and their hosts." *Trends Microbiol* 12 (10):458–65. doi:10.1016/j.tim.2004.08.005.

Shah, S. D., J. Doorbar, and R. A. Goldstein. 2010. "Analysis of host–parasite incongruence in papillomavirus evolution using importance sampling." *Mol Biol Evol.* doi:10.1093/molbev/msq015.

Shanta, M. Zimmer, and S. Burke Donald. 2009. "Historical perspective—Emergence of influenza A (H1N1) viruses." *New Engl J Med.* doi:10.1056/NEJMra0904322.

Sharp, P. M., and B. H. Hahn. 2010. "The evolution of HIV-1 and the origin of AIDS." *P Roy Soc Lond B Bio* 365 (1552):2487–94. doi:10.1098/rstb.2010.0031.

Sharp, P. A. 1985. "On the origin of RNA splicing and introns." *Cell* 42 (2):397–400.

Sharp, P. M., and B. H. Hahn. 2011. "Origins of HIV and the AIDS pandemic." *Cold Spring Harb Perspect Med* 1 (1):a006841. doi:10.1101/cshperspect.a006841.

Shchelkunov, S. N. 2013. "An increasing danger of zoonotic orthopoxvirus infections." *PLOS Pathog* 9 (12):e1003756. doi:10.1371/journal.ppat.1003756.

Shchelkunov, S. N. 2009. "How long ago did smallpox virus emerge?" *Arch Virol* 154 (12):1865–71. doi:10.1007/s00705–009–0536–0.

Sheehy, A. M., N. C. Gaddis, J. D. Choi, and M. H. Malim. 2002. "Isolation of a human gene that inhibits HIV-1 infection and is suppressed by the viral Vif protein." *Nature* 418 (6898):646–50. doi:10.1038/nature00939.

Shope, R. E. 1936. "The incidence of neutralizing antibodies for swine influenza virus in the sera of human beings of different ages." *J Exp Med* 63 (5):669–84.

Short, S. M. 2012. "The ecology of viruses that infect eukaryotic algae." *Environ Microbiol* 14 (9):2253–71. doi:10.1111/j.1462–2920.2012.02706.x.

Shortridge, K. F., R. G. Webster, W. K. Butterfield, and C. H. Campbell. 1977. "Persistence of Hong Kong influenza virus variants in pigs." *Science* 196 (4297):1454–55.

Silvestri, G., D. L. Sodora, R. A. Koup, M. Paiardini, S. P. O'Neil, H. M. McClure, S. I. Staprans, and M. B. Feinberg. 2003. "Nonpathogenic SIV infection of sooty mangabeys is characterized by limited bystander immunopathology despite chronic high-level viremia." *Immunity* 18 (3):441–52.

Simmons, G., D. Clarke, J. McKee, P. Young, and J. Meers. 2014. "Discovery of a novel retrovirus sequence in an Australian native rodent (Melomys burtoni): A putative link between gibbon ape leukemia virus and koala retrovirus." *PLOS ONE* 9 (9):e106954. doi:10.1371/journal.pone.0106954.

Simmons, G. S., P. R. Young, J. J. Hanger, K. Jones, D. Clarke, J. J. McKee, and J. Meers. 2012. "Prevalence of koala retrovirus in geographically diverse populations in Australia." *Aust Vet J* 90 (10):404–9. doi:10.1111/j.1751–0813.2012.00964.x.

Simon-Loriere, E., O. Faye, O. Faye, L. Koivogui, N. Magassouba, S. Keita, J. M. Thiberge, L. Diancourt, et al. 2015. "Distinct lineages of Ebola virus in Guinea during the 2014 West African epidemic." *Nature* 524 (7563):102–4. doi:10.1038/nature14612.

Smith, D. J., A. S. Lapedes, J. C. de Jong, T. M. Bestebroer, G. F. Rimmelzwaan, A. D. Osterhaus, and R. A. Fouchier. 2004. "Mapping the antigenic and genetic evolution of influenza virus." *Science* 305 (5682):371–76. doi:10.1126/science.1097211.

Smith, E. C., J. B. Case, H. Blanc, O. Isakov, N. Shomron, M. Vignuzzi, and M. R. Denison. 2015. "Mutations in coronavirus nonstructural protein 10 decrease virus replication fidelity." *J Virol* 89 (12):6418–26. doi:10.1128/JVI.00110–15.

Smith, G. J., D. Vijaykrishna, J. Bahl, S. J. Lycett, M. Worobey, O. G. Pybus, S. K. Ma, et al. 2009. "Origins and evolutionary genomics of the 2009 swine-origin H1N1 influenza A epidemic." *Nature* 459 (7250):1122–25. doi:10.1038/nature08182.

Sobey, W. R. 1969. "Selection for resistance to myxomatosis in domestic rabbits (Oryctolagus cuniculus)." *J Hyg (Lond)* 67 (4):743–54.

Spencer, J. V., K. M. Lockridge, P. A. Barry, G. Lin, M. Tsang, M. E. Penfold, and T. J. Schall. 2002. "Potent immunosuppressive activities of cytomegalovirus-encoded interleukin-10." *J Virol* 76 (3):1285–92.

Stanford, M. M., S. J. Werden, and G. McFadden. 2007. "Myxoma virus in the European rabbit: interactions between the virus and its susceptible host." *Vet Res* 38 (2):299–318. doi:10.1051/vetres:2006054.

Stanley, M. 2010. "HPV—immune response to infection and vaccination." *Infect Agents Cancer* 5 (1):19. doi:10.1186/1750–9378–5–19.

Steinberg, K. M., and B. R. Levin. 2007. "Grazing protozoa and the evolution of the Escherichia coli O157:H7 Shiga toxin-encoding prophage." *P Roy Soc B-Biol Sci* 274 (1621):1921–29. doi:10.1098/rspb.2007.0245.

Stephens, C. R., and H. Waelbroeck. 1999. "Codon bias and mutability in HIV sequences." *J Mol Evol* 48 (4):390–97. doi:10.1007/PL00006483.

Stocking, C., and C. A. Kozak. 2008. "Murine endogenous retroviruses." *Cell Mol Life Sci* 65 (21):3383–98. doi:10.1007/s00018–008–8497–0.

Stoye, J. P. 2006. "Koala retrovirus: a genome invasion in real time." *Genome Biol* 7 (11):241. doi:10.1186/gb-2006–7–11–241.

Stoye, J. P. 2012. "Studies of endogenous retroviruses reveal a continuing evolutionary saga." *Nat Rev Microbiol* 10 (6):395–406. doi:10.1038/nrmicro2783.

Streicker, D. G., A. S. Turmelle, M. J. Vonhof, I. V. Kuzmin, G. F. McCracken, and C. E. Rupprecht. 2010. "Host phylogeny constrains cross-species emergence and establishment of rabies virus in bats." *Science* 329 (5992):676–79. doi:10.1126/science.1188836.

Stremlau, M., M. Perron, M. Lee, Y. Li, B. Song, H. Javanbakht, F. Diaz-Griffero, D. J. Anderson, W. I. Sundquist, and J. Sodroski. 2006. "Specific recognition and accelerated uncoating of retroviral capsids by the TRIM5alpha restriction factor." *Proc Natl Acad Sci USA* 103 (14):5514–19. doi:10.1073/pnas.0509996103.

Suarez, D. L., M. L. Perdue, N. Cox, T. Rowe, C. Bender, J. Huang, and D. E. Swayne. 1998. "Comparisons of highly virulent H5N1 influenza A viruses isolated from humans and chickens from Hong Kong." *J Virol* 72 (8):6678–88.

Sullivan, M. B., D. Lindell, J. A. Lee, L. R. Thompson, J. P. Bielawski, and S. W. Chisholm. 2006. "Prevalence and evolution of core photosystem II genes in marine cyanobacterial viruses and their hosts." *PLOS Biol* 4 (8). doi:10.1371/journal.pbio.0040234.

Suttle, C. A. 2007. "Marine viruses—major players in the global ecosystem." *Nat Rev Microbiol* 5 (10):801–12. doi:10.1038/nrmicro1750.

Suzan-Monti, M., B. La Scola, L. Barrassi, L. Espinosa, and D. Raoult. 2007. "Ultrastructural characterization of the giant volcano-like virus factory of Acanthamoeba polyphaga Mimivirus." *PLOS ONE* 2 (3). doi:10.1371/journal.pone.0000328.

Sykes, A., M. R. Edwards, J. Macintyre, A. del Rosario, E. Bakhsoliani, M. B. Trujillo-Torralbo, O. M. Kon, P. Mallia, M. McHale, and S. L. Johnston. 2012. "Rhinovirus 16-induced IFN-alpha and IFN-beta are deficient in bronchoalveolar lavage cells in asthmatic patients." *J Allergy Clin Immunol* 129 (6):1506–14 e6. doi:10.1016/j.jaci.2012.03.044.

Tabler, M., and M. Tsagris. 2004. "Viroids: petite RNA pathogens with distinguished talents." *Trends Plant Sci* 9 (7):339–48. doi:10.1016/j.tplants.2004.05.007.

Tapparel, C., S. Cordey, T,.Junier, L. Farinelli, S. V. Belle, P. M. Soccal, J. D. Aubert, E. M. Zdobnov, and L. Kaiser. 2011. "Rhinovirus genome variation during chronic upper and lower respiratory tract infections." *PLOS ONE* 6 (6). doi:10.1371/journal.pone.0021163.

Tarlinton, R., J. Meers, and P. Young. 2008. "Biology and evolution of the endogenous koala retrovirus." *Cell Mol Life Sci* 65 (21):3413–21. doi:10.1007/s00018–008–8499-y.

Tarlinton, R. E., J. Meers, and P. R. Young. 2006. "Retroviral invasion of the koala genome." *Nature* 442 (7098):79–81. doi:10.1038/nature04841.

Taubenberger, J. K., and D. M. Morens. 2006. "1918 influenza: The mother of all pandemics." *Emerg Infect Dis.* doi:10.3201/eid1201.050979.

Taylor, D. J., R. W. Leach, and J. Bruenn. 2010. "Filoviruses are ancient and integrated into mammalian genomes." *BMC Evol Biol* 10:193. doi:10.1186/1471-2148-10-193.

Tormo-Más, M. Á., I. MirI, A. Shrestha, S. M. Tallent, S. Campoy, Í. Lasa, J. Barbé, R. P. Novick, G. E. Christie, and J. Penadés, R. 2010. "Moonlighting bacteriophage proteins derepress staphylococcal pathogenicity islands." *Nature* 465 (7299):779–82. doi:10.1038/nature09065.

Tsagris, E. M., A. E. Martinez de Alba, M. Gozmanova, and K. Kalantidis. 2008. "Viroids." *Cell Microbiol* 10 (11):2168–79. doi:10.1111/j.1462-5822.2008.01231.x.

Tsetsarkin, K. A., R. Chen, M. B. Sherman, and S. C. Weaver. 2011. "Chikungunya virus: Evolution and genetic determinants of emergence." *Curr Opin Virol* 1 (4):310–7. doi:10.1016/j.coviro.2011.07.004.

Tumpey, T. M., C. F. Basler, P. V. Aguilar, H. Zeng, A. Solorzano, D. E. Swayne, N. J. Cox, et al. 2005. "Characterization of the reconstructed 1918 Spanish influenza pandemic virus." *Science* 310 (5745):77–80. doi:10.1126/science.1119392.

Twort, F. W. 1915. "An investigation on the nature or ultra-microscopic viruses." *Lancet* 2:1241–43.

Ubeda, C., E. Maiques, E. Knecht, I. Lasa, R. P. Novick, and J. R. Penadés. 2005. "Antibiotic-induced SOS response promotes horizontal dissemination of pathogenicity island-encoded virulence factors in staphylococci." *Mol Microbiol* 56 (3):836–44. doi:10.1111/j.1365-2958.2005.04584.x.

Van Doorslaer, K. 2013. "Evolution of the papillomaviridae." *Virology* 445 (1–2):11–20. doi:10.1016/j.virol.2013.05.012.

Van Heuverswyn, F., Y. Li, E. Bailes, C. Neel, B. Lafay, B. F. Keele, K. S. Shaw, et al. 2007. "Genetic diversity and phylogeographic clustering of SIVcpzPtt in wild chimpanzees in Cameroon." *Virology* 368 (1):155–71. doi:10.1016/j.virol.2007.06.018.

Van Valen, L. 1973. "A New Evolutionary Law." *Evol. Theory* 1:1–30.

Vignuzzi, M., J. K. Stone, J. J. Arnold, C. E. Cameron, and R. Andino. 2006. "Quasispecies diversity determines pathogenesis through cooperative interactions in a viral population." *Nature* 439 (7074):344–48. doi:10.1038/nature04388.

Virgin, H. W. 2014. "The virome in mammalian physiology and disease." *Cell* 157 (1):142–50. doi:10.1016/j.cell.2014.02.032.

Volchkov, V. E., V. A. Volchkova, E. Muhlberger, L. V. Kolesnikova, M. Weik, O. Dolnik, and H. D. Klenk. 2001. "Recovery of infectious Ebola virus from complementary DNA: RNA editing of the GP gene and viral cytotoxicity." *Science* 291 (5510):1965–69. doi:10.1126/science.1057269.

Vrancken, B., A. Rambaut, M. A. Suchard, A. Drummond, G. Baele, I. Derdelinckx, E. Van Wijngaerden, A. M. Vandamme, K. Van Laethem, and P. Lemey. 2014. "The genealogical population dynamics of HIV-1 in a large transmission chain: bridging within and among host evolutionary rates." *PLOS Comput Biol* 10 (4):e1003505. doi:10.1371/journal.pcbi.1003505.

Wagner, P. L., D. W. K. Acheson, and M. K. Waldor. 2001. "Human neutrophils and their products induce Shiga toxin production by enterohemorrhagic Escherichia coli." *Infect Immun* 69 (3):1934–37.

Wagner, P. L., M. N. Neely, X. Zhang, D. W. K. Acheson, M. K. Waldor, and D. I. Friedman. 2001. "Role for a phage promoter in Shiga toxin 2 expression from a pathogenic Escherichia coli strain." *J Bacteriol* 183 (6):2081–85. doi:10.1128/JB.183.6.2081–2085.2001.

Wagner, Patrick L., and Matthew K. Waldor. 2002. "Bacteriophage control of bacterial virulence." *Infect Immun* 70 (8):3985–93. doi:10.1128/IAI.70.8.3985–3993.2002.

Waldor, M. K., and J. J. Mekalanos. 1996. "Lysogenic conversion by a filamentous phage encoding cholera toxin." *Science* 272 (5270):1910–14.

Walsh, P. D., K. A. Abernethy, M. Bermejo, R. Beyers, P. De Wachter, M. E. Akou, B. Huijbregts, et al. 2003. "Catastrophic ape decline in western equatorial Africa." *Nature* 422 (6932):611–14. doi:10.1038/nature01566.

Warshauer, D. M., E. C. Dick, A. D. Mandel, T. C. Flynn, and R. S. Jerde. 1989. "Rhinovirus infections in an isolated antarctic station. Transmission of the viruses and susceptibility of the population." *Am J Epidemiol* 129 (2):319–40.

Watson, J. D. 1993. "Early speculations and facts about RNA templates." In *The RNA World*, edited by R. S. Gesteland and J. F. Atkins (xv–xxiii). Woodbury, NY: Cold Spring Harbor Laboratory Press.

Webby, R. J., and R. G. Webster. 2003. "Are we ready for pandemic influenza?" *Science* 302 (5650):1519–22. doi:10.1126/science.1090350.

Webster, R. G., W. J. Bean, O. T. Gorman, T. M. Chambers, and Y. Kawaoka. 1992. "Evolution and ecology of influenza A viruses." *Microbiol Rev* 56 (1): 152–79.

Webster, R. G. 2002. "The importance of animal influenza for human disease." *Vaccine* 20. doi:10.1016/S0264–410X(02)00123–8.

Webster, R. G., and E. A. Govorkova. 2014. "Continuing challenges in influenza." *Ann NY Acad Sci* 1323:115–39. doi:10.1111/nyas.12462.

Wei, X., J. M. Decker, S. Wang, H. Hui, J. C. Kappes, X. Wu, J. F. Salazar-Gonzalez, et al. 2003. "Antibody neutralization and escape by HIV-1." *Nature* 422 (6929):307–12. doi:10.1038/nature01470.

Weinbauer, M. G., and F. Rassoulzadegan. 2003. "Are viruses driving microbial diversification and diversity?" *Environ Microbiol.* doi:10.1046/j.1462-2920.2003.00539.x.

Weissmann, C. 2012. "The end of the road." *Prion* 6 (2):97–104. doi:10.4161/pri.19778.

Wertheim, J. O., D. K. Chu, J. S. Peiris, S. L. Kosakovsky Pond, and L. L. Poon. 2013. "A case for the ancient origin of coronaviruses." *J Virol* 87 (12):7039–45. doi:10.1128/JVI.03273-12.

Wertheim, J. O., M. D. Smith, D. M. Smith, K. Scheffler, and S. L. Kosakovsky Pond. 2014. "Evolutionary origins of human herpes simplex viruses 1 and 2." *Mol Biol Evol* 31 (9):2356–64. doi:10.1093/molbev/msu185.

Wertheim, J. O., and M. Worobey. 2009. "Dating the age of the SIV lineages that gave rise to HIV-1 and HIV-2." *PLOS Comput Biol* 5 (5). doi:10.1371/journal.pcbi.1000377.

Whitley, R. J., and B. Roizman. 2001. "Herpes simplex virus infections." *Lancet* 357 (9267):1513–18. doi:10.1016/S0140-6736(00)04638-9.

WHO. 2009. "WHO experts consultation on Ebola Reston pathogenicity in humans." Geneva: World Health Organization. http://www.who.int/csr/resources/publications/WHO_HSE_EPR_2009_2/en/.

WHO. 2014. "Influenza (seasonal) fact sheet." Geneva: World Health Organization. http://www.who.int/mediacentre/factsheets/fs211/en/.

WHO. 2015a. "Nipah virus (NiV) infection." Geneva: World Health Organization. http://www.who.int/csr/disease/nipah/en/.

WHO. 2015b. "Poliomyelitis (polio)." Geneva: World Health Organization. http://www.who.int/topics/poliomyelitis/en/.

WHO. 2015c. "Situation Reports: Ebola Response Roadmap." Geneva: World Health Organization. Accessed 2015. http://www.afro.who.int/en/clusters-a-programmes/dpc/epidemic-a-pandemic-alert-and-response/sitreps/4333-situation-reports-ebola-response-roadmap.html.

WHO. 2015d. "Surveillance and Montioring."Geneva: World Health Organization. http://www.who.int/influenza/surveillance_monitoring/en/.

WHO. 2016. "Work stream 1—Priority global disease threats." Geneva: World Health Organization. http://www.who.int/csr/research-and-development/workstream1-prioritize-pathogens/en/.

Williamson, S. 2003. "Adaptation in the *env* gene of HIV-1 and evolutionary theories of disease progression." *Mol Biol Evol.* doi:10.1093/molbev/msg144.

Williamson, S. J., D. B. Rusch, S. Yooseph, A. L. Halpern, K. B. Heidelberg, J. I. Glass, C. Andrews-Pfannkoch, et al. 2008. "The Sorcerer II Global Ocean Sampling expedition: Metagenomic characterization of viruses within aquatic microbial samples." *PLOS ONE* 3 (1):e1456. doi:10.1371/journal.pone.0001456.

Williamson, S. J., S. C. Cary, K. E. Williamson, R. R. Helton, S. R. Bench, D. Winget, and K. E. Wommack. 2008. "Lysogenic virus-host interactions predominate at deep-sea diffuse-flow hydrothermal vents." *ISME J* 2 (11):1112–21. doi:10.1038/ismej.2008.73.

Wilson, A. C., and I. Mohr. 2012. "A cultured affair: HSV latency and reactivation in neurons." *Trends Microbiol* 20 (12):604–11. doi:10.1016/j.tim.2012.08.005.

Wittmann, T. J., R. Biek, A. Hassanin, P. Rouquet, P. Reed, P. Yaba, X. Pourrut, L. A. Real, J. P. Gonzalez, and E. M. Leroy. 2007. "Isolates of Zaire ebolavirus from wild apes reveal genetic lineage and recombinants." *P Natl Acad Sci USA* 104 (43):17123–27. doi:10.1073/pnas.0704076104.

Woelk, C. H., and E. C. Holmes. 2002. "Reduced positive selection in vector-borne RNA viruses." *Mol Biol Evol* 19 (12):2333–36.

Woese, C. R. 2002. "On the evolution of cells." *P Natl Acad Sci USA* 99 (13):8742–47. doi:10.1073/pnas.132266999.

Wong, C. S., S. Jelacic, R. L. Habeeb, S. L. Watkins, and P. I. Tarr. 2000. "The risk of the hemolytic–uremic syndrome after antibiotic treatment of Escherichia coli O157:H7 infections." *New Engl J Med* 342 (26):1930–36. doi:10.1056/NEJM200006293422601.

Woo, P. C., S. K. Lau, C. S. Lam, C. C. Lau, A. K. Tsang, J. H. Lau, R. Bai, et al. 2012. "Discovery of seven novel Mammalian and avian coronaviruses in the genus deltacoronavirus supports bat coronaviruses as the gene source of alphacoronavirus and betacoronavirus and avian coronaviruses as the gene source of gammacoronavirus and deltacoronavirus." *J Virol* 86 (7):3995–4008. doi:10.1128/JVI.06540-11.

Woolhouse, M., F. Scott, and Z. Hudson. 2012. "Human viruses: Discovery and emergence." *Philos T Roy Soc Lond B Bio* 367 (1604):2864–71. doi:10.1098/rstb.2011.0354.

Woolhouse, M. E., R. Howey, E. Gaunt, L. Reilly, M. Chase-Topping, and N. Savill. 2008. "Temporal trends in the discovery of human viruses." *Proc Biol Sci* 275 (1647):2111–15. doi:10.1098/rspb.2008.0294.

Woolhouse, M. E., and S. Gowtage-Sequeria. 2005. "Host range and emerging and reemerging pathogens." *Emerg Infect Dis* 11 (12):1842–47. doi:10.3201/eid1112.050997.

Worobey, M., M. Gemmel, D. E. Teuwen, T. Haselkorn, K. Kunstman, M. Bunce, J-J. J. Muyembe, et al. 2008. "Direct evidence of extensive diversity of HIV-1 in Kinshasa by 1960." *Nature* 455 (7213):661–64. doi:10.1038/nature07390.

Worobey, M., P. Telfer, S. Souquière, M. Hunter, C. A. Coleman, M. J. Metzger, P. Reed, et al. 2010. "Island biogeography reveals the deep history of SIV." *Science* 329 (5998):1487. doi:10.1126/science.1193550.

Wright, G. D. 2007. "The antibiotic resistome: the nexus of chemical and genetic diversity." *Nat Rev Microbiol* 5 (3):175–86. doi:10.1038/nrmicro1614.

Wu, Z., L. Yang, F. Yang, X. Ren, J. Jiang, J. Dong, L. Sun, Y. Zhu, H. Zhou, and Q. Jin. 2014. "Novel Henipa-like virus, Mojiang Paramyxovirus, in rats, China, 2012." *Emerg Infect Dis* 20 (6):1064–66. doi:10.3201/eid2006.131022.

Xu, X., Subbarao, N. J. Cox, and Y. Guo. 1999. "Genetic characterization of the pathogenic influenza A/Goose/Guangdong/1/96 (H5N1) virus: Similarity of its hemagglutinin gene to those of H5N1 viruses from the 1997 outbreaks in Hong Kong." *Virology* 261 (1):15–19.

Yap, M., W., E. Colbeck, S. A. Ellis, and J. Stoye, P. 2014. "Evolution of the retroviral restriction gene Fv1: Inhibition of non-MLV retroviruses." *PLOS Pathog* 10 (3). doi:10.1371/journal.ppat.1003968.

Yin, X., X. Yin, B. Rao, C. Xie, P. Zhang, X. Qi, P. Wei, and H. Liu. 2014. "Antibodies against avian-like A (H1N1) swine influenza virus among swine farm residents in eastern China." *J Med Virol* 86 (4):592–96. doi:10.1002/jmv.23842.

Young, V. B., and T. M. Schmidt. 2004. "Antibiotic-associated diarrhea accompanied by large-scale alterations in the composition of the fecal microbiota." *J Clin Microbiol* 42 (3):1203–6.

Yue, L., K. J. Pfafferott, J. Baalwa, K. Conrod, C. C. Dong, C. Chui, R. Rong, D. T. Claiborne, et al. 2015. "Transmitted virus fitness and host T cell responses collectively define divergent infection outcomes in two HIV-1 recipients." *PLOS Pathog* 11 (1). doi:10.1371/journal.ppat.1004565.

Yutin, N., P. Colson, D. Raoult, and E. V. Koonin. 2013. "Mimiviridae: Clusters of orthologous genes, reconstruction of gene repertoire evolution and proposed expansion of the giant virus family." *Virol J* 10:106. doi:10.1186/1743–422X-10–106.

Yutin, N., D. Raoult, and E. V. Koonin. 2013. "Virophages, polintons, and transpovirons: A complex evolutionary network of diverse selfish genetic elements with different reproduction strategies." *Virol J* 10:158. doi:10.1186/1743–422X-10–158.

Yutin, N., Y. I. Wolf, D. Raoult, and E. V. Koonin. 2009. "Eukaryotic large nucleo-cytoplasmic DNA viruses: Clusters of orthologous genes and reconstruction of viral genome evolution." *Virol J* 6:223. doi:10.1186/1743–422X-6–223.

Zaitlin, M. 1998. "The discovery of the causal agent of the tobacco mosaic disease." In *Discoveries in Plant Biology*, edited by S. D. and Yang Kung, S. F. (105–10). Hong Kong: World Publishing Co.

Zampieri, C. A., N. J. Sullivan, and G. J. Nabel. 2007. "Immunopathology of highly virulent pathogens: insights from Ebola virus." *Nat Immunol* 8 (11):1159–64. doi:10.1038/ni1519.

Zhang, X., A. D. McDaniel, L. E. Wolf, G. T. Keusch, M. K. Waldor, and D. W. Acheson. 2000. "Quinolone antibiotics induce Shiga toxin-encoding bacteriophages, toxin production, and death in mice." *J Infect Dis* 181 (2):664–70. doi:10.1086/315239.

Zhang, X., M. Hasoksuz, D. Spiro, R. Halpin, S. Wang, S. Stollar, D. Janies, et al. 2006. "Complete genomic sequences, a key residue in the spike protein and deletions in nonstructural protein 3b of US strains of the virulent and attenuated coronaviruses, transmissible gastroenteritis virus and porcine respiratory coronavirus." *Virology* 358 (2). doi:10.1016/j.virol.2006.08.051.

Zhao, K., Y. Ishida, T. K. Oleksyk, C. A. Winkler, and A. L. Roca. 2012. "Evidence for selection at HIV host susceptibility genes in a West Central African human population." *BMC Evol Biol* 12:237. doi:10.1186/1471–2148–12–237.

Zhuang, J., A. E. Jetzt, G. Sun, H. Yu, G. Klarmann, Y. Ron, B. D. Preston, and J. P. Dougherty. 2002. "Human immunodeficiency virus type 1 recombination: rate, fidelity, and putative hot spots." *J Virol* 76 (22):11273–82.

Zimmer, S., M., and D. S. Burke. 2009. "Historical Perspective—Emergence of Influenza A (H1N1) Viruses." *New Engl J Med.* doi:10.1056/NEJMra0904322.

ACKNOWLEDGMENTS

I am indebted to several friends and colleagues who have enthusiastically supported this project—in particular, Jacques Archambault, Pierre Bonneau, and Peter White for their interest, encouragement, and critical feedback and for many stimulating discussions and inspiration. I would also like to thank Michael Fisher at Harvard University Press, who first saw the potential of the project, and my editor, Janice Audet, for her enthusiasm and gentle steering of a disorganized manuscript into its current form. I owe a debt of gratitude to my teachers, mentors, and the many inspirational scientists on whose research the book is based. They have all, in different ways, sustained my enduring enthusiasm for the biomedical research enterprise. Above all I am deeply grateful to my wife, Janet, without whose confidence and unwavering support I would have achieved far less and this book would not have been written.

INDEX